Rational Extended Thermodynamics beyond the Monatomic Gas

Prof. Masaru Sugiyama (*left*) and Prof. Tommaso Ruggeri (*right*). Oberwolfach, March 2014 (©MFO; source http://owpdb.mfo.de/detail?photo_id=18572)

Tommaso Ruggeri • Masaru Sugiyama

Rational Extended Thermodynamics beyond the Monatomic Gas

Springer

Tommaso Ruggeri
Dept. of Mathematics and Res. Center of Applied Mathematics AM^2
University of Bologna
Bologna, Italy

Masaru Sugiyama
Graduate School of Engineering
Nagoya Institute of Technology
Nagoya, Japan

ISBN 978-3-319-13340-9 ISBN 978-3-319-13341-6 (eBook)
DOI 10.1007/978-3-319-13341-6

Library of Congress Control Number: 2015947291

Springer Cham Heidelberg New York Dordrecht London

Printed on acid-free paper

Springer International Publishing AG Switzerland is part of Springer Science+Business Media (www.springer.com)

To Ester and Tomoko

Preface

This book aims to explain recent developments in Rational Extended Thermodynamics (RET), in particular those that have occurred since the publication of the book Rational Extended Thermodynamics, Second Edition (Springer, New York, 1998) by Ingo Müller and Tommaso Ruggeri.

RET is a phenomenological field theory capable of describing nonequilibrium phenomena with steep gradients and rapid changes in space-time out of local equilibrium. Classical thermodynamics of irreversible processes (TIP) relies essentially on the assumption of local equilibrium. Therefore, the validity range of RET is wider than that of TIP. Moreover, RET can predict the finite speed of disturbances because its basic system of field equations is hyperbolic. In contrast, TIP predicts the infinite speed of disturbances because of its parabolic character, which is fatal in a relativistic framework.

RET was strongly motivated by—and is in perfect agreement with—the kinetic theory, in particular, the system of moment equations derived from the Boltzmann equation. In RET, the differential system is closed by the universal principles: the objectivity principle, the entropy principle, and the principle of causality and stability. This permits an intimate connection between RET and the mathematical theory of hyperbolic systems with convex extension (symmetric systems). It is, therefore, possible to give a qualitative analysis, and the Cauchy problem is well posed. For example, a well-known theory of viscous heat-conducting fluids based on TIP is the classical Navier-Stokes-Fourier theory with five independent field variables: the mass density, the velocity and the temperature. On the other hand, RET adopts more independent field variables by incorporating nonequilibrium variables such as viscous stress and heat flux into the theory.

The limitation of the previous RET is, however, that its validity range has been restricted to rarefied monatomic gases. The present book presents the recent results that have overcome this limitation, that is, the results concerned with polyatomic gases, moderately dense gases, and mixtures of gases with multi-temperature.

The features of the book may be summarized as follows:

- We firstly explain the results of RET in the case of monatomic gases briefly, which are necessary to understand the progress in the new approach of RET.
- We present the hyperbolic theory of polyatomic and moderately dense gases with 14 fields, which, in the parabolic limit, reduces to the Navier-Stokes-Fourier theory. The singular limit of the theory of monatomic gases with 13 fields is also considered.
- We present some typical applications of the theory: sound wave, light scattering, shock wave, heat conduction, fluctuation. We compare the theoretical results with experimental data.
- The 14-field theory gives us a complete phenomenological model, but its differential system is rather complex. For this reason, we have constructed a simplified theory with six fields. This simplified theory preserves the main physical properties of the more complex theory of 14 variables, in particular when the bulk viscosity plays a more important role than the shear viscosity and the heat conductivity. This situation is observed in many gases such as rarefied hydrogen gases and carbon dioxide gases at some temperature ranges. This model is particularly interesting because it is also valid in a situation far from equilibrium.
- We present a theory of *molecular* RET with an arbitrary number of field variables by using the method of closure based on both the maximum entropy principle and the entropy principle. And we prove that two closures are equivalent.
- Recent results in respect of mixtures of gases with multi-temperature are presented together with a natural definition of the average temperature.
- Qualitative analysis of the differential system is done by taking into account the fact that, due to the convexity of the entropy, there exists a privileged field (*main field*) such that the system becomes symmetric hyperbolic. The existence of the global smooth solution and the convergence to equilibrium are also studied.
- We summarize open problems and try to provide an outlook on future studies.

This book is designed for applied mathematicians, physicists, and engineers. We hope that the methodology presented can offer powerful models for possible applications to, say, re-entry of a satellite into the atmosphere of a planet, semiconductors, and nano-scale phenomena.

Bologna, Italy
Nagoya, Japan
May 2015

Tommaso Ruggeri
Masaru Sugiyama

Acknowledgements

We are particularly indebted to our collaborators, Takashi Arima and Shigeru Taniguchi, who read the manuscript, offered critical comments, and continuously helped us to improve almost all parts of the book.

Our perspectives on Extended Thermodynamics have been strongly influenced by the pioneering works of Ingo Müller, who introduced both of us to this beautiful research field.

The topic addressed in this book spans three different research areas: thermomechanics of continuous media, the kinetic theory of gases, and the hyperbolic systems of balance laws. Therefore, many colleagues, co-authors, and collaborators have affected our research, either directly through joint papers and discussions or indirectly by way of their papers. We sincerely thank all of them. Although the list of their names is too long to record here, we would especially like to mention Guy Boillat, who produced fundamental works on the mathematical problems that are intimately related to this book.

We are also grateful to the Mathematical Research Institute of Oberwolfach, which gave us the opportunity to work together on the book under the Research in Pairs program in 2014.

Finally, we would like to thank Springer, and in particular the Executive Editor, Dr. Francesca Bonadei, and Dr. Francesca Ferrari, for their help.

Contents

Part II Survey of Rational Extended Thermodynamics of Monatomic Gas

Part III RET of Polyatomic Gas and Dense Gas with 14 Fields

List of Symbols

$A_{(ij)}$	Symmetric part of a tensor
$A_{<ij>}$	Deviatoric tensor
$A_{[ij]}$	Antisymmetric part of a tensor
$D = 3 + f^i$	Degrees of freedom of a molecule
$F^{\alpha_1 \alpha_2 \dots \alpha_N}$	Relativistic moment tensor of order N
F_A, F_{iA}, P_A	Densities, Fluxes and Productions with a multi-index
$F_{k_1 k_2 \dots k_N}$	Momentum-like field (or moment) of order N
$G(t)$	Acceleration amplitude in an acceleration wave
$G_{kkk_1 k_2 \dots k_M}$	Energy-like field (or moment) of order M
I	Extra variable in the distribution function of a polyatomic gas
$J_{\alpha\beta}, J_{i\alpha\beta}$	Symmetric matrices appearing in the MEP procedure
K_n	Knudsen number
M_0	Mach number
$P^{\alpha_1 \alpha_2 \dots \alpha_N}$	Relativistic production tensor of order N
$P_{k_1 k_2 \dots k_N}$	Momentum-like production of order N
Q	Collision term or quadratic form for convexity of entropy
$Q_{kkk_1 k_2 \dots k_M}$	Energy-like production term of order M
S	Dynamic structure factor, Entropy density in a Mixture
T	Absolute temperature
$T^{\alpha\beta}$	Energy-momentum tensor
Z, Ω	Partition function
Γ	Function that appears in convexity of entropy or Gamma function
Π	Dynamic pressure (nonequilibrium pressure)
Π_θ	Dynamical pressure due to the difference of temperatures
$\Pi_{k_1 k_2 \dots k_M}$	Difference between moments of order M that vanishes in the monatomic case (when $M = 0$ coincides with Π)
Σ	Entropy production
Θ_α	Diffusion temperature
α	Attenuation factor in linear waves
$\alpha = \frac{D-5}{2}$	Power exponent related to the degrees of freedom of a molecule

$\bar{\Sigma}$	Entropy production of a principal subsystem
$\bar{h}^\alpha$	Entropy four-vector of a principal subsystem
$\boldsymbol{\Phi}^i$	Non-convective flux vector
$\bullet = \frac{d}{dt}$	Material time derivative
χ	Variable of the distribution function in the MEP procedure
δ_{ij}	Kronecker's delta
$\frac{\partial}{\partial t}$	Time derivative
$\frac{\partial}{\partial x_i}$	Spatial derivative
γ	Ratio of specific heats
$\hat{\boldsymbol{\Phi}}^i$	Non-convective flux at zero velocity
$\hat{\mathbf{F}}^0$	Density vector at zero velocity
$\hat{\mathbf{F}}^i$	Flux vector at zero velocity
$\hat{\mathbf{f}}$	Production at zero velocity
$\hat{c}_v$	Dimensionless specific heat
$\hat{h}^\alpha$	Entropy four-vector at zero velocity
$\hat{h}'^\alpha$	Potential four-vector at zero velocity
κ	Thermal conductivity (heat conductivity)
λ	Characteristic velocity
$\lambda, \lambda_i, \lambda_{ij}, \mu_i$	Lagrange multipliers
$\lambda^{(k)}$	k-characteristic velocity
λ_{max}	Maximum characteristic velocity
$\mathbf{A}^D$	Deviatoric part of a tensor
$\mathbf{A}^T$	Transpose of a tensor
$\mathbf{A}^r$	Matrices that appear in the Galilean invariance
$\mathbf{C} \equiv (C_i)$	Peculiar velocity
$\mathbf{D}$	Strain velocity tensor
$\mathbf{F}^0$	Density vector
$\mathbf{F}^i$	Flux vector
$\mathbf{J}_\alpha$	Diffusion flux
$\mathbf{X}$	Matrix that dictates the velocity dependence in the Galilean invariance
$\mathbf{c} \equiv (c_i)$	Microscopic velocity
$\mathbf{d}$	Right characteristic eigenvector
$\mathbf{f}$	Production vector
$\mathbf{l}$	Left characteristic eigenvector
$\mathbf{q} \equiv (q_i)$	Heat flux
$\mathbf{t} \equiv (t_{ij})$	Stress tensor
$\mathbf{u}'$	Main field
$\mathbf{v} \equiv (v_i)$	Velocity
$\mathbf{v}'$	First block of the main field vector corresponding to conservation laws
$\mathbf{w}'$	Second block of the main field vector corresponding to balance laws
$\mathbf{x} \equiv (x_i)$	Position vector
$\mathscr{A}$	Amplitude of an acceleration wave, Affinity in the Meixner theory

$\mathscr{D}$	Thermal diffusion
$\mathscr{E}$	Internal energy in the Meixner theory
$\mathscr{G}$	Chemical potential in nonequilibrium
$\mathscr{P}$	Pressure in the Meixner theory
$\mathscr{S}$	Entropy density in the Meixner theory
$\mathscr{T}$	Temperature in nonequilibrium
$\mathfrak{r}$, $\mathfrak{r}_{\langle ij\rangle}$, $\mathfrak{s}_i$	Random forces
μ	Shear viscosity or Lagrange multiplier
$\mu^{(\bar{k})}$	$\bar{k}$-characteristic velocity of the equilibrium subsystem
∇	Nabla vector
ν	Bulk viscosity
ω	Frequency
ρ	Mass density
σ^{rr}	Radial viscous stress component
σ_{ij}	Viscous stress tensor
τ	Relaxation time
τ_S	Relaxation time for the shear (deviatoric viscous) stress tensor
τ_Π	Relaxation time for the dynamical pressure
$\tau_\alpha, \mathbf{m}_\alpha, e_\alpha$	Production terms due to the interaction between constituents
τ_q	Relaxation time for the heat flux
ε	Internal energy density
φ_i or $\hat{h}^i$	Entropy flux at zero velocity
$\|\| \cdot \|\|_2$	L^2 Norm
ξ	Internal variable in the Mexiner theory
c_0	Equilibrium sound velocity
c_v	Specific heat at constant volume
f	Distribution function
f^i	Internal degrees of freedom of a molecule
f_E	Equilibrium distribution function for a polyatomic gas
f_M	Maxwellian distribution function
f_{G}	Grad distribution function
g	Chemical potential
$g_{\alpha\beta}$	Metric tensor
h^0 or h	Entropy density
h^α	Entropy four-vector
h^i	Entropy flux
h'^α	Potential four-vector
k	Nonequilibrium entropy density or complex wave number
k_B	Boltzmann constant
m	Atomic mass
p	Equilibrium pressure
p_α	Four-momentum
q^r	Radial heat flux component
s	Shock velocity or Entropy density in equilibrium
t	Time

t_{cr}	Critical time
$u', u'_{\alpha}, u'_{\alpha\beta}$	Relativistic main field
u_{α}	Diffusion velocity
v_{ph}	Phase velocity
$[[\cdot]]$	Jump across wave front

Chapter 1
Introduction

Abstract Rational extended thermodynamics (RET), which is compatible with the kinetic theory of gases and is also closely related to the mathematical theory of hyperbolic systems, has been a successful theory of rarefied monatomic gases to explain nonequilibrium phenomena such as light scattering, sound dispersion, shock wave structure, nonequilibrium radiation. These subjects were treated in detail in the book "Rational Extended Thermodynamics" by Ingo Müller and Tommaso Ruggeri.

The other subjects remained to be explored in RET were those of polyatomic gases, of dense gases in general, and of mixtures of gases with multi-temperature. This is the aim of the present book to discuss such new RET.

In this chapter, before going into the details, we give some introductory perspective on these subjects starting with a short history of nonequilibrium thermodynamics.

The new RET theory includes the 14-field theory of dense gases that reduces to the classical Navier-Stokes Fourier theory in the parabolic limit (*Maxwellian iteration*), to the singular limit of a monatomic gas with 13 fields, and to the subsystem with 6 fields. The 6-field theory is the minimal dissipative system, where the dissipation is only due to the dynamic pressure, after the Euler system of perfect fluids. The concept of nonequilibrium temperature is also discussed.

For rarefied polyatomic gases, we discuss a theory of *molecular* RET with arbitrary number of field variables by using the methods of closure based on both the maximum entropy principle and the entropy principle. It can be proved that the two methods are equivalent to each other.

Several applications of the new RET theory are reviewed as well.

Moreover we discuss the theory of a mixture of gases with multi-temperature, i.e., a mixture in which each constituent has its own temperature.

In the new approach, the qualitative analysis of the differential system is also done by taking into account the fact that, due to the convexity of the entropy, there exists a privileged field (*main field*) such that the system becomes symmetric hyperbolic. Existence of global smooth solutions and convergence to equilibrium are also discussed.

T. Ruggeri, M. Sugiyama, *Rational Extended Thermodynamics beyond the Monatomic Gas*, DOI 10.1007/978-3-319-13341-6_1

1.1 Thermodynamics of Irreversible Processes and the Laws of Navier-Stokes and Fourier

1.1.1 Dawn of Thermodynamics

The nineteenth century witnessed the birth of thermodynamics and its great developments. Thermodynamics was expected to address the demands of the new industrial mode of production with revolutionary new technology such as a steam engine that transforms heat into work. Many scientists made fundamental contributions to this new discipline: Carnot, Mayer, Joule, Helmholtz, Clausius, Kelvin (Thomson), Maxwell, Gibbs, Duhem, Nernst, and many others.

Three fundamental laws of thermodynamics have been established. The first is the law of conservation of energy for thermodynamic systems. The second law selects appropriate evolution of a system by introducing a physical quantity called entropy. The third law is concerned with the value of the entropy at zero absolute temperature. On the basis of these laws, theory of thermodynamics has been successfully constructed, and has been utilized in various theoretical fields and diverse practical applications. See, for examples the text books [1–4].

Thermodynamics, in particular, the second law of thermodynamics raised the following questions: Can thermodynamics be traced back, in some way, to mechanics? Thermodynamics predicts the so-called arrow of time (irreversibility), but the equations of motion for molecules in mechanics have a strict symmetry with respect to the time reversal (reversibility). If the thermodynamics is a branch of mechanics, how does the irreversibility come out? Boltzmann tried to answer to this fundamental problem by using the newly developed H-theorem in the kinetic theory. However, it was revealed that Boltzmann had introduced implicitly an assumption, sometimes called assumption of molecular chaos, into his proof of the H-theorem and had broken the time reversal symmetry. From a mathematical point of view, the study of the limiting processes has been made in order to deduce the laws of motion of continua from the atomistic level. This is the 6th Hilbert problem, to which some authors tried to give answers [5–7].

Thermodynamic studies focusing especially on irreversible processes were made also in the nineteenth century. For example, Fourier's law of heat conduction, Navier-Stokes' law of viscous flow, Fick's law of mass diffusion, Ohm's law of electrical conduction, and thermo-electrical coupling effects such as the Seebeck effect and the Peltier effect were found empirically.

From the pioneering works of Onsager, Eckart, Meixner, Prigogine and others, thermodynamics of irreversible processes (TIP) emerged as a systematic nonequilibrium thermodynamic theory in the middle of the last century. For its details, see the monumental book by de Groot and Mazur [8]. The Navier-Stokes Fourier theory for viscous and heat-conducting fluids [8, 9], for example, can be regarded as one of the typical TIP theories. TIP has been useful in various practical situations involving nonequilibrium processes such as mass diffusion, viscous flow, heat conduction, chemical reaction, electrical conduction. If we want to analyze nonequilibrium phenomena keeping its validity range in mind, TIP continues to be useful.

In this section, we review briefly the theoretical structure of TIP through studying one-component viscous and heat-conducting fluids. The empirical Navier-Stokes and Fourier laws are naturally derived from TIP. We also point out its validity range.

1.1.2 TIP of One-Component Viscous and Heat-Conducting Fluids

In thermodynamics of continuous media, time-evolution of relevant densities is expressed in the *balance form*: Let $\mathbf{F}^0(\mathbf{x}, t)$ be a R^N-vector of the densities depending on the space variable $\mathbf{x} \equiv (x_i) \in \Omega \subset R^3$ and the time $t \in R^+$, then we have

$$\frac{d}{dt}\int_\Omega \mathbf{F}^0 \, d\Omega = -\int_\Sigma \boldsymbol{\Phi}^i n_i d\Sigma + \int_\Omega \mathbf{f} d\Omega, \tag{1.1}$$

where the first integral on the right-hand side represents the fluxes $\boldsymbol{\Phi}^i \in R^N$ $(i = 1, 2, 3)$ through the surface Σ with unit outward normal vector $\mathbf{n} \equiv (n_i)$, while the second integral represents the productions.[1] Under suitable regularity assumption, the system (1.1) can be put in the local form:

$$\frac{\partial \mathbf{F}^0}{\partial t} + \frac{\partial \mathbf{F}^i}{\partial x^i} = \mathbf{f}, \qquad \mathbf{F}^i = \mathbf{F}^0 v_i + \boldsymbol{\Phi}^i, \tag{1.2}$$

where $\mathbf{v} \equiv (v_i)$ is the velocity.

Hereafter, let us study a one-component fluid as a typical example. In this case, five independent fields: the mass density $\rho(\mathbf{x}, t)$, the velocity $\mathbf{v}(\mathbf{x}, t)$, and the temperature $T(\mathbf{x}, t)$ are the unknowns that we want to determine. These fields should satisfy the conservation laws of mass, momentum, and energy:

$$\begin{aligned}
&\frac{\partial \rho}{\partial t} + \frac{\partial \rho v_i}{\partial x_i} = 0, \\
&\frac{\partial \rho v_j}{\partial t} + \frac{\partial}{\partial x_i}\left(\rho v_i v_j - t_{ij}\right) = 0, \\
&\frac{\partial}{\partial t}\left(\frac{\rho v^2}{2} + \rho\varepsilon\right) + \frac{\partial}{\partial x_i}\left\{\left(\frac{\rho v^2}{2} + \rho\varepsilon\right) v_i - t_{ik} v_k + q_i\right\} = 0.
\end{aligned} \tag{1.3}$$

[1] We adopt the summation convention, i.e., we take summation over repeated indices. $i, j = 1, 2, 3$.

This is a particular case of (1.2) with

$$\mathbf{F}^0 = \begin{pmatrix} \rho \\ \rho v_j \\ \frac{1}{2}\rho v^2 + \rho\varepsilon \end{pmatrix}, \quad \mathbf{F}^i = \begin{pmatrix} \rho v_i \\ \rho v_i v_j - t_{ij} \\ \left(\frac{1}{2}\rho v^2 + \rho\varepsilon\right) v_i - t_{ik}v_k + q_i \end{pmatrix}, \quad \mathbf{f} = \begin{pmatrix} 0 \\ 0_j \\ 0 \end{pmatrix}, \tag{1.4}$$

where $i,j = 1,2,3$. And t_{ij}, ε and q_i are, respectively, the symmetric stress tensor, the specific internal energy, and the heat flux.

As usual the stress tensor $t_{ij} = -p\delta_{ij} + \sigma_{ij}$ can be decomposed into an isotropic part and a deviatoric part (symmetric traceless part) denoted by the brackets $\langle\ \rangle$:

$$t_{ij} = -(p + \Pi)\delta_{ij} + \sigma_{\langle ij\rangle}, \tag{1.5}$$

where p is the equilibrium pressure, δ_{ij} is the Kronecker's delta, σ_{ij} is the viscous stress tensor and $\Pi = -\sigma_{ll}/3$ is the so-called dynamic pressure (nonequilibrium pressure). The deviatoric tensor $\sigma_{\langle ij\rangle}$ is called the shear stress tensor. As is well known the previous system (1.3) can be rewritten, for classical solutions, in the form:

$$\begin{aligned} \dot{\rho} + \rho\frac{\partial v_j}{\partial x_j} &= 0, \\ \rho\dot{v}_i - \frac{\partial t_{ij}}{\partial x_j} &= 0, \\ \rho\dot{\varepsilon} - t_{ij}\frac{\partial v_i}{\partial x_j} + \frac{\partial q_i}{\partial x_i} &= 0, \end{aligned} \tag{1.6}$$

where a dot on a quantity denotes the material time derivative operator:

$$\frac{\partial}{\partial t} + v_i\frac{\partial}{\partial x_i}.$$

The specific internal energy ε and the pressure p are considered to be functions of ρ and T that are prescribed by the thermal and caloric equations of state of equilibrium thermodynamics: $p = p(\rho, T), \varepsilon = \varepsilon(\rho, T)$.

In order to close the system (1.6), we need the constitutive relations of σ_{ij} and q_i in terms of the independent fields ρ, v_i and T. In TIP, such relations are derived in a heuristic manner from the entropy balance equation that is based on the Gibbs equation of equilibrium thermodynamics[2]:

$$T\,ds = d\varepsilon - \frac{p}{\rho^2}\,d\rho \quad \Longleftrightarrow \quad \dot{s} = \frac{1}{T}\left(\dot{\varepsilon} - \frac{p}{\rho^2}\dot{\rho}\right), \tag{1.7}$$

[2]The adoption of an equilibrium thermodynamic relation in a small volume element is sometimes called " assumption of local equilibrium". As for the validity criterion of the assumption, see, for example, Pottier [10].

where s is the specific entropy. Elimination of $\dot{\varepsilon}$ and $\dot{\rho}$ by using (1.6) gives, after some calculations, the relation:

$$\rho\dot{s} + \frac{\partial}{\partial x_i}\left(\frac{q_i}{T}\right) = \frac{1}{T}\sigma_{\langle ij\rangle}\frac{\partial v_{\langle i}}{\partial x_{j\rangle}} - \frac{1}{T}\Pi\frac{\partial v_i}{\partial x_i} - \frac{1}{T^2}q_i\frac{\partial T}{\partial x_i}, \tag{1.8}$$

or equivalently:

$$\frac{\partial \rho s}{\partial t} + \frac{\partial}{\partial x_i}\left(\rho s v_i + \frac{q_i}{T}\right) = \frac{1}{T}\sigma_{\langle ij\rangle}\frac{\partial v_{\langle i}}{\partial x_{j\rangle}} - \frac{1}{T}\Pi\frac{\partial v_i}{\partial x_i} - \frac{1}{T^2}q_i\frac{\partial T}{\partial x_i}, \tag{1.9}$$

which can be seen as a balance equation of the entropy. Then we may have the following interpretation:

(intrinsic) entropy flux; $\varphi_i = \frac{q_i}{T}$,
entropy production : $\Sigma = \frac{1}{T}\sigma_{\langle ij\rangle}\frac{\partial v_{\langle i}}{\partial x_{j\rangle}} - \frac{1}{T}\Pi\frac{\partial v_i}{\partial x_i} - \frac{1}{T^2}q_i\frac{\partial T}{\partial x_i}$.

The entropy production is a sum of products of:

Dissipative fluxes	**Thermodynamic forces**	
shear stress $\sigma_{\langle ij\rangle}$	deviatoric velocity gradient	$\frac{1}{T}\frac{\partial v_{<i}}{\partial x_{j>}}$,
dynamic pressure Π	divergence of velocity	$-\frac{1}{T}\frac{\partial v_i}{\partial x_i}$,
heat flux q_i	temperature gradient	$-\frac{1}{T^2}\frac{\partial T}{\partial x_i}$.

1.1.3 Laws of Navier-Stokes and Fourier

From the second law of thermodynamics, the entropy production must be non-negative. Assuming linear relations between the dissipative fluxes and the thermodynamic forces, we have the constitutive equations (phenomenological equations) of the type:

$$\begin{aligned} \sigma_{\langle ij\rangle} &= 2\mu\frac{\partial v_{<i}}{\partial x_{j>}} && \mu \geq 0, \\ \Pi &= -\nu\frac{\partial v_i}{\partial x_i} && \nu \geq 0, \\ q_i &= -\kappa\frac{\partial T}{\partial x_i} && \kappa \geq 0. \end{aligned} \tag{1.10}$$

These are known as the laws of Navier-Stokes and Fourier with μ and ν being the shear and bulk viscosities and κ the thermal conductivity. All of these coefficients may be functions of ρ and T.

Along with the thermal and caloric equations of state, (1.10) is adopted as the constitutive equations of TIP, and the differential system (1.3) is closed, i.e., five equations for five unknowns.

1.1.4 Parabolic Structure and the Prediction of Infinite Speed of Waves in TIP

TIP raises the problem of infinite speed of waves. Historically, in order to avoid this unphysical prediction of TIP, extended thermodynamics (ET) [11] was conceived.

Let us, therefore, discuss firstly this problem through studying heat conduction in a fluid at rest with a constant mass density ρ (heat conduction in a rigid heat conductor). Inserting the Fourier law $(1.10)_3$ into the energy balance equation $(1.6)_3$, we obtain the equation of heat conduction:

$$\frac{\partial T}{\partial t} = \mathscr{D} \Delta T, \tag{1.11}$$

where

$$\mathscr{D} = \frac{\kappa}{\rho c_v} \quad \text{and} \quad c_v = \left(\frac{\partial \varepsilon}{\partial T}\right)_\rho$$

are, respectively, the thermal diffusion coefficient and the specific heat at constant volume. We assume, for simplicity, that $\mathscr{D}$ is constant.

The solution of (1.11) in an unbounded domain with an initial value $T(\mathbf{x}, 0)$ is given by

$$T(\mathbf{x}, t) = \frac{1}{(4\pi \mathscr{D} t)^{3/2}} \int_{-\infty}^{\infty} T(\mathbf{y}, 0) e^{-\frac{(\mathbf{y}-\mathbf{x})^2}{4\mathscr{D} t}} d\mathbf{y}. \tag{1.12}$$

We notice that $T(\mathbf{x}, t)$ is nonzero for any $\mathbf{x}$ if $t > 0$ even though the initial value $T(\mathbf{x}, 0)$ is nonzero only in a bounded domain. This phenomenon has been sometimes called a paradox, because the temperature propagates with infinitely large speed. From a mathematical point of view, this is due to the parabolic character of the basic equation (1.11).

The assertion of infinite speed is, of course, beyond the validity range of TIP. We cannot describe properly such a rapid change by TIP because TIP is based on the local equilibrium assumption. On the other hand, it is well known that equations (1.11) and (1.12) have been utilized quite successfully in various practical situations.

Indeed, if we take carefully its validity range into account, and if we do not care about its unphysical predictions, we would have useful results from a practical point of view.

There is, however, a situation where the infinite speed should be avoided strictly. It is a relativistic thermodynamic case where propagation speed of a wave should be less or equal to the light speed.

How can we construct a new thermodynamic theory that predicts only finite speed of waves by generalizing TIP?

1.1.5 Cattaneo Equation

1.1.5.1 Classical Cattaneo Equation

A heuristic argument made by Cattaneo [12] gives us an interesting suggestion for the question above.

The heat flux vector $\mathbf{q}$ is assumed, by Cattaneo, to be proportional not only to the gradient of temperature ∇T but also to the gradient of the time derivative of the temperature $\nabla \dot{T}$ in such a way that

$$\mathbf{q} = -\kappa \left(\nabla T - \tau \nabla \dot{T} \right).$$

Assuming that the relaxation time τ is very small, he somehow arrived at the following equation, which is known as the (classical) Cattaneo equation:

$$\tau \dot{\mathbf{q}} + \mathbf{q} = -\kappa \nabla T, \tag{1.13}$$

which reduces to the Fourier law when $\tau \to 0$. Combining (1.13) with the energy equation $(1.6)_3$, we obtain

$$\tau \ddot{T} + \dot{T} = \mathscr{D} \Delta T. \tag{1.14}$$

This equation, called *telegraph equation*, is hyperbolic provided $\tau > 0$. It predicts the propagation of heat pulses with finite speed:

$$V = \sqrt{\frac{\mathscr{D}}{\tau}}.$$

If the relaxation time τ is negligibly small, the Eq. (1.14) reduces to the heat conduction equation (1.11). However, we should be careful about this because it is a singular limit process from hyperbolic to parabolic equation.

What points should we learn from Cattaneo's heuristic argument for constructing the new thermodynamic theory ET? The essential points are the following two:

1. Dissipative fluxes, such as the heat flux, should be introduced as independent variables into the theory in addition to the usual thermodynamic quantities in TIP.
2. Relaxation processes of these dissipative fluxes should be properly taken into consideration in order to make the theory hyperbolic.

1.1.5.2 Generalized Cattaneo Equation and the Second Sound

Although the Cattaneo equation was obtained by the heuristic argument, it has been a popular model in applications concerning rigid heat conductors. Huge literature exists on this subject. See, for example, the review papers of Joseph and Preziosi [13, 14] and the recent book of Straughan [15].

In reality, however, it is simple to verify that the classical Cattaneo equation with internal energy depending only on the temperature T is compatible with the entropy principle if and only if the ratio between the relaxation time and the heat conductivity is proportional to T^2. In fact, Coleman, Fabrizio and Owen [16] noticed for the first time that, in order to be compatible with the entropy principle for a generic relaxation time and the heat conductivity, the internal energy should have the dependence not only on T but also on the square of $\mathbf{q}$. This assertion was, however, criticized by Morro and Ruggeri [17], who revisited a previous paper of Ruggeri [18]. And they proposed a more general model than the Cattaneo model by adopting a system of equations of balance type. This adoption overcomes the difficulty and is compatible with the internal energy depending only on the temperature T. Therefore the Cattaneo equation is physically meaningful only near equilibrium where only linear terms in nonequilibrium variables are present in the field equations.

The importance of hyperbolic equations has its experimental evidence in the so-called second sound, i.e., a heat wave. The second sound was observed first in liquid helium at low temperatures [19] and then was expected in crystals [20]. Recently sophisticated experiments were made in [21] observing the propagation of a heat wave in an ultra-cold quantum gas. In the generalized Cattaneo model, we can discover interesting new phenomena in crystals such as the passage from hot to cold shocks at a critical temperature [22, 23], the behavior of a simple wave and shock formation [24]. Finally we recall that, starting from the pioneering papers of Guyer and Krumhansl [25, 26], an interesting approach to this problem on the basis of the phonon-gas theory was taken by Dreyer and Struchtrup [27] and by Larecki and Banach [28, 29].

1.2 First Tentative of Extended Thermodynamics and Rational Extended Thermodynamics

The first approach to ET, made by Müller [30] in a classical framework and by Müller (PHD thesis) and independently by Israel [31] in a relativistic context, is based on the modification of the Gibbs equation (1.7) incorporating also the effects of dissipative fluxes. This point of view has been adopted by several authors and is the starting point of Extended Irreversible Thermodynamics (EIT), which has gained popularity through the book of Jou et al. [32].

The introduction of dissipative fluxes as independent variables together with the usual thermodynamic quantities into the theory has a deep implication. It is evident that ET has gone beyond the local equilibrium assumption because dissipative fluxes play an essential role to characterize a nonequilibrium state in the theory.

The applicability range of ET becomes, in this way, to be wider than that of TIP. In other words, ET is applicable to highly nonequilibrium phenomena where the local equilibrium assumption is no longer valid. Examples of such phenomena are shock waves, micro- and nano-flows, second sounds, light scattering and so on. Thus ET is expected to be useful not only for relativistic cases but also for such non-relativistic cases.

Such an approach was, however, criticized by Ruggeri [33] because the entropy production depends strongly on the choice of the entropy flux, and there appear different field equations for a different entropy flux. Moreover the differential system is not a priori in the form of balance laws. This implies, from a mathematical point of view, that it is not possible to define weak solutions and therefore impossible to study, in particular, shock waves.

Later a revision of ET was proposed by Liu and Müller [34] in a classic context, and by Liu, Müller and Ruggeri in a relativistic framework [35]. This new approach was named Rational Extended Thermodynamics (RET) and the main results obtained at that time were summarized in the two editions of the book by Müller and Ruggeri [11, 36].

The mathematical framework of RET consists of a hierarchy of balance laws. The same hierarchical structure can be seen in the system of moment equations in the kinetic theory with the truncation at some arbitrary order of moments. The RET theory, in this respect, resembles to the theory of moment equations. However, the closure of RET is achieved by means of the universal principles of physics: *objectivity principle*, *entropy principle*, and *principle of causality and stability*.

1.3 Rational Thermodynamics and the Entropy Principle

Before we present the details of Rational Extended Thermodynamics, we need to recall the approach of the so-called *Rational Thermodynamics* (RT). RT is mainly due to the Truesdell school [37] and has the starting point in the fundamental paper of Coleman and Noll [38] where they reinterpreted the second law of thermodynamics as a selection rule for constitutive equations. We have seen that the basic system of equations of continuous media is in the form of balance laws (1.1), (1.2). To close the system, constitutive equations are necessary. The idea of Coleman and Noll is to regard the entropy law:

$$\frac{\partial \rho s}{\partial t} + \frac{\partial}{\partial x_i}\left(\rho s v_i + \frac{q_i}{T}\right) \geqslant 0 \tag{1.15}$$

as a constraint for the admissible physical constitutive equations. In this way, the prescription of the arrow of time that characterizes the irreversibility is made indirectly. In fact, if we pick up some constitutive equations, we are not sure whether the second law is certainly satisfied or not. Therefore, according with this new idea, we select at the beginning the class of admissible constitutive equations that satisfy the inequality for any initial or boundary data. This was a very important observation and very useful as we will see in Sect. 2.2 where we will characterize, in more precise way, the entropy principle. We have seen before that, from (1.9), how the Navier-Stokes Fourier constitutive equations (1.10) are compatible with the entropy principle.

The limitation of the idea of Coleman and Noll is, however, in the postulation that the entropy flux is in the form of Clausius, i.e., the ratio of heat flux to temperature, q_i/T. Müller noticed with the help of kinetic considerations that this requirement is too much restrictive and he proposed to extend the entropy principle with a general entropy flux as a constitutive quantity [39].

An interesting book based on the framework of mathematical methods of Rational Mechanics and Thermodynamics is the one due to Šilhavý [40].

1.4 Other Approaches

We observe that, in addition to *Classical Irreversible Thermodynamics* (CIT or TIP), *Rational Thermodynamics* (RT), *Extended Irreversible Thermodynamics* (EIT) that we briefly summarized before and *Rational Extended Thermodynamics* (RET), there are other approaches in nonequilibrium thermodynamics with exotic acronyms.

The most popular and interesting one is GENERIC, an acronym for *General Equation for Non-Equilibrium Reversible-Irreversible Coupling* proposed by Grmela and Öttinger [41–43]. It is the general form of dynamic equation for a system with both reversible and irreversible dynamics.

Another approach proposed by Müller, Reitebuch and Weiss is the so-called *Consistent-Order Extended Thermodynamics* (COET) [44]. In this approach every nonequilibrium variable has a certain order of magnitude and the set of variables taken into account contains only variables up to a chosen order.

A different point of view of nonequilibrium thermodynamics based on a revisit of the entropy principle and the introduction of the so-called *calortropy* is the subject of a book of Eu [45].

Connection between nonequilibrium thermodynamics and variational principles is the subject of the book of Gyarmati [46].

A first partial tentative to give a comparison among different approaches of nonequilibrium theories was recently presented by Cimmelli, Jou, Ruggeri and Ván (see [47] and references cited therein).

An interesting review of nonequilibrium thermodynamics, elucidating different pathways to macroscopic equations, was published by Müller and Weiss [48].

Finally we recall that there exist parabolic-type approaches to nonequilibrium processes by using kinetic considerations: the works by Torrilon and Struchtrup [49, 50] via regularization of moment-equations and the works by Bobylev and Windfäll [51] via revisited Chapman-Enskog method.

1.5 Rational Extended Thermodynamics and the Kinetic Theory

The study of nonequilibrium phenomena in gases is particularly important. We have two complementary approaches to rarefied gases, namely the *continuum approach* and the *kinetic approach*.

The continuum model consists in the description of the system by means of macroscopic equations (e.g., fluid-dynamic equations) obtained on the basis of conservation laws and appropriate constitutive equations. A typical example is TIP. The applicability of this classical macroscopic theory is, however, inherently restricted to a nonequilibrium state characterized by a *small* Knudsen number K_n, which is a measure to what extent the gas is rarefied:

$$K_n = \frac{\text{mean free path of molecule}}{\text{macroscopic characteristic length}}.$$

The transport coefficients associated to dissipation processes are not provided by the theory except for the sign. Usually we require experimental data on the coefficients.

The approach based on the kinetic theory [52–54] postulates that the state of a gas can be described by the velocity distribution function. The evolution of the distribution function is governed by the Boltzmann equation. The kinetic theory is applicable to a nonequilibrium state characterized by a *large* K_n, and the transport coefficients naturally emerge from the theory itself. Therefore the range of the applicability of the Boltzmann equation has been limited to rarefied gases.

The RET theory, a generalization of the TIP theory, also belongs to the continuum approach but is applicable to a nonequilibrium state with larger K_n. In a sense, RET is a sort of bridge between TIP and the kinetic theory. An interesting point to be noticed is that, in the case of rarefied gases, there exists a common applicability range of the RET theory and the kinetic theory. Therefore, in such a range, the results from the two theories should be consistent with each other. Because of this, we can expect that the kinetic-theoretical considerations can motivate us to establish the mathematical structure of the RET theory.

1.5.1 Boltzmann Equation and the Moments

The kinetic theory describes a state of a rarefied gas by using the phase density (velocity distribution function) $f(\mathbf{x}, t, \mathbf{c})$, where $f(\mathbf{x}, t, \mathbf{c})d\mathbf{c}$ is the number density of (monatomic) molecules at the point $\mathbf{x}$ and time t that have velocities between $\mathbf{c}$ and $\mathbf{c} + d\mathbf{c}$. Time-evolution of the phase density is governed by the Boltzmann equation:

$$\partial_t f + c_i \, \partial_i f = Q, \tag{1.16}$$

where the right-hand side, the collision term, describes the effect of collisions between molecules. Here $\partial_t \equiv \frac{\partial}{\partial t}$ and $\partial_i \equiv \frac{\partial}{\partial x_i}$.

As is well known, almost all macroscopic thermodynamic quantities are identified as moments of the phase density:

$$F_{k_1 k_2 \cdots k_j} = \int_{\mathbb{R}^3} m f c_{k_1} c_{k_2} \cdots c_{k_j} d\mathbf{c}, \tag{1.17}$$

where m is the mass of a molecule. The moments satisfy a hierarchy of the balance laws in which the *flux* in one equation becomes the *density* in the next one:

$$\begin{aligned}
&\partial_t F + \partial_i F_i = 0 \\
&\qquad \swarrow \\
&\partial_t F_{k_1} + \partial_i F_{ik_1} = 0 \\
&\qquad \swarrow \\
&\partial_t F_{k_1 k_2} + \partial_i F_{ik_1 k_2} = P_{k_1 k_2} \\
&\qquad \swarrow \\
&\partial_t F_{k_1 k_2 k_3} + \partial_i F_{ik_1 k_2 k_3} = P_{k_1 k_2 k_3} \\
&\vdots \\
&\partial_t F_{k_1 k_2 \ldots k_n} + \partial_i F_{ik_1 k_2 \ldots k_n} = P_{k_1 k_2 \ldots k_n} \\
&\vdots
\end{aligned} \tag{1.18}$$

where

$$P_{k_1 k_2 \cdots k_j} = \int_{\mathbb{R}^3} m Q c_{k_1} c_{k_2} \cdots c_{k_j} d\mathbf{c}. \tag{1.19}$$

Taking $P_{kk} = 0$ into account, we notice that the first five equations are exactly the conservation laws, and correspond to the conservation laws (1.3) of mass, momentum and energy, respectively.

Due to the structure of the hierarchy (1.18), we obtain, from $(1.18)_2$ and the trace of $(1.18)_3$, the relation $3(p + \Pi) = 2\rho\varepsilon$. As Π is a nonequilibrium variable vanishing in equilibrium, we have

$$p = \frac{2}{3}\rho\varepsilon \quad \text{and} \quad \Pi \equiv 0. \tag{1.20}$$

Then the gas under consideration is indeed monatomic, and the dynamic pressure vanishes identically.

For the quantities defined by

$$h^0 = -k_B \int_{\mathbb{R}^3} f \log f \, d\mathbf{c}, \quad h^i = -k_B \int_{\mathbb{R}^3} f \log f \, c^i \, d\mathbf{c} \tag{1.21}$$

with k_B being the Boltzmann constant, it is possible to prove the famous H-theorem:

$$\partial_t h^0 + \partial_i h^i = \Sigma \geqslant 0. \tag{1.22}$$

This represents the balance law of entropy if we identify h^0 , h^i and Σ as the entropy density, the entropy flux, and the entropy production, respectively.

1.5.2 Closure of RET

If we truncate the hierarchy at the density with the tensorial order n, we encounter the *problem of closure* because the last flux and the productions are not in the list of the densities. The first idea of RET [11] was to consider the truncated system as a phenomenological system of continuum mechanics and then to consider the quantities out of the list as constitutive functions:

$$\begin{aligned} F_{k_1 k_2 \dots k_n k_{n+1}} &\equiv F_{k_1 k_2 \dots k_n k_{n+1}} \left(F, F_{k_1}, F_{k_1 k_2}, \dots F_{k_1 k_2 \dots k_n}\right), \\ P_{k_1 k_2 \dots k_j} &\equiv P_{k_1 k_2 \dots k_j} \left(F, F_{k_1}, F_{k_1 k_2}, \dots F_{k_1 k_2 \dots k_n}\right), \quad 2 \leqslant j \leqslant n. \end{aligned} \tag{1.23}$$

The constitutive quantities at one point and time depend on the independent fields at the same point and time, i.e., local and instantaneous. As mentioned above, according to the continuum theory, the restrictions on the constitutive equations come from the *universal principles* (see Chap. 2).

1.5.3 Macroscopic Approach of RET with 13 Fields

The first attempt of RET was the 13-field case. Thirteen is a special number because, by comparison between the first five moments and conservation laws (1.25) below, it is possible to relate, in a unique way, the 13 moments $\{F, F_i, F_{ij}, F_{lli}\}$ to the 13 fields $\{\rho, v_i, \sigma_{\langle ij\rangle}, q_i\}$. Therefore, in the case of rarefied monatomic gases, the first 13 moments have concrete physical meanings.

The balance laws in this case are given by

$$\begin{aligned}
&\partial_t F + \partial_i F_i = 0,\\
&\partial_t F_{k_1} + \partial_i F_{ik_1} = 0,\\
&\partial_t F_{k_1k_2} + \partial_i F_{ik_1k_2} = P_{k_1k_2},\\
&\partial_t F_{kkj} + \partial_i F_{kkij} = P_{kkj}.
\end{aligned} \tag{1.24}$$

The constitutive quantities are $F_{i<k_1k_2>}$, F_{ik_1kk}, $P_{<k_1k_2>}$ and P_{kkj}.

The restrictions by the universal principles—in particular the entropy principle—are so strong that, at least for processes not too far from equilibrium, the system can be completely closed. In this case, the closed system is given, by employing the usual symbols, as [11, 34]

$$\begin{aligned}
&\frac{\partial \rho}{\partial t} + \frac{\partial}{\partial x_k}(\rho v_k) = 0,\\
&\frac{\partial \rho v_i}{\partial t} + \frac{\partial}{\partial x_k}\left\{\rho v_i v_k + p\delta_{ik} - \sigma_{\langle ik\rangle}\right\} = 0,\\
&\frac{\partial}{\partial t}(\rho v^2 + 2\rho\varepsilon) + \frac{\partial}{\partial x_k}\left\{\rho v^2 v_k + 2(\rho\varepsilon + p)v_k - 2\sigma_{\langle kl\rangle}v_l + 2q_k\right\} = 0,\\
&\frac{\partial}{\partial t}\left\{\rho v_i v_j + p\delta_{ij} - \sigma_{\langle ij\rangle}\right\} +\\
&\quad + \frac{\partial}{\partial x_k}\left\{\rho v_i v_j v_k + p(v_i\delta_{jk} + v_j\delta_{ki} + v_k\delta_{ij}) - \sigma_{\langle ij\rangle}v_k - \sigma_{\langle jk\rangle}v_i - \sigma_{\langle ki\rangle}v_j\right.\\
&\qquad\left. + \frac{2}{5}(q_i\delta_{jk} + q_j\delta_{ki} + q_k\delta_{ij})\right\} = \frac{\sigma_{\langle ij\rangle}}{\tau_S},\\
&\frac{\partial}{\partial t}\left\{\rho v^2 v_i + 2\left(\rho\varepsilon + p\right)v_i - \sigma_{\langle li\rangle}v_l + 2q_i\right\} +\\
&+ \frac{\partial}{\partial x_k}\left\{\rho v^2 v_i v_k + 2\rho\varepsilon v_i v_k + p(v^2\delta_{ik} + 4v_i v_k) - \sigma_{\langle ik\rangle}v^2 - 2\sigma_{\langle li\rangle}v_l v_k - 2\sigma_{\langle lk\rangle}v_l v_i\right.\\
&\qquad\left. + \frac{4}{5}q_l v_l\delta_{ik} + \frac{14}{5}q_i v_k + \frac{14}{5}q_k v_i + 5\frac{k_B}{m}Tp\delta_{ik} - 7\frac{k_B}{m}T\sigma_{\langle ik\rangle}\right\}\\
&= -2\frac{q_i}{\tau_q} + 2\frac{\sigma_{\langle ij\rangle}}{\tau_S}v_j,
\end{aligned} \tag{1.25}$$

where τ_S and τ_q are the relaxation times, which are related to the viscosity and the heat conductivity, respectively.

The first five equations in (1.25) are the usual conservation laws of mass, momentum and energy, while the last two blocks are the balance laws for the viscous stress σ_{ij} (in the present case of monatomic gases, $\sigma_{ij} = \sigma_{\langle ij \rangle}$) and for the heat flux q_i, respectively. They reduce to the Navier-Stokes and Fourier constitutive equations when the relaxation times are small [11].

We emphasize that we have assumed the system (1.24) motivated by the kinetic theory but, after that, our procedure has been completely macroscopic. In particular, we have not relied on the fact that the $F's$ are related to the distribution function of the kinetic theory.

1.5.4 Grad Distribution

It is interesting to observe that the macroscopic universal principles give us the result that is in perfect agreement with the kinetic-theoretical result derived from the Grad distribution function [55]. The idea of Grad is to use a perturbation method based on the expansion of the distribution function in terms of the Hermite polynomials around the Maxwellian distribution f_M:

$$f_M = \frac{\rho}{m} \sqrt{\frac{m}{2\pi k_B T}}^3 e^{-\frac{m(C_1^2+C_2^2+C_3^2)}{2k_B T}},$$

where $C_i = c_i - v_i$ is the peculiar velocity. He obtained the distribution function:

$$f_{\mathrm{G}} = f_M \left\{ 1 - \frac{1}{2p} \sigma_{\langle ij \rangle} \left(\frac{m}{k_B T} C_i C_j - \delta_{ij} \right) - \frac{1}{p} \frac{m}{k_B T} q_i C_i \left(1 - 5 \frac{m}{k_B T} C^2 \right) \right\} . \tag{1.26}$$

Inserting f_{G} into the definition of the moments (1.17), we can accomplish the closure, and have the same result as that derived from the method of RET (1.25). This is the first success of RET because it is proved clearly that both the macroscopic approach and the kinetic Grad approach give the same result!

1.5.5 Closure via the Maximum Entropy Principle and Molecular RET of Monatomic Gases

The 13-moment theory has extensively shown its superiority over the Navier-Stokes Fourier theory. However, in some situations, for examples, high-frequency sound waves, light scattering with large scattering angle, shock waves of large Mach number, even the 13-moment theory can not provide satisfactory results. In order to remedy this difficulty a larger number of moments are required. RET can provide a coherent theoretical framework for building *theories with larger number of moments* that give results in excellent agreement with experiments. In this sense, RET is the theory of theories. But, usually, it is too difficult to adopt the pure continuum approach for a system with such a large number of field variables.

Therefore it is necessary to recall that the field variables are the moments of a distribution function truncated at some order. And then the closure of the balance equations of the moments, which is known as the *maximum entropy principle* (MEP), should be introduced. This is the procedure of the so-called *molecular extended thermodynamics* (molecular RET). The principle of maximum entropy has its root in statistical mechanics. It is developed by Jaynes in the context of the theory of information basing on the Shannon entropy [56, 57].

Nowadays the importance of MEP is recognized fully due to the numerous applications in many fields, for example, in the field of computer graphics.

MEP states that the probability distribution that represents the current state of knowledge in the best way is the one with the largest entropy. Another way of stating this is as follows: take precisely stated prior data or testable information about a probability distribution function. Then consider the set of all trial probability distributions that would encode the prior data. Of those, one with maximal information entropy is the proper distribution, according to this principle.

Concerning the applicability of MEP in nonequilibrium thermodynamics, this was originally motivated by the following two reasons: One is the similarity between the field equations in ET and the moment equations. The other is the observation made by Kogan [58] that Grad's distribution function maximizes the entropy. The MEP was proposed in RET for the first time by Dreyer [59].

In this way the 13-moment theory can be obtained in three different ways: RET, Grad, and MEP. A remarkable point is that all closures are equivalent to each other!

The MEP procedure was then generalized by Müller and Ruggeri to the case of any number of moments [36], and later proposed again and popularized by Levermore [60]. The complete equivalence between the entropy principle and the MEP was finally proved by Boillat and Ruggeri [61]. Later MEP was formulated also in a quantum-mechanical context [62].

A comprehensive review of the state-of-the-art of the maximum entropy principle in both classical (MEP) and quantal (QMEP) formulations was presented by Trovato and Reggiani [63]. Interesting approach of MEP to radiation was given by Larecki and Banach [64, 65]. Some mathematical delicate questions concerning the domain of validity of the MEP was given in a series of papers by Junk and co-workers [66–68].

The details on MEP in RET will be given in Chap. 4.

1.6 New Approach to Polyatomic Gas and Dense Gas

Unfortunately the previous RET theory, being strictly connected with the kinetic theory, suffers from nearly the same limitations as the Boltzmann equation. Indeed, the previous RET is valid only for rarefied monatomic gases, where the specific internal energy ε and the pressure p are connected by the relation $2\rho\varepsilon = 3p$, and the dynamic pressure Π vanishes identically.

In the case of polyatomic gases, on the other hand, the rotational and vibrational degrees of freedom of a molecule, which are not present in monatomic gases, come into play [69], and in the case of dense gases, as the average distance between the constituent molecules is finite, the interaction between the molecules cannot be neglected. From a mathematical standpoint, these effects are responsible for intrinsic changes in the structure of the system of field equations. A simple hierarchy of field equations as in the case of monatomic gases is no longer valid. In particular, the internal specific energy is no longer related to the pressure in a simple way.

1.6.1 Macroscopic Approach with 14 Fields

After some pioneering works [70, 71], a *14-field* RET theory for dense gases and for rarefied polyatomic ones has recently been developed by Arima, Taniguchi, Ruggeri and Sugiyama [72]. This theory adopts two parallel hierarchies (binary hierarchy) for the independent fields: the mass density, the velocity, the internal energy, the shear stress, the dynamic pressure and the heat flux. One hierarchy consists of balance equations for the mass density, the momentum density and the momentum flux (*momentum-like* hierarchy), and the other one consists of balance equations for the energy density and the energy flux (*energy-like* hierarchy):

$$\begin{aligned}
&\partial_t F + \partial_i F_i = 0, && \\
&\partial_t F_{k_1} + \partial_i F_{ik_1} = 0, && \\
&\partial_t F_{k_1k_2} + \partial_i F_{ik_1k_2} = P_{k_1k_2}, && \partial_t G_{kk} + \partial_i G_{ikk} = 0, \\
& && \partial_t G_{kkj} + \partial_i G_{kkij} = Q_{kkj}.
\end{aligned} \tag{1.27}$$

These hierarchies cannot merge with each other in contrast to the case of rarefied monatomic gases because the specific internal energy (the intrinsic part of the energy density) is no longer related to the pressure (one of the intrinsic parts of the momentum flux).

By means of the closure procedure of the RET theory, the constitutive equations are determined explicitly by the thermal and caloric equations of state. For example, let us consider the particular case of rarefied polyatomic gases with the thermal and caloric equations of state given by

$$p = \frac{k_B}{m}\rho T \quad \text{and} \quad \varepsilon = \frac{D}{2}\frac{k_B}{m}T, \qquad (D = 3 + f^i) \tag{1.28}$$

where D is related to the degrees of freedom of a molecule given by the sum of the space dimension 3 for the translational motion and the contribution from the internal degrees of freedom $f^i (\geqslant 0)$. For monatomic gases, $D = 3$.

Then, as will be explained in Chap. 5, we obtain the closed system as follows:

$$
\begin{aligned}
&\frac{\partial \rho}{\partial t} + \frac{\partial}{\partial x_k}(\rho v_k) = 0, \\
&\frac{\partial \rho v_i}{\partial t} + \frac{\partial}{\partial x_k}\left\{\rho v_i v_k + (p+\Pi)\delta_{ik} - \sigma_{\langle ik\rangle}\right\} = 0, \\
&\frac{\partial}{\partial t}(\rho v^2 + 2\rho\varepsilon) + \frac{\partial}{\partial x_k}\left\{\rho v^2 v_k + 2(\rho\varepsilon + p + \Pi)v_k - 2\sigma_{\langle kl\rangle}v_l + 2q_k\right\} = 0, \\
&\frac{\partial}{\partial t}\left\{\rho v_i v_j + (p+\Pi)\delta_{ij} - \sigma_{\langle ij\rangle}\right\} + \\
&\quad + \frac{\partial}{\partial x_k}\left\{\rho v_i v_j v_k + (p+\Pi)(v_i\delta_{jk} + v_j\delta_{ki} + v_k\delta_{ij}) - \sigma_{\langle ij\rangle}v_k - \sigma_{\langle jk\rangle}v_i - \sigma_{\langle ki\rangle}v_j + \right. \\
&\qquad \left. + \frac{2}{D+2}(q_i\delta_{jk} + q_j\delta_{ki} + q_k\delta_{ij})\right\} = -\frac{\Pi\delta_{ij}}{\tau_\Pi} + \frac{\sigma_{\langle ij\rangle}}{\tau_S}, \\
&\frac{\partial}{\partial t}\left\{\rho v^2 v_i + 2\left(\rho\varepsilon + p + \Pi\right)v_i - 2\sigma_{\langle li\rangle}v_l + 2q_i\right\} + \\
&\quad + \frac{\partial}{\partial x_k}\left\{\rho v^2 v_i v_k + 2\rho\varepsilon v_i v_k + (p+\Pi)(v^2\delta_{ik} + 4v_i v_k) - \sigma_{\langle ik\rangle}v^2 \right. \\
&\qquad - 2\sigma_{\langle li\rangle}v_l v_k - 2\sigma_{\langle lk\rangle}v_l v_i \\
&\qquad + \frac{4}{D+2}q_l v_l \delta_{ik} + \frac{2D+8}{D+2}q_i v_k + \frac{2D+8}{D+2}q_k v_i + \frac{k_B}{m}T\left[(D+2)p\right. \\
&\qquad \left. + (D+4)\Pi\right]\delta_{ik} - \\
&\qquad \left. - \frac{k_B}{m}T(D+4)\sigma_{\langle ik\rangle}\right\} = -2\frac{q_i}{\tau_q} - 2\left(\frac{\Pi\delta_{ij}}{\tau_\Pi} - \frac{\sigma_{\langle ij\rangle}}{\tau_S}\right)v_j,
\end{aligned}
\tag{1.29}
$$

where τ_S, τ_Π, and τ_q are the relaxation times . The Navier-Stokes Fourier theory is contained in the present theory as a limit of small relaxation times (the Maxwellian iteration [11, 73]) as shown in [72] (see also [74]). Details will be explained in Chap. 5, Sect. 5.3.6.

The validity of the theory has been confirmed by comparing its predictions to experimental data for rarefied polyatomic gases in the cases of ultrasonic waves [75, 76], light scattering [77] and shock waves [78, 79]. The application of the theory to heat conduction was considered as well [80]. See Part IV for details.

1.6.2 Singular Limit from Polyatomic to Monatomic Gas

Let us consider the limiting process from polyatomic to monatomic rarefied gases when we let D approach 3 (i.e., $D \to 3$) from above, where D is assumed to be a continuous variable. The limit is a singular in the sense that the system for rarefied polyatomic gases with 14 independent variables seems to converge to the system with only 13 independent variables for rarefied monatomic gases. What happened to the remaining one equation?

It is possible to prove that the 14 equations and the solutions of the system (1.29) converge to the 13 equations and solutions of monatomic gas (1.25) with $\Pi = 0$, respectively, provided that the initial data are chosen in such a way that the data are compatible with a monatomic gas, i.e., $\Pi(\mathbf{x}, 0) = 0$ [81]. Details will be given in Sect. 5.6.

1.6.3 MEP Closure and the Molecular Approach for the **14**-*Moment Theory*

Concerning the kinetic counterpart, a crucial step towards the development of the theory of rarefied polyatomic gases was made by Borgnakke and Larsen [82]. The distribution function is assumed to depend on an additional continuous variable representing the energy of the internal modes of a molecule in order to take into account the exchange of energy (other than translational one) in binary collisions. This model was initially used for Monte Carlo simulations of polyatomic gases, and later it was applied to the derivation of the generalized Boltzmann equation by Bourgat, Desvillettes, Le Tallec and Perthame [83].

As a consequence of the introduction of one additional parameter, the velocity distribution function $f(t, \mathbf{x}, \boldsymbol{c}, I)$ is defined on the extended domain $[0, \infty) \times R^3 \times R^3 \times [0, \infty)$. Its rate of change is determined by the Boltzmann equation which has the same form as the one of monatomic gases (1.16) but the collision integral $Q(f)$ takes into account the influence of the internal degrees of freedom through the collisional cross section.

Recently Pavić, Ruggeri and Simić proved [84] that, by means of the maximum entropy principle, the kinetic model for rarefied polyatomic gases presented in [82] and [83] yields appropriate macroscopic balance laws. This is a natural generalization of the classical procedure of MEP from monatomic gases to polyatomic gases. They considered the case of 14 moments, and showed the complete agreement with the binary hierarchy (1.27). The moments are defined by

$$\begin{pmatrix} F \\ F_{i_1} \\ F_{i_1 i_2} \end{pmatrix} = \int_{\mathbb{R}^3} \int_0^\infty m \begin{pmatrix} 1 \\ c_{i_1} \\ c_{i_1} c_{i_2} \end{pmatrix} f(t, \mathbf{x}, \boldsymbol{c}, I)\, \varphi(I)\, dI\, d\boldsymbol{c},$$

$$\begin{pmatrix} G_{pp} \\ G_{ppk_1} \end{pmatrix} = \int_{\mathbb{R}^3} \int_0^\infty m \begin{pmatrix} c^2 + 2\frac{I}{m} \\ \left(c^2 + 2\frac{I}{m}\right) c_{k_1} \end{pmatrix} f(t, \mathbf{x}, \boldsymbol{c}, I)\, \varphi(I)\, dI\, d\boldsymbol{c}. \tag{1.30}$$

The weighting function $\varphi(I)$ is determined in such a way that it recovers the caloric equation of state in equilibrium for polyatomic gases. It can be shown that $\varphi(I) = I^\alpha$ leads to an appropriate caloric equation (1.28) provided that

$$\alpha = \frac{D-5}{2}. \tag{1.31}$$

The important feature of the nonequilibrium distribution obtained with MEP is its compatibility with the distribution obtained by Mallinger using the Grad procedure for diatomic molecules ($\alpha = 0$) [85]. Therefore, also for rarefied polyatomic gases, the three closure procedures (RET, MEP and Grad) give the same result!

1.6.4 Applications of the 14-Field RET Theory

Some applications of the 14-field theory to specific problems have been made. In this section we review briefly some of them. Their detailed exposition will be made in Part IV.

1.6.4.1 Dispersion Relation for Sound in Rarefied Diatomic Gases

The dispersion relation for sound in rarefied diatomic gases; hydrogen, deuterium and hydrogen deuteride gases basing on the 14-field RET theory was recently studied in detail [75]. The relation was compared with those obtained in experiments and by the Navier-Stokes Fourier (NSF) theory. As is expected, the applicable frequency-range of the RET theory was shown to be much wider than that of the NSF theory. The values of the bulk viscosity and the relaxation times involved in nonequilibrium processes were evaluated. It was found that the relaxation time related to the dynamic pressure has a possibility to become much larger than the other relaxation times related to the shear stress and the heat flux. The isotope effects on sound propagation were also clarified [75, 76]. The analysis was made in the temperature range where the rotational and vibrational modes in a molecule play an important role. The RET theory can be applied to many other rarefied polyatomic gases in a wider temperature range where the rotational and/or vibrational modes in a molecule play a role.

1.6.4.2 Shock Wave Structure in a Rarefied Polyatomic Gas

The shock wave structure in a rarefied polyatomic gas is, under some conditions, quite different from the shock wave structure in a rarefied monatomic gas due to the presence of the microscopic internal modes in a polyatomic molecule such as the rotational and vibrational modes [86, 87]. For examples: (1) The shock wave

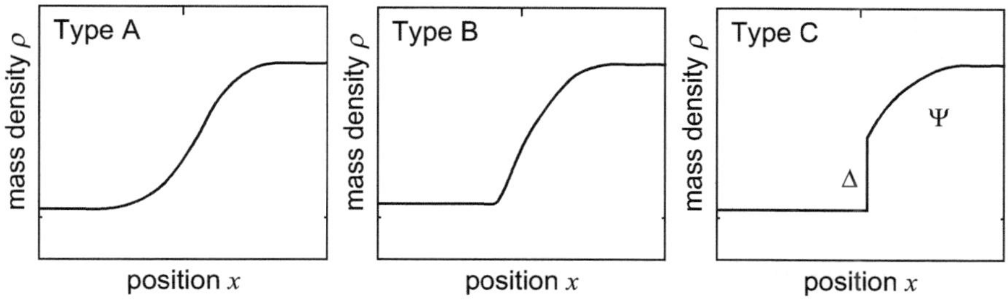

Fig. 1.1 Schematic representation of three types of the shock wave structure in a rarefied polyatomic gas, where ρ and x are the mass density and the position, respectively. As the Mach number increases from unity, the profile of the shock wave structure changes from Type A to Type B, and then to Type C that consists of the thin layer Δ and the thick layer Ψ

thickness in a rarefied monatomic gas is of the order of the mean free path. On the other hand, owing to the slow relaxation process involving the internal modes, the thickness of a shock wave in a rarefied polyatomic gas is several orders larger than the mean free path. (2) As the Mach number increases from unity, the profile of the shock wave structure in a polyatomic rarefied gas changes from the nearly symmetric profile (Type A) to the asymmetric profile (Type B), and then changes further to the profile composed of thin and thick layers (Type C) [88–93]. Schematic profiles of the mass density are shown in Fig. 1.1. Such change of the shock wave profile with the Mach number cannot be observed in a monatomic gas.

In order to explain the shock wave structure in a rarefied polyatomic gas, there have been two well-known approaches. One was proposed by Bethe and Teller [94] and the other is proposed by Gilbarg and Paolucci [95]. Although the Bethe-Teller theory can describe qualitatively the shock wave structure of Type C, its theoretical basis is not clear enough. The Gilbarg-Paolucci theory, on the other hand, cannot explain asymmetric shock wave structure (Type B) nor thin layer (Type C).

Recently it was shown that the 14-field RET theory can describe the shock wave structure of all Types A to C in a rarefied polyatomic gas [78, 79]. In other words the 14-field RET theory has overcome the difficulties encountered in the previous two approaches. This new approach indicates clearly the usefulness of the RET theory for the analysis of shock wave phenomena. These basic studies will be useful for various practical applications, for example, the re-entry of a satellite into the atmosphere of a planet.

1.6.4.3 Some Other Applications

Some of the possible applications of the ET14 theory of rarefied polyatomic gases are as follows:

- Light scattering.

- Heat conduction in a gas at rest confined between two infinite parallel plates, two coaxial cylinders, and two concentric spheres.
- Fluctuation hydrodynamics.

These topics will be explained briefly in Chap. 9.

1.6.5 Molecular RET of Polyatomic Gases

For polyatomic gases, we can also consider the closure of the system with generic number of moments. In this case, we use the same idea introduced in [84]. We consider a distribution function depending on the additional parameter I that takes into account the internal degrees of freedom of a molecule and adopt the generalized hierarchy structure (1.30).

We can define a binary hierarchy of the moments in the following way [84, 96]: We consider now the same binary hierarchy of 14 moments but for a generic number of moments truncated, for the F-series, at the index of truncation N and, for the G-series, at the index M:

$$\begin{array}{ll}
\partial_t F + \partial_i F_i = 0, & \\
\partial_t F_{k_1} + \partial_i F_{ik_1} = 0, & \\
\partial_t F_{k_1k_2} + \partial_i F_{ik_1k_2} = P_{k_1k_2}, & \partial_t G_{kk} + \partial_i G_{ikk} = 0, \\
\vdots & \partial_t G_{kkj_1} + \partial_i G_{kkij_1} = Q_{kkj_1}, \\
\vdots & \vdots \\
\partial_t F_{k_1k_2\ldots k_N} + \partial_i F_{ik_1k_2\ldots k_N} = P_{k_1k_2\ldots k_N}, & \vdots \\
 & \partial_t G_{kkj_1j_2\ldots j_M} + \partial_i G_{kkij_1j_2\ldots j_M} = Q_{kkj_1j_2\ldots j_M}.
\end{array} \tag{1.32}$$

The truncation order N of the F-hierarchy (*momentum-type* hierarchy) and the order M of the G-hierarchy (*energy-type* hierarchy) are a priori independent of each other. It is worth noting that the first and the second equations of the F-hierarchy represent the conservation laws of mass and momentum, respectively ($P \equiv 0, P_i \equiv 0$), while the first equation of the G-hierarchy represents the conservation law of energy ($Q_{ll} \equiv 0$), and that, in each of the two hierarchies, the flux in one equation appears as the density in the following equation—a feature in common with the single hierarchy of monatomic gases.

The Euler 5-moment system is a particular case of (1.32) with $N = 1, M = 0$. And the 14-moment system (1.27) is another particular case of (1.32) with $N = 2, M = 1$.

It was proved that the truncation indices N and M of the two hierarchies are, in reality, not independent of each other because of the physical reasons; (1) Galilean invariance of field equations and (2) the fact that the characteristic velocities depend on the degrees of freedom of a molecule. We have arrived at the conclusion that the relation $M = N - 1$ should be satisfied [96].

The closure of the system is achieved by means of the maximum entropy principle. And it was also proved that, in the present case, this closure is equivalent to the closure by the entropy principle with concave entropy density. In this way, the system becomes to be symmetric when it is written in terms of the *main field* components, definition of which will be shown in Chap. 2.

The characteristic velocities in an equilibrium state are analyzed, which play an important role in the following cases: the propagation of acceleration waves [97, 98], the determination of the phase velocity of linear waves in the high-frequency limit [99, 100], and the *subshock* formation [101].

With regard to this, it will be discussed how the characteristic velocities of the system depend on the internal degrees of freedom and on the order of the truncation of hierarchies. In particular, the two limit cases of monatomic gases and of a gas with infinite internal degrees of freedom will be investigated.

Finally, using the convexity arguments and the sub-characteristic conditions for the principal subsystems, the lower-bound estimate for the maximum characteristic velocity was obtained as follows:

$$\frac{\lambda_{(N)}^{E,\max}}{c_0} \geqslant \sqrt{\frac{6}{5}\left(N - \frac{1}{2}\right)}, \qquad \left(c_0 = \sqrt{\frac{5}{3}\frac{k_B}{m}T}\right).$$

This is independent of the degrees of freedom of a molecule. Noteworthy point is that this estimate is exactly the same as that for monatomic gases established by Boillat and Ruggeri [61]. This is used recently by Slemrod in his analysis of the hydrodynamic limit of the Boltzmann equation and Hilbert's sixth problem [6]. Therefore, also for polyatomic gases, the maximum characteristic velocity tends to be unbounded when the order of the hierarchies tends to infinity!

As in the case of 14 fields [81], it is possible to prove [102] that, in the limit $D \to 3$, the solutions of the system for rarefied polyatomic gases, which is composed by $\frac{1}{6}(N + 1)(N + 2)(2N + 3)$ equations, converge to those of the system of rarefied monatomic gases, which is composed by $\frac{1}{6}(N + 1)(N^2 + 8N + 6)$ equations. In fact, in this limit, the differences between the trace part of the F-hierarchy and the G-hierarchy:

$$\Pi_{k_1 k_2 \ldots k_\alpha} \equiv \lim_{D\to 3}(F_{kkk_1 k_2 \ldots k_\alpha} - G_{kkk_1 k_2 \ldots k_\alpha}), \qquad (0 \leqslant \alpha \leqslant N - 2),$$

are governed by the following balance equations:

$$\partial_t \Pi_{k_1 k_2 \ldots k_\alpha} + \partial_i \Pi_{ik_1 k_2 \ldots k_\alpha} = \lim_{D\to 3}(P_{kkk_1 k_2 \ldots k_\alpha} - Q_{kkk_1 k_2 \ldots k_\alpha}). \tag{1.33}$$

At least in the BGK approximation, it was proved in [102] that $\Pi_{ik_1k_2\ldots k_\alpha}$ and the right-hand side terms are linear combination of $\Pi_{k_1k_2\ldots k_\alpha}$ and therefore the system (1.33) composed by $\frac{1}{6}(N-1)N(N+1)$ equations admits only the zero solution provided that the initial data are compatible with a monatomic gas, i.e., $\Pi_{k_1k_2\ldots k_\alpha}(\mathbf{x},0)=0$.

1.6.6 6-Field RET Theory and Comparison with the Meixner Theory of Relaxation Processes

The 14-field RET theory (ET14) gives us a complete phenomenological model but its differential system is rather complex. For this reason we construct a simplified RET theory with 6 fields (ET6). This simplified theory preserves the main physical properties of the more complex ET14 theory, when the bulk viscosity plays more important role than the shear viscosity and the heat conductivity. This situation is observed in many polyatomic gases at some temperature ranges. This model is particularly interesting because, as seen later, it is also valid in a situation far from equilibrium.

In the 14-field RET theory, there exist three relaxation times τ_S, τ_Π, and τ_q that characterize the relaxation of the shear stress, the dynamic pressure, and the heat flux, respectively. The relaxation times depend on the mass density and the temperature, and their magnitudes are usually comparable with each other. As mentioned above, the shear and bulk viscosities and the heat conductivity in the Navier-Stokes Fourier theory can be expressed in terms of the relaxation times.

It was revealed recently, however, by studying the dispersion relation of linear harmonic waves [75] and the shock wave structure [78, 103] that, in an appropriate temperature range of some polyatomic gases such as a hydrogen gas or a carbon dioxide gas, the relaxation time τ_Π is several orders larger than the other two relaxation times τ_S and τ_q. In such a situation, the dynamic pressure relaxes very slowly compared with the relaxation of the shear stress and the heat flux. And the effect of the shear stress and the heat flux on the relaxation process is negligibly small.

In order to focus our attention on such slow relaxation phenomena, a simplified version of the 14-field RET theory, that is, a RET theory with six independent fields of the mass density, the velocity, the temperature, and the dynamic pressure has been proposed [104, 105]. It was shown that the well-known Meixner theory of relaxation processes [106, 107] is equivalent to the 6-field RET theory.

Due to the relative simplicity of the RET6 theory, a theory that is valid also far from equilibrium was constructed [108]. Details of the 6-field RET theory will be presented in Chaps. 11–14.

1.6.7 Nonequilibrium Temperature

If our study is restricted within nonequilibrium thermodynamics under the local equilibrium assumption, there exists no conceptual difficulty in the temperature in nonequilibrium. Therefore, TIP does no suffer from such a difficulty.

However, if we go beyond the local equilibrium assumption to study highly nonequilibrium phenomena, we encounter an extremely difficult problem. The establishment of a suitable definition of the nonequilibrium temperature in such a situation has always been a big challenge. In RET (and also the kinetic theory) of monatomic gases, the so-called kinetic temperature, which is defined by the thermal average of the kinetic energy of a molecule, has usually been adopted as a nonequilibrium temperature.[3] While, in the papers by Barbera et al. [109] and Au et al. [110] for example, the so-called thermodynamic temperature was also introduced in the framework of RET. The thermodynamic temperature is defined by the zeroth law of thermodynamics, that is, the continuity conditions of the heat flux and of the entropy flux, especially, at the boundary of a system [3].

For a survey about the nonequilibrium temperature in more general context, see the review paper [111].

In Chap. 15, we will study the temperature and also the chemical potential in polyatomic gases in nonequilibrium. In addition to the kinetic temperature, we introduce another well-defined nonequilibrium temperature and chemical potential on the basis of the generalized Gibbs relation in ET where the *main field* plays an essential role. And subsequently these quantities are examined explicitly in the RET theories with 6 and 14 fields. In the 6-field RET theory, in particular, it will be shown in Chap. 11 that their definitions correspond exactly to the definitions of the temperature and chemical potential in the Meixner theory of relaxation processes [106, 107]. An example in the case of shock waves will also be shown.

1.7 Mixture of Gases with Multi-Temperature

In Part VII, we will study some models of a mixture of compressible fluids. In particular, we will discuss the most general model of a mixture in which each constituent has its own temperature (multi-temperature, or MT).

We will firstly compare the solutions of this model with those with a unique common temperature (single temperature or ST) [112]. In the case of Eulerian fluids, it will be shown that the corresponding ST differential system is a principal subsystem of the MT system [112]. Global behavior of smooth solutions for large time for both systems will also be discussed by applying the Shizuta Kawashima K-condition (see Sect. 2.6.2).

[3] In computer simulations by the molecular-dynamics method, the kinetic temperature has also been exclusively adopted as the temperature in nonequilibrium.

Secondly, we introduce the concept of the average temperature of a mixture based on the consideration that the internal energy of a mixture with multi-temperature is the same as that of a single-temperature mixture [113, 114].

As a consequence, it is shown that the entropy of a mixture reaches a local maximum in equilibrium. Through the procedure of the Maxwellian iteration new constitutive equations for nonequilibrium temperatures of the constituents are obtained in a classical limit, together with Fick's law for the diffusion flux.

In order to justify the Maxwellian iteration, we will present, for dissipative fluids, a possible approach to a classical theory of a mixture with multi-temperature. We will prove that the differences of the temperatures between the constituents imply the existence of the dynamic pressure even if the fluids have zero bulk viscosity [113]. Shock structure in a mixture is the subject of Chap. 17.

1.8 Qualitative Analysis

In RET, the differential system is closed by the universal principles: the objectivity principle, the entropy principle, and the principle of causality and stability. This permits an intimate connection between RET and the mathematical theory of hyperbolic systems with convex extension. Then there exists a privileged field (*main field*) such that the differential system becomes to be a symmetric hyperbolic system with the well-posedness of the local (in time) Cauchy problem [115–117].

Moreover the requirement that the balance laws are invariant with respect to the Galilean transformation permits to fix, in a unique way, the velocity dependence in the field equations. The entropy principle becomes a constraint only for the constitutive equations [117].

If the system is nonlinear and hyperbolic, global smooth solutions can exist due to the interrelationship between the first five conservation laws and the remaining dissipative ones. In fact, for generic hyperbolic systems of balance laws, endowed with a convex entropy law and dissipation, the Kawashima-Shizuta condition (K-condition) [118] becomes a sufficient condition for the existence of global smooth solutions, provided that the initial data are sufficiently smooth [119–122]. Recently Lou and Ruggeri [123] observed that there exists a weaker K-condition that is a necessary (but unfortunately not sufficient) condition for the global existence of smooth solutions. It was proved that the assumptions of the previous theorems are fulfilled in both classical [124] and relativistic [125, 126] monatomic RET, and also in the case of mixtures of gases with multi-temperature [112]. The same property exists also in the case of polyatomic rarefied gases. Details will be explained in Chap. 14 for polyatomic gas with six fields and in Chap. 16 for the model of mixture with multi-temperature. In Chap. 18, we will discuss the parabolic limit of the ET theories via the Maxwellian iteration.

1.9 About this Book

The book is composed of eight Parts I–VIII, with two additional chapters; "Introduction" at the beginning and "Open Problems" at the end.

In Part I, we present the mathematical structure of RET and its intimate connection with hyperbolic nonlinear system of first order. In particular, basing on the entropy principle and the convexity of entropy, we can rewrite the original system in a symmetric form using the main field. The main field is crucial also for clarifying the nesting structure of RET. Moreover symmetric hyperbolic systems have good properties concerning the well-posedness of the Cauchy problem. We discuss recent theorems that guarantee the existence of global smooth solutions for all time and, as is expected from a physical point of view, the asymptotic tendency to an equilibrium state. The Galilean invariance dictates the explicit dependence of the field equations on the velocity, while the entropy principle gives a selection rule for admissible constitutive equations. One chapter is devoted to explain the general properties of wave propagation phenomena like linear waves, acceleration waves, shock waves, and shock wave structure. These are important to understand RET deeply.

The Part II is a brief summary of the main results of ET of monatomic gases, details of which are explained in the book of Müller and Ruggeri [11]. This Part is necessary to understand the new progress that will be explained in the present book.

From Part III, we start to explain the newly obtained results after the book of Müller and Ruggeri. In Part III, we study the ET theory of 14 independent fields that is applicable to rarefied polyatomic gases and also to dense gases. The finding that the system of field equations should have the structure of the binary hierarchy is the main breakthrough in the new progress. We also study the molecular RET with the maximum entropy principle. This is compared with the above phenomenological approach, and the consistency between two approaches is shown.

Part IV deals with some applications of the ET14 theory. The linear waves and shock waves are studied. Comparison of the theoretical results with experimental data shows the remarkable superiority of the ET14 theory to the classical Navier-Stokes Fourier theory. Some other possible applications are also briefly discussed.

In Part V, the molecular ET theory of rarefied polyatomic gases is explained. We grasp the mathematical structure of the ET theory with binary hierarchy of arbitrary number of independent densities. Nesting theory of many moments is studied.

Part VI is devoted to the study of the theory that is applicable to the phenomena far from equilibrium within the ET6 theory. The correspondence relation between the ET6 theory and the Meixner theory is established. Then the ET6 theory is applied to the shock wave structure to solve the pending question about the change of the shock profile with the Mach number. Acceleration waves are also studied in this context. Lastly the concept of the nonequilibrium temperature is revisited on the basis of the results obtained.

In Part VII, multi-temperature mixture of Euler fluids is studied. Then we study the shock structure in a mixture. We study, in particular, the temperature overshoot in a shock wave,

In Part VIII, the parabolic limit of the hyperbolic ET systems obtained via Maxwellian iteration is analyzed. In particular we study the question whether the entropy principle is preserved in this limit and we discuss the range of validity of the so-called regularized system.

References

1. E. Fermi, *Thermodynamics*. Dover Books on Physics (Dover, New York, 1956). ISBN-13 978-0486603612
2. I. Müller, *A History of Thermodynamics The Doctrine of Energy and Entropy* (Springer, Berlin, 2007). ISBN 978-3-540-46226-2
3. I. Müller, *Thermodynamics* (Pitman, Boston, 1985). ISBN 0-273-08577-8
4. H.B. Callen, *Thermodynamics* (Wiley, New York, 1960)
5. L. Saint-Raymond, *Hydrodynamic Limits of the Boltzmann Equation*. Lecture Notes in Mathematics, vol. 1971 (Springer, Berlin, 2009)
6. M. Slemrod, From Boltzmann to Euler: Hilbert's 6th problem revisited. Comput. Math. Appl. **65**, 1497 (2013)
7. A.N. Gorban, I. Karlin, Hilbert's 6th Problem: exact and approximate hydrodynamic manifolds for kinetic equations. Bull. Am. Math. Soc. **51**, 186 (2014)
8. S.R. de Groot, P. Mazur, *Non-equilibrium Thermodynamics* (North-Holland, Amsterdam, 1963)
9. L.D. Landau, E.M. Lifshitz, *Fluid Mechanics* (Pergamon Press, London, 1958)
10. N. Pottier, *Nonequilibrium Statistical Physics* (Oxford University Press, Oxford, 2010)
11. I. Müller, T. Ruggeri, *Rational Extended Thermodynamics*, 2nd edn. (Springer, New York, 1998)
12. C. Cattaneo, Sulla conduzione del calore Atti Sem. Mat. Fis. Univ. Modena **3**, 83 (1948)
13. D.D. Joseph, L. Preziosi, Heat waves. Rev. Mod. Phys. **61**, 41 (1989)
14. D.D. Joseph, L. Preziosi, Addendum to the paper heat waves. Rev. Mod. Phys. **62**, 375 (1990)
15. B. Straughan, *Heat Waves* (Springer, New York, 2011)
16. B.D. Coleman, M. Fabrizio, D.R. Owen, On the thermodynamics of second sound in dielectric crystals. Arch. Ration. Mech. Anal. **80**, 135–158 (1983)
17. A. Morro, T. Ruggeri, Non equilibrium properties of solids through second sound measurements. J. Phys. C Solid State Phys. **21**, 1743 (1988)
18. T. Ruggeri, Struttura dei sistemi alle derivate parziali compatibili con un principio di entropia e termodinamica estesa. Suppl. Boll. UMI **4**(5), 261 (1985)
19. L. Landau, Theory of the superfluidity of Helium II. J. Phys. USSR **5**, 71 (1941)
20. V. Peshkov, Second sound in Helium II. J. Phys. USSR **8**, 381 (1944)
21. L.A. Sidorenkov, M.K. Tey, R. Grimm, Y.-H. Hou, L. Pitaevskii, S. Stringari, Second sound and the superfluid fraction in a Fermi gas with resonant interactions. Nature **498**, 78 (2013)
22. T. Ruggeri, A. Muracchini, L. Seccia, Shock waves and second sound in a rigid heat conductor: a critical temperature for NaF and Bi. Phys. Rev. Lett. **64**, 2640 (1990)
23. T. Ruggeri, A. Muracchini, L. Seccia, Continuum approach to phonon gas and shape changes of second sound via shock wave theory. Nuovo Cimento D **16**(1), 15 (1994)
24. T. Ruggeri, A. Muracchini, L. Seccia, Second sound and characteristic temperature in solids. Phys. Rev. B **54**(1), 332 (1996)

25. R.A. Guyer, J.A. Krumhansl, Solution of the linearized phonon Boltzmann equation. Phys. Rev. **148**, 766 (1966)
26. R.A. Guyer, J.A. Krumhansl, Thermal conductivity, second sound and phonon hydrodynamic phenomena in nonmetallic crystals. Phys. Rev. **148**, 778 (1966)
27. W. Dreyer, H. Struchtrup, Heat pulse experiments revisited. Contin. Mech. Thermodyn. **5**, 3 (1993)
28. W. Larecki, Z. Banach, Influence of nonlinearity of the phonon dispersion relation on wave velocities in the four-moment maximum entropy phonon hydrodynamics. Physica D **266**, 65 (2014)
29. W. Larecki, Z. Banach, Consistency of the phenomenological theories of wave-type heat transport with the hydrodynamics of a phonon gas. J. Phys. A Math. Theor. **43**, 385501 (2010)
30. I. Müller, Zum Paradoxon der Wermeleitungstheorie. Zeitschrift für Physik **198**, 329 (1967)
31. W. Israel, Nonstationary irreversible thermodynamics: a causal relativistic theory. Ann. Phys. **100**, 310 (1976)
32. D. Jou, J. Casas-Vázquez, G. Lebon, *Extended Irreversible Thermodynamics* (Springer, Berlin, 1996)
33. T. Ruggeri, Symmetric hyperbolic system of conservative equations for a viscous conducting fluid. Acta Mech. **47**, 167 (1983)
34. I.-S. Liu, I. Müller, Extended thermodynamics of classical and degenerate ideal gases. Arch. Ration. Mech. Anal. **83**, 285 (1983)
35. I.-S. Liu, I. Müller, T. Ruggeri, Relativistic thermodynamics of gases. Ann. Phys. **169**, 191 (1986)
36. I. Müller, T. Ruggeri, *Extended Thermodynamics*. Springer Tracts in Natural Philosophy, vol. 37, 1st edn. (Springer, New York, 1993)
37. C. Truesdell, *Rational Thermodynamics* (McGraw-Hill, New York, 1969)
38. B.D. Coleman, W. Noll, The thermodynamics of elastic materials with heat conduction and viscosity. Arch. Ration. Mech. Anal. **13**, 167 (1963)
39. I. Müller, On the entropy inequality. Arch. Ration. Mech. Anal. **26**, 118 (1967)
40. M. Šilhavý, *The Mechanics and Thermodynamics of Continuous Media* (Springer, Berlin, 1997). ISBN 3-540-58378-5
41. M. Grmela, H.C. Öttinger, Dynamics and thermodynamics of complex fluids. I. Development of a general formalism. Phys. Rev. E **56**, 6620 (1997)
42. H.C. Öttinger, M. Grmela, Dynamics and thermodynamics of complex fluids. II. Illustrations of a general formalism. Phys. Rev. E **56**, 6633 (1997)
43. H.C. Öttinger, *Beyond Equilibrium Thermodynamics* (Wiley, Hoboken, 2004)
44. I. Müller, D. Reitebuch, W. Weiss, Extended thermodynamics – consistent in order of magnitude. Contin. Mech. Thermodyn. **15**, 113 (2003)
45. B.C. Eu, *Generalized Thermodynamics. The Thermodynamics of Irreversible Processes and Generalized Hydrodynamics* (Kluwer Academic, Dordrecht, 2002). ISBN 1-4020-0788-4
46. I. Gyarmati, *Non-equilibrium Thermodynamics. Field Theory and Variational Principles* (Springer, Berlin, 1970)
47. V.A. Cimmelli, D. Jou, T. Ruggeri, P. Ván, Entropy principle and recent results in non-equilibrium theories. Entropy **16**, 1756 (2014)
48. I. Müller, W. Weiss, Thermodynamics of irreversible processes – past and present. Eur. Phys. J. H. **37**, 139 (2012)
49. M. Torrilon, H. Struchtrup, Regularized 13 moment equations: shock structure calculations and comparison to Burnett models. J. Fluid Mech. **513**, 171 (2004)
50. H. Struchtrup, *Macroscopic Transport Equations for Rarefied Gas Flow, Approximation Methods in Kinetic Theory* (Springer, Berlin/Heidelberg, 2005)
51. A. Bobylev, Å. Windfäll, Boltzmann equation and hydrodynamics at the Burnett level. Kinet. Relat. Models **5**(2), 237 (2012)
52. C. Cercignani, *The Boltzmann Equation and Its Applications* (Springer, New York, 1988)
53. Y. Sone, *Kinetic Theory and Fluid Dynamics* (Birkhäuser, Boston, 2002)

54. Y. Sone, *Molecular Gas Dynamics, Theory, Techniques, and Applications* (Birkhäuser, Boston, 2007)
55. H. Grad, On the kinetic theory of rarefied gases. Commun. Pure Appl. Math. **2**(4), 331 (1949)
56. E.T. Jaynes, Information theory and statistical mechanics. Phys. Rev. **106**, 620 (1957); Information theory and statistical mechanics II. Phys. Rev. **108**, 171 (1957)
57. J.N. Kapur, *Maximum Entropy Models in Science and Engineering* (Wiley, New York, 1989)
58. M.N. Kogan, On the principle of maximum entropy, in *Rarefied Gas Dynamics*, vol. 1 (Academic, New York, 1967), pp. 359–368
59. W. Dreyer, Maximization of the entropy in non-equilibrium. J. Phys. A Math. Gen. **20**, 6505 (1987)
60. C.D. Levermore, Moment closure hierarchies for kinetic theories. J. Stat. Phys. **83**, 1021 (1996)
61. G. Boillat, T. Ruggeri, Moment equations in the kinetic theory of gases and wave velocities. Contin. Mech. Thermodyn. **9**, 205 (1997)
62. P. Degond, C. Ringhofer, Quantum moment hydrodynamics and the entropy principle. J. Stat. Phys. **112**, 587 (2003)
63. M. Trovato, L. Reggiani, Maximum entropy principle and hydrodynamic models in statistical mechanics. Riv. Nuovo Cimento Soc. Ital. Fis. **35**, 99 (2012)
64. W. Larecki, Z. Banach, Entropic derivation of the spectral Eddington factors. J. Quant. Spectrosc. Radiat. Transf. **11**, 22486 (2011)
65. W. Larecki, Z. Banach, One-dimensional maximum entropy radiation hydrodynamics: three-moment theory. J. Phys. A Math. Theor. **45**, 385501 (2012)
66. M. Junk, Domain of definition of Levermore's five-moment system. J. Stat. Phys. **93**, 1143 (1988)
67. M. Junk, Maximum entropy for reduced moment problems. Math. Models Methods Appl. Sci. **10**, 1001 (2000)
68. M. Junk, A. Unterreiter, Maximum entropy moment systems and Galilean invariance. Contin. Mech. Thermodyn. **14**, 563 (2002)
69. G.M. Kremer, *An Introduction to the Boltzmann Equation and Transport Processes in Gases* (Springer, Berlin, 2010)
70. I.-S. Liu, Extended thermodynamics of fluids and virial equations of state. Arch. Ration. Mech. Anal. **88**, 1 (1985)
71. G.M. Kremer, Extended thermodynamics of non-ideal gases. Physica A **144**, 156 (1987)
72. T. Arima, S. Taniguchi, T. Ruggeri, M. Sugiyama, Extended thermodynamics of dense gases. Contin. Mech. Thermodyn. **24**, 271 (2011)
73. E. Ikenberry, C. Truesdell, On the pressure and the flux of energy in a gas according to Maxwell's kinetic theory. J. Ration. Mech. Anal. **5**, 1 (1956)
74. T. Ruggeri, Can constitutive relations be represented by non-local equations? Q. Appl. Math. **70**, 597 (2012)
75. T. Arima, S. Taniguchi, T. Ruggeri, M. Sugiyama, Dispersion relation for sound in rarefied polyatomic gases based on extended thermodynamics. Contin. Mech. Thermodyn. **25**, 727 (2013)
76. T. Arima, S. Taniguchi, T. Ruggeri, M. Sugiyama, A study of linear waves based on extended thermodynamics for rarefied polyatomic gases. Acta Applicandae Mathematicae **132**, 15 (2014)
77. T. Arima, S. Taniguchi, M. Sugiyama, Light scattering in rarefied polyatomic gases based on extended thermodynamics, in *Proceedings of the 34th Symposium on Ultrasonic Electronics* (2013), pp. 15–16
78. S. Taniguchi, T. Arima, T. Ruggeri, M. Sugiyama, Thermodynamic theory of the shock wave structure in a rarefied polyatomic gas: beyond the Bethe-Teller theory. Phys. Rev. E **89**, 013025 (2014)
79. S. Taniguchi, T. Arima, T. Ruggeri, M. Sugiyama, Shock wave structure in a rarefied polyatomic gas based on extended thermodynamics. Acta Applicandae Mathematicae **132**, 583 (2014)

80. E. Barbera, F. Brini, M. Sugiyama, Heat transfer problem in a van der Waals gas. Acta Applicandae Mathematicae **132**, 41 (2014)
81. T. Arima, S. Taniguchi, T. Ruggeri, M. Sugiyama, Monatomic rarefied gas as a singular limit of poyatomic gas in extended thermodynamics. Phys. Lett. A **377**, 2136 (2013)
82. C. Borgnakke, P. S. Larsen, Statistical collision model for monte carlo simulation of polyatomic gas mixture. J. Comput. Phys. **18**, 405 (1975)
83. J.-F. Bourgat, L. Desvillettes, P. Le Tallec, B. Perthame, Microreversible collisions for polyatomic gases. Eur. J. Mech. B. Fluids **13**, 237 (1994)
84. M. Pavić, T. Ruggeri, S. Simić, Maximum entropy principle for rarefied polyatomic gases. Physica A **392**, 1302 (2013)
85. F. Mallinger, Generalization of the Grad theory to polyatomic gases. INRIA Research Report (1998), p. 3581
86. W.G. Vincenti, C.H. Kruger Jr., *Introduction to Physical Gas Dynamics* (Wiley, New York, 1965)
87. Ya.B. Zel'dovich, Yu.P. Raizer, *Physics of Shock Waves and High-Temperature Hydrodynamic Phenomena* (Dover, Mineola, 2002)
88. E.F. Smiley, E.H. Winkler, Z.I. Slawsky, Measurement of the vibrational relaxation effect in CO_2 by means of shock tube interferograms. J. Chem. Phys. **20**, 923 (1952)
89. E.F. Smiley, E.H. Winkler, Shock-tube measurements of vibrational relaxation. J. Chem. Phys. **22**, 2018 (1954)
90. W.C. Griffith, W. Bleakney, Shock waves in gases. Am. J. Phys. **22**, 597 (1954)
91. W. Griffith, D. Brickl, V. Blackman, Structure of shock waves in polyatomic gases. Phys. Rev. **102**, 1209 (1956)
92. N.H. Johannesen, H.K. Zienkiewicz, P.A. Blythe, J.H. Gerrard, Experimental and theoretical analysis of vibrational relaxation regions in carbon dioxide. J. Fluid Mech. **13**, 213 (1962)
93. W.C. Griffith, A. Kenny, On fully-dispersed shock waves in carbon dioxide. J. Fluid Mech. **3**, 286 (1957)
94. H.A. Bethe, E. Teller, Deviations from thermal equilibrium in shock waves, Reprinted by Engineering Research Institute. University of Michigan
95. D. Gilbarg, D. Paolucci, The structure of shock waves in the continuum theory of fluids. J. Ration. Mech. Anal. **2**, 617 (1953)
96. T. Arima, A. Mentrelli, T. Ruggeri, Molecular extended thermodynamics of rarefied polyatomic gases and wave velocities for increasing number of moments. Ann. Phys. **345**, 111 (2014)
97. G. Boillat, T. Ruggeri, On the evolution law of the weak discontinuities for hyperbolic quasilinear systems. Wave Motion **1**(2), 149 (1979)
98. T. Ruggeri, Stability and discontinuity waves for symmetric hyperbolic systems, in *Non-linear Wave Motion*, ed. by A. Jeffrey (Longman, New York, 1989), pp. 148–161
99. A. Muracchini, T. Ruggeri, L. Seccia, Dispersion relation in the high frequency limit and non linear wave stability for hyperbolic dissipative systems. Wave Motion **15**(2), 143 (1992)
100. Z. Banach, W. Larecki, T. Ruggeri, Dispersion relation in the limit of high frequency for a hyperbolic system with multiple eigenvalues. Wave Motion **51**, 955 (2014)
101. G. Boillat, T. Ruggeri, On the shock structure problem for hyperbolic system of balance laws and convex entropy. Contin. Mech. Thermodyn. **10**, 285 (1998)
102. T. Arima, T. Ruggeri, M. Sugiyama, S. Taniguchi, A singular limit of rarefied polyatomic gas to monatomic gas in molecular extended thermodynamics, submitted
103. S. Taniguchi, T. Arima, T. Ruggeri, M. Sugiyama, Effect of dynamic pressure on the shock wave structure in a rarefied polyatomic gas. Phys. Fluids **26**, 016103 (2014)
104. T. Arima, S. Taniguchi, T. Ruggeri, M. Sugiyama, Extended thermodynamics of real gases with dynamic pressure: an extension of Meixner's theory. Phys. Lett. A **376**, 2799 (2012)
105. T. Arima, T. Ruggeri, M. Sugiyama, S. Taniguchi, On the six-field model of fluids based on extended thermodynamics. Meccanica **49**, 2181 (2014)
106. J. Meixner, Absorption und dispersion des schalles in gasen mit chemisch reagierenden und anregbaren komponenten. I. Teil. Ann. Phys. **43**, 470 (1943)

107. J. Meixner, Allgemeine theorie der schallabsorption in gasen und flussigkeiten unter berucksichtigung der transporterscheinungen. Acoustica **2**, 101 (1952)
108. T. Arima, S. Taniguchi, T. Ruggeri, M. Sugiyama, Nonlinear extended thermodynamics of real gases with 6 fields. Int. J. Non Linear Mech. **72**, 6 (2015)
109. E. Barbera, I. Müller, M. Sugiyama, On the temperature of a rarefied gas in non-equilibrium. Meccanica **34**, 103 (1999)
110. J. Au, I. Müller, T. Ruggeri, Temperature jumps at the boundary of a rarefied gas. Contin. Mech. Thermodyn. **12**(1), 19 (2000)
111. J. Casas-Vázquez, D. Jou, Temperature in non-equilibrium states: a review of open problems and current proposals. Rep. Prog. Phys. **66**, 1937 (2003)
112. T. Ruggeri, S. Simić, On the hyperbolic system of a mixture of eulerian fluids: a comparison between single and multi-temperature models. Math. Methods Appl. Sci. **30**, 827 (2007)
113. H. Gouin, T. Ruggeri, Identification of an average temperature and a dynamical pressure in a multitemperature mixture of fluids. Phys. Rev. E **78**, 016303 (2008)
114. T. Ruggeri, S. Simić, Average temperature and maxwellian iteration in multitemperature mixtures of fluids. Phys. Rev. E **80**, 026317 (2009)
115. G. Boillat, Sur l'existence et la recherche d'équations de conservation supplémentaires pour les systémes hyperboliques. C. R. Acad. Sci. Paris A **278**, 909 (1974)
116. T. Ruggeri, A. Strumia, Main field and convex covariant density for quasi-linear hyperbolic systems. Relativistic fluid dynamics. Ann. Inst. H. Poincaré Sect. A **34**, 65 (1981)
117. T. Ruggeri, Galilean invariance and entropy principle for systems of balance Laws. The structure of the extended thermodynamics. Contin. Mech. Thermodyn. **1**, 3 (1989)
118. Y. Shizuta, S. Kawashima, Systems of equations of hyperbolic-parabolic type with applications to the discrete Boltzmann equation. Hokkaido Math. J. **14**, 249 (1985)
119. B. Hanouzet, R. Natalini, Global existence of smooth solutions for partially dissipative hyperbolic systems with a convex entropy. Arch. Ration. Mech. Anal. **169**, 89 (2003)
120. W.-A. Yong, Entropy and global existence for hyperbolic balance laws. Arch. Ration. Mech. Anal. **172**, 247 (2004)
121. S. Bianchini, B. Hanouzet, R. Natalini, Asymptotic behavior of smooth solutions for partially dissipative hyperbolic systems with a convex entropy. Commun. Pure Appl. Math. **60**, 1559 (2007)
122. T. Ruggeri, D. Serre, Stability of constant equilibrium state for dissipative balance laws system with a convex entropy. Q. Appl. Math. **62**, 163 (2004)
123. J. Lou, T. Ruggeri, Acceleration waves and weak Shizuta-Kawashima condition. Suppl. Rend. Circ. Mat. Palermo "Non Linear Hyperbolic Fields and Waves. A Tribute to Guy Boillat" **78**, 187 (2006)
124. T. Ruggeri, Global existence of smooth solutions and stability of the constant state for dissipative hyperbolic systems with applications to extended thermodynamics, in *Trends and Applications of Mathematics to Mechanics*. STAMM 2002 (Springer, Berlin, 2005)
125. T. Ruggeri, Entropy principle and relativistic extended thermodynamics: global existence of smooth solutions and stability of equilibrium state. Nuovo Cimento Soc. Ital. Fis. B **119**, 809 (2004); Lecture Notes of the International Conference in Honour of Y. Choquet-Bruhat: Analysis, Manifolds and Geometric Structures in Physics, ed. by G. Ferrarese T. Ruggeri, Elba (2004)
126. T. Ruggeri, Extended relativistic thermodynamics, in *General Relativity and the Einstein Equations*, ed. by Y.C. Bruhat (Oxford University Press, Oxford, 2009), pp. 334–340

Part I
Mathematical Structure and Waves

Chapter 2
Mathematical Structure

Abstract In this chapter, we give a survey on the mathematical structure of the system of RET, which is strictly related to the mathematical problems of hyperbolic systems in balance form with a convex entropy density. We summarize the main results: The proof of the existence of the *main field* in terms of which a system becomes symmetric, and several properties derived from the qualitative analysis concerning symmetric hyperbolic systems. In particular, the Cauchy problem is well-posed locally in time, and if the so-called *K-condition* is satisfied, there exist global smooth solutions provided that the initial data are sufficiently small. Moreover the main field permits to identify natural subsystems and in this way we have a structure of nesting theories. The main property of these subsystems is that the characteristic velocities satisfy the so-called *sub-characteristic* conditions that imply, in particular, that the maximum characteristic velocity does not decrease when the number of equations increases. Another beautiful general property is the compatibility of the balance laws with the Galilean invariance that dictates the precise dependence of the field equations on the velocity.

2.1 System of Balance Laws

The approach of RET [1] is basing on the assumption that the density $\mathbf{F}^0$, the flux $\mathbf{F}^i$ and the production $\mathbf{f}$ in the balance-law system (1.2) depend locally on the field variable $\mathbf{u}(x,t) \in R^N$, and that the quasi-linear dissipative system of balance laws:

$$\frac{\partial \mathbf{F}^0(\mathbf{u})}{\partial t} + \frac{\partial \mathbf{F}^i(\mathbf{u})}{\partial x^i} = \mathbf{f}(\mathbf{u}) \tag{2.1}$$

is hyperbolic with respect to the t-direction. The variable $\mathbf{u}(\mathbf{x}, t)$ is the unknown field vector to be determined in a problem under consideration.

It is convenient to rewrite the system (2.1). The most compact form of the hyperbolic system in the space-time is to use the relativistic notation: $x^0 = t$; $\partial_\alpha = \partial/\partial x^\alpha$ $(\alpha = 0, 1, 2, 3)$. Then we have

$$\partial_\alpha \mathbf{F}^\alpha(\mathbf{u}) = \mathbf{f}(\mathbf{u}). \tag{2.2}$$

T. Ruggeri, M. Sugiyama, *Rational Extended Thermodynamics beyond the Monatomic Gas*, DOI 10.1007/978-3-319-13341-6_2

2.1.1 Hyperbolicity in the t-Direction

The system (2.2) is a particular case of the quasi-linear first order system of PDE's:

$$\mathbf{A}^{\alpha}(\mathbf{u})\partial_{\alpha}\mathbf{u} = \mathbf{F}(\mathbf{u}) \quad \text{with} \quad \mathbf{A}^{\alpha} = \frac{\partial \mathbf{F}^{\alpha}}{\partial \mathbf{u}}. \tag{2.3}$$

The system (2.3) is called *hyperbolic in the t-direction*, if it has the following two properties:

- $\det \mathbf{A}^0 \neq 0$.
- For all unit vectors $\mathbf{n} \equiv (n_i)$, the eigenvalue problem:

$$\left(\mathbf{A}^i n_i - \lambda \mathbf{A}^0\right) \mathbf{d} = 0 \tag{2.4}$$

  admits only real eigenvalues $\lambda^{(i)}$ $(i = 1, 2, \ldots N)$ and a set of linearly independent right eigenvectors $\mathbf{d}^{(i)}$.

The λ's are called *characteristic velocities* and the polynomial

$$\det\left(\mathbf{A}^i n_i - \lambda \mathbf{A}^0\right) = 0 \tag{2.5}$$

is called *characteristic polynomial*. The left eigenvectors $\mathbf{l}^{(i)}$ $(i = 1, 2, \ldots N)$ are defined by

$$\mathbf{l}\left(\mathbf{A}^i n_i - \lambda \mathbf{A}^0\right) = 0, \tag{2.6}$$

and may be chosen in such a way that

$$\mathbf{l}^{(i)} \cdot \mathbf{d}^{(j)} = \delta^{ij}, \qquad \text{for all} \qquad i,j = 1, 2, \ldots N. \tag{2.7}$$

2.1.2 Symmetric Hyperbolic System

The system (2.3) is called *symmetric hyperbolic* (in the t-direction), or briefly symmetric—by the definition of Friedrichs—, if:

- The matrices $\mathbf{A}^{\alpha}$ are symmetric.
- The matrix $\mathbf{A}^0$ is positive definite.

By linear algebra every symmetric system is hyperbolic, but the reverse statement is not true.

Symmetric systems play an important role in extended thermodynamics, because the theory uses hyperbolic equations, and the entropy principle, together with the convexity of the entropy density, ensures that they form a *symmetric* hyperbolic system.

2.2 Axioms of Rational Extended Thermodynamics

The main axioms of RET are summarized as follows:

- The balance laws (2.2) must satisfy the relativity principle, i.e., the balance laws are invariant under a Galilean transformation (or a Lorentz transformation in a relativistic case) and the proper constitutive equations are invariant under any change of observer.
- The entropy principle requires that constitutive equations must be selected so that all thermodynamic processes are compatible with the second law of thermodynamics, where a thermodynamic process is defined as a solution of the total system (balance laws plus constitutive equations). More precisely, the principle consists in the following axioms:

  (i) *There exists an additive and objective scalar, which we call entropy.*
  (ii) *The entropy density and the flux of the entropy are constitutive functions to be determined.*
  (iii) *The entropy production is non-negative for all thermodynamic processes.*

  The axiom (i) says that the body possesses an additive quantity called entropy, like mass and energy. As for any additive quantity, there exists a balance law of entropy:

$$\partial_\alpha h^\alpha(\mathbf{u}) = \Sigma(\mathbf{u}), \tag{2.8}$$

  where h^0, h^i and Σ are, respectively, the entropy density, the entropy flux and the entropy production. The entropy production depends also on $\mathbf{u}$. The second axiom (ii) says that both h^0 and h^i are constitutive quantities to be determined as functions of the field variables. Finally, the third axiom (iii) requires that the entropy production is non-negative for all thermodynamic processes:

$$\Sigma \geqslant 0. \tag{2.9}$$

  We have seen, in Sect. 1.3, the genesis of this entropy principle. In the context of hyperbolic systems of conservation laws this formulation is due to Friedrichs and Lax [2].
- The requirement of the causality and the thermodynamic stability, i.e., the entropy density must be a convex function of the density fields. When $\mathbf{u} \equiv \mathbf{F}^0$, we have the condition:

$$\frac{\partial^2 h^0}{\partial \mathbf{F}^0 \partial \mathbf{F}^0} \quad : \text{ negative definite.} \tag{2.10}$$

In the mathematical community, the sign of the entropy density h^0 is usually opposite to the one adopted in physics. Therefore following the tradition we still use the word, convexity, instead of the more appropriate word, concavity.

2.3 Entropy Principle and Symmetric Systems

2.3.1 General Discussions

First we want to prove the following important theorem:

Theorem 2.1 (Symmetric Form) *By selecting the main field* $\mathbf{u}'$ *as the field variables, the original system (2.2) can be written in a symmetric form with Hessian matrices:*

$$\partial_\alpha \left(\frac{\partial h'^\alpha}{\partial \mathbf{u}'} \right) = \mathbf{f} \quad \Longleftrightarrow \quad \frac{\partial^2 h'^\alpha}{\partial \mathbf{u}' \partial \mathbf{u}'} \partial_\alpha \mathbf{u}' = \mathbf{f} \tag{2.11}$$

provided that h^0 *is a convex function of* $\mathbf{u} \equiv \mathbf{F}^0$ *(or equivalently the Legendre transform* h'^0 *is a convex function of the dual field* $\mathbf{u}'$*).*

Proof The compatibility between (2.2) and (2.8), taking into account that both are quasi-linear equations, implies the existence of a *main field* $\mathbf{u}' \equiv \mathbf{u}'(\mathbf{u})$ such that [3]:

$$\partial_\alpha h^\alpha - \Sigma \;\equiv\; \mathbf{u}' \cdot (\partial_\alpha \mathbf{F}^\alpha - \mathbf{f}) \,. \tag{2.12}$$

As a consequence of this identity, we have

$$dh^\alpha = \mathbf{u}' \cdot d\mathbf{F}^\alpha, \qquad \Sigma = \mathbf{u}' \cdot \mathbf{f} \geqslant 0. \tag{2.13}$$

Boillat [4] and later, in a covariant formulation, Ruggeri and Strumia [3] introduced four potentials h'^α:

$$h'^\alpha = \mathbf{u}' \cdot \mathbf{F}^\alpha - h^\alpha. \tag{2.14}$$

By choosing $\mathbf{u}'$ as a field, we obtain from $(2.13)_1$ and (2.14):

$$\mathbf{F}^\alpha = \frac{\partial h'^\alpha}{\partial \mathbf{u}'}. \tag{2.15}$$

Inserting (2.15) into (2.2) we obtain the form (2.11). As the density vector $\mathbf{F}^0$ is a possible field then the change between the density $\mathbf{F}^0$ and the main field $\mathbf{u}'$ is ensured by the convexity condition of the entropy. In fact h'^0 is the Legendre transform of h^0 and the convexity condition (2.10) imply the convexity of h'^0 with respect the dual field $\mathbf{u}'$. Therefore the system (2.11) is symmetric and the proof is given. It is easy to verify that the vice versa is also true. Every system of the form (2.11) satisfies automatically the entropy principle (2.8), (2.9) with (2.14) provided that the production term $\mathbf{f}$ as the function of $\mathbf{u}'$ satisfies the inequality $(2.13)_2$.

We notice, from (2.15), that the convexity condition (2.10) can be rewritten as

$$Q = \delta \mathbf{u}' \cdot \frac{\partial h'^0}{\partial \mathbf{u}' \partial \mathbf{u}'} \delta \mathbf{u}' = \delta \mathbf{u}' \cdot \delta \mathbf{F}^0 < 0 \qquad \forall\, \delta \mathbf{u}' \neq 0. \tag{2.16}$$

Historical Remarks: The history concerning the symmetrization seems to be not so clear and it sometimes does stir up debate. We here want to list some remarks on it to our knowledge:

- In 1961, Godunov wrote a short paper "An interesting class of quasi-linear systems" [5] in which he was able to prove, in the case of Euler fluids, that if we choose, as variables, $\left(\left(-g + \frac{v^2}{2}\right)/T, -\mathbf{v}/T, 1/T\right)$ (g is the chemical potential), the original system becomes symmetric. Moreover he proved that all systems that come from a variational principle can be put in a symmetric form.
- In 1971, Friedrichs and Lax proved [2] that all systems that are compatible with the entropy principle are symmetrizable. This means that, for an original system that is not symmetric, a new one after a pre-multiplication of a matrix $\mathbf{H}(\mathbf{u})$ becomes to be symmetric. As a consequence the symmetric system is different with respect to the previous system. There is the disadvantage that weak solutions of the original system are not weak solutions of the new one. It seems that they ignored completely Godunov's work!
- In 1974, Boillat [4] introduced the field $\mathbf{u}' = \partial h^0 / \partial \mathbf{F}^0$, and he realized that the original system can be put in a symmetric form. Therefore, to our knowledge, he is the first person who symmetrized the original hyperbolic system that is compatible with the entropy principle. At that time, he did not know Godunov's paper, but after he discovered the paper of Godunov, he has called this system *Godunov system* (and Godunov always quoted Boillat). The reader can find more details in the Lecture Notes in Mathematics n. 1640 (1986) of a Cime Course by Boillat, Dafermos, Lax and Liu. In the part written by Boillat [6], there are several physical examples of symmetrization by using this technique: non-linear elasticity, Born-Infeld non-linear electro-dynamics, magneto-fluid dynamics, etc. in which the field introduced by Boillat was used to symmetrize these systems.
- In 1981, Ruggeri and Strumia [3] were interested in extending this technique to a relativistic case by using a covariant formulation. But they realized that it is impossible to define $\mathbf{u}'$ in the same manner as before because h^0 is the temporal part of the four vector h^α and therefore is not a scalar invariant. Also $\mathbf{F}^0$ is not a vector. Therefore they introduced $\mathbf{u}'$ as a multiplier such that if we multiply the balance laws by $\mathbf{u}'$ we obtain identically the supplementary entropy law [see (2.12)]. This has a benefit that all components of $\mathbf{u}'$ are tensors and, in the classical case, reduce to the Boillat field. They realized the importance of this change of variables and for this reason they proposed for the first time the

name *main field*. The technique of Lagrange multipliers to explore the entropy principle is similar to the one used by Ruggeri and Strumia and was given first by I-Shi Liu [7]. However, it should be noticed that the Lagrange multipliers that define the main field are obtained only when the system is written in a balance form. While, if we use a differential system in a different form (for example, the system expressed by using the material derivative), the Lagrange multipliers are not independent variables (see for more details Sect. 2.7).

- In 1982, Boillat extended the symmetrization also to the case with constraints [8]. This problem was considered also by Dafermos [9].
- In 1983, in the first tentative to construct ET, Ruggeri realized that it is possible to construct a symmetrization also for parabolic systems and he wrote down for the first time the expression of the main field for Navier-Stokes Fourier fluids [10].
- In 1989, in the paper [11], it was proved that symmetrization is compatible with the Galilean invariance.
- In a relativistic case, there is also a subtle point because the entropy depends on the choice of the temporal congruence. In 1990, a discussion about the convexity and the symmetrization and a choice of temporal congruence was the subject of the paper [12].

2.3.2 Symmetric System of Euler Fluids

As an example of the previous general results, let us consider the simple case of Euler fluids. We have

$$\mathbf{F}^0 = \begin{pmatrix} \rho \\ \rho v_k \\ \frac{1}{2}\rho v^2 + \rho\varepsilon \end{pmatrix}, \quad \mathbf{F}^i = \begin{pmatrix} \rho v_i \\ \rho v_i v_k + p\delta_{ik} \\ (\frac{1}{2}\rho v^2 + \rho\varepsilon + p)v_i \end{pmatrix}, \quad \mathbf{f} = \begin{pmatrix} 0 \\ 0_k \\ 0 \end{pmatrix}, \tag{2.17}$$

$(i, k = 1, 2, 3)$. From (2.13), (2.14), and the Gibbs equation (1.7), we have the main field, which coincides with the one derived by Godunov [5]:

$$\mathbf{u}' = \frac{1}{T}\begin{pmatrix} -g + \frac{v^2}{2} \\ -v_k \\ 1 \end{pmatrix}, \tag{2.18}$$

and the potentials:

$$h'^0 = -\frac{p}{T}, \quad h'^i = -v_i\frac{p}{T}. \tag{2.19}$$

The convexity condition is always satisfied under the usual thermodynamic stability condition, that is, the positivity of the specific heat and of the compressibility,

$$\left(\frac{\partial \varepsilon}{\partial T}\right)_\rho > 0, \qquad \left(\frac{\partial p}{\partial \rho}\right)_T > 0. \tag{2.20}$$

2.4 Principal Subsystems

Importance of the main field was again recognized when Boillat and Ruggeri discovered that this field permits to define *nesting theories* through the concept of principal subsystems [13].

Let us split the main field $\mathbf{u}' \in R^N$ into two parts $\mathbf{u}' \equiv (\mathbf{v}', \mathbf{w}')$, where $\mathbf{v}' \in R^M$, $\mathbf{w}' \in R^{N-M}$, $(0 < M < N)$, then the system (2.11) with $\mathbf{f} \equiv (\mathbf{r}, \mathbf{g})$ reads:

$$\partial_\alpha \left(\frac{\partial h'^\alpha(\mathbf{v}', \mathbf{w}')}{\partial \mathbf{v}'}\right) = \mathbf{r}(\mathbf{v}', \mathbf{w}'), \tag{2.21}$$

$$\partial_\alpha \left(\frac{\partial h'^\alpha(\mathbf{v}', \mathbf{w}')}{\partial \mathbf{w}'}\right) = \mathbf{g}(\mathbf{v}', \mathbf{w}'). \tag{2.22}$$

Definition 2.1 (Principal Subsystem) Given some assigned constant value $\mathbf{w}'_*$ of $\mathbf{w}'$, we call the system[1]:

$$\partial_\alpha \left(\frac{\partial h'^\alpha(\mathbf{v}', \mathbf{w}'_*)}{\partial \mathbf{v}'}\right) = \mathbf{r}(\mathbf{v}', \mathbf{w}'_*) \tag{2.23}$$

principal subsystem of (2.11). In other words, a principal subsystem (there are $2^N - 2$ such subsystems) coincides with the first block of the system putting $\mathbf{w}' = \mathbf{w}'_*$.

The principal subsystems have two important properties: they admit also a convex subentropy law and the spectrum of the characteristic velocities is contained in the spectrum of the full system (subcharacteristic conditions). In fact it is possible to prove the following theorems [13]:

Theorem 2.2 *Solutions of a principal subsystem satisfy also a supplementary law (subentropy law):*

$$\partial_\alpha \overline{h}^\alpha = \overline{\Sigma}, \tag{2.24}$$

[1]The definition and the properties remain valid for prescribed values of $\mathbf{w}'_*$ that depend on x^α in an arbitrary manner. In this case the principal subsystem is not autonomous [13].

where the entropy four-vector $\overline{h}^{\alpha}(\mathbf{v}', \mathbf{w}'_*)$ *and the entropy production* $\overline{\Sigma}$ *are related to the restrictions of the entropy four-vector* $h^{\alpha}(\mathbf{v}', \mathbf{w}'_*)$ *and of the entropy production* $\Sigma(\mathbf{v}', \mathbf{w}'_*)$ *of the full system:*

$$\overline{h}^{\alpha}(\mathbf{v}', \mathbf{w}'_*) = h^{\alpha}(\mathbf{v}', \mathbf{w}'_*) - \mathbf{w}'_* \cdot \left(\frac{\partial h'^{\alpha}}{\partial \mathbf{w}'}\right)_{\mathbf{w}' \equiv \mathbf{w}'_*},$$

$$\overline{\Sigma} = \Sigma(\mathbf{v}', \mathbf{w}'_*) - \mathbf{w}'_* \cdot \mathbf{g}(\mathbf{v}', \mathbf{w}'_*).$$

The subentropy is convex and therefore every principal subsystem are also symmetric hyperbolic.

Let $\lambda^{(k)}(\mathbf{v}', \mathbf{w}', \mathbf{n})$ and $\overline{\lambda}^{(\overline{k})}(\mathbf{v}', \mathbf{w}'_*, \mathbf{n})$ be the characteristic velocities of the total system and of the subsystem, respectively, where $\mathbf{n}$ is the unit normal to the wave front. In general, solutions of the subsystem are not particular solutions of the system (for $\mathbf{w}' = \mathbf{w}'_*$) and the spectrum of the $\overline{\lambda}$'s is not part of the spectrum of the λ's. However let us define

$$\lambda_{\max} = \max_{k=1,2,\ldots,N} \lambda^{(k)}, \qquad \overline{\lambda}_{\max} = \max_{\overline{k}=1,2,\ldots,M} \overline{\lambda}^{(\overline{k})}$$

and similarly for the minima, then we have a theorem:

Theorem 2.3 (Subcharacteristic Conditions) *Under the assumption that* h^0 *is a convex function, the following subcharacteristic conditions hold for every principal subsystem:*

$$\lambda_{\max}(\mathbf{v}', \mathbf{w}'_*, \mathbf{n}) \geqslant \overline{\lambda}_{\max}(\mathbf{v}', \mathbf{w}'_*, \mathbf{n}); \qquad \lambda_{\min}(\mathbf{v}', \mathbf{w}'_*, \mathbf{n}) \leqslant \overline{\lambda}_{\min}(\mathbf{v}', \mathbf{w}'_*, \mathbf{n}), \tag{2.25}$$

$\forall\ \mathbf{v}' \in R^M$ *and* $\forall\ \mathbf{n} \in R^3 : \|\ \mathbf{n}\ \| = 1$.

The proof of the theorems is given in [13].

2.5 Conservation and Balance Laws, and Equilibrium Subsystem

A particular case of (2.21) and (2.22) is the case where the first M equations are conservation laws, i.e., $\mathbf{r} \equiv 0$. This is the case of all extended thermodynamic theories. Then the block of conservation laws is expressed by

$$\partial_{\alpha}\mathbf{V}^{\alpha} = 0, \quad \Longleftrightarrow \quad \partial_{\alpha}\left(\frac{\partial h'^{\alpha}(\mathbf{v}', \mathbf{w}')}{\partial \mathbf{v}'}\right) = 0, \tag{2.26}$$

and a block of balance laws by

$$\partial_\alpha \mathbf{W}^\alpha = \mathbf{g}, \quad \Longleftrightarrow \quad \partial_\alpha \left(\frac{\partial h'^\alpha(\mathbf{v}', \mathbf{w}')}{\partial \mathbf{w}'} \right) = \mathbf{g}(\mathbf{v}', \mathbf{w}'). \tag{2.27}$$

In this case, it is possible to define, as in usual thermodynamics, the equilibrium state:

Definition 2.2 (Equilibrium State) An equilibrium state is a state for which the entropy production $\Sigma|_E$ vanishes and hence attains its minimum value.

It is possible to prove the following theorem [13, 14]:

Theorem 2.4 (Equilibrium Manifold) *In an equilibrium state, under the assumption of dissipative productions i.e., the assumption that*

$$\mathbf{D} = \frac{1}{2} \left\{ \frac{\partial \mathbf{g}}{\partial \mathbf{w}'} + \left(\frac{\partial \mathbf{g}}{\partial \mathbf{w}'} \right)^T \right\} \Bigg|_E \quad \textit{is negative definite,} \tag{2.28}$$

the production vanishes and the main field components vanish except for the first M components. Thus

$$\mathbf{g}|_E = 0, \qquad \mathbf{w}'|_E = 0. \tag{2.29}$$

Therefore, in an equilibrium state, all the components of the main field corresponding to the Lagrange multipliers of the balance laws (2.27) vanish, and only the Lagrange multipliers corresponding to the conservation laws (2.26) survive. This confirms again the importance of the main field!

We have another important characteristic property of the equilibrium state [15, 16]:

Theorem 2.5 (Maximum of the Entropy) *In equilibrium, the entropy density h^0 is maximal, i.e.,*

$$h^0 < h^0|_E \quad \forall\, \mathbf{u}' \neq \mathbf{u}'|_E, \quad \textit{where} \quad h^0|_E = h^0 \left(\mathbf{v}'|_E, 0 \right),$$

and $\mathbf{v}'|_E$ denotes the restriction in equilibrium of the components of the main field corresponding to the Lagrange multipliers of conservation laws.

Therefore we find also at this general level the well-known property of the entropy in thermodynamics, i.e., the property of maximal entropy in equilibrium.

If we limit our study within one-dimensional space, the system (2.21) and (2.22) assumes the form:

$$\begin{cases} \mathbf{V}_t + \left(k'_{\mathbf{v}'} \right)_x = 0, \\ \mathbf{W}_t + \left(k'_{\mathbf{w}'} \right)_x = -\mathbf{G}\left(\mathbf{v}', \mathbf{w}' \right) \mathbf{w}' \end{cases} \tag{2.30}$$

with $\mathbf{V} = h'_{\mathbf{v}'}$, $\mathbf{W} = h'_{\mathbf{w}'}$ and $h' = h'^{0}, k' = h'^{1}$. The matrix $\mathbf{G}$ is a positive definite $(N-M) \times (N-M)$ matrix.

2.6 Qualitative Analysis

In this section, we discuss the importance of the entropy principle for the qualitative analysis of the Cauchy problem.

2.6.1 *Competition Between Hyperbolicity and Dissipation*

In the general theory of hyperbolic conservation laws and hyperbolic-parabolic conservation laws, the existence of a strictly convex entropy function is a basic condition for the well-posedness. In fact, if the fluxes $\mathbf{F}^i$ and the production $\mathbf{f}$ are smooth enough in a suitable convex open set $\in R^N$, it is well known that system (2.2) has a unique local (in time) smooth solution for smooth initial data [2, 17, 18].

However, in a general case, even for arbitrarily small and smooth initial data, there is no global continuation for these smooth solutions, which may develop singularities, shocks, or blowup, in a finite time, see for instance [19, 20].

On the other hand, in many physical examples, thanks to the interplay between the source term and the hyperbolicity, there exist global smooth solutions for a suitable set of initial data. This is the case, for example, of the isentropic Euler system with damping. Roughly speaking, for such a system, the relaxation term induces a dissipative effect. This effect then competes with the hyperbolicity. If the dissipation is sufficiently strong so as to dominate the hyperbolicity, the system is *dissipative*, and we expect that the classical solution exists for all time and converges to a constant state. While, if the dissipation and the hyperbolicity are equally important, we expect that only part of the perturbation diffuses. In the latter case the system is called *of composite type* by Zeng [21].

2.6.1.1 A Simple Example: Burgers' Equation

The simplest example of the problem is represented by Burgers' equation:

$$u_t + uu_x = 0, \qquad u(x,0) = u_0(x), \tag{2.31}$$

which, by using the method of characteristics, can be rewritten as

$$\begin{cases} \dfrac{du}{dt} = 0, & u(0) = u_0(x_0), \\ \dfrac{dx}{dt} = u, & x(0) = x_0, \end{cases}$$

and therefore admits the general solution:

$$\begin{cases} u(x,t) = u_0(x_0), \\ x = x_0 + u_0(x_0)t. \end{cases} \tag{2.32}$$

Here u is a function of (x, t) through the parameter x_0.

The invertibility for x_0 as a function of (x, t) is lost for each value of x_0 at the time

$$t_c(x_0) = -\frac{1}{u_0'(x_0)}.$$

The critical time is defined by the smallest positive value of $t_c(x_0)$:

$$t_{cr} = \inf_{x_0} \{t_c(x_0) > 0\}.$$

If, instead of (2.31), we take into account a dissipative production term like

$$u_t + uu_x = -\nu u, \qquad u(x,0) = u_0(x)$$

with ν=constant > 0, then we have

$$\begin{cases} \dfrac{du}{dt} = -\nu u, & u(0) = u_0(x_0), \\ \dfrac{dx}{dt} = u, & x(0) = x_0, \end{cases}$$

which admits the solution:

$$\begin{cases} u(x,t) = u_0(x_0)e^{-\nu t}, \\ x = x_0 + \dfrac{u_0(x_0)}{\nu}\left(1 - e^{-\nu t}\right), \end{cases} \tag{2.33}$$

and we have

$$t_c(x_0) = -\frac{1}{\nu}\log\left(1 + \frac{\nu}{u_0'(x_0)}\right), \qquad t_{cr} = \inf_{x_0}\{t_c(x_0) > 0\}.$$

Therefore if

$$\nu > \nu^* \tag{2.34}$$

with $\nu^* = \max_{x_0} |u_0'(x_0)|$, the classical solution exists any time: *the dissipation wins over the hyperbolicity*. While, if

$$\nu \leq \nu*,$$

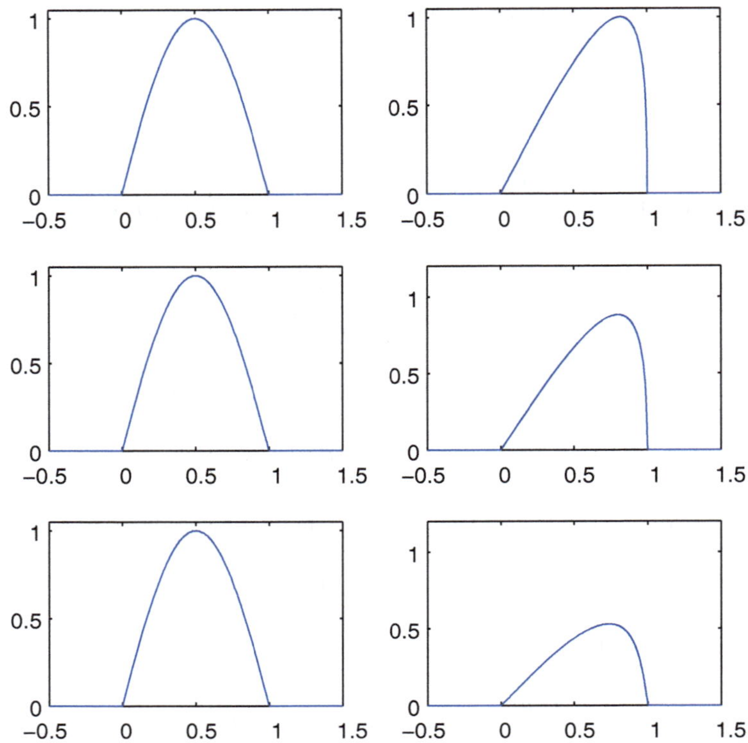

Fig. 2.1 Burgers' equation: evolution of the profile from the initial time (*left*) to the critical time (*right*) for $\nu = 0$, $0 < \nu \leqslant \nu^*$ and for $\nu > \nu^*$ (respectively, from *top* to *bottom*)

the hyperbolicity dominates the dissipation and, in general, we do not have global existence of a smooth solution. See for these cases Fig. 2.1.

2.6.2 Shizuta Kawashima K-Condition

In general, there are several ways to identify whether a hyperbolic system with relaxation is of dissipative type or of composite type. One way, which is completely parallel to the case of the hyperbolic-parabolic system, was discussed first by Shizuta and Kawashima [22]. It is known in the literature as *K- condition* or *genuine coupling*.

Definition 2.3 (K-Condition) A system (2.3) satisfies the K-condition if, in the equilibrium manifold, any right characteristic eigenvectors $\mathbf{d}$ of (2.4) are not in the null space of $\nabla\mathbf{f}$, where $\nabla \equiv \partial/\partial\mathbf{u}$:

$$\nabla\mathbf{f}\ \mathbf{d}|_E \neq 0 \quad \forall\mathbf{d}. \tag{2.35}$$

2.6.3 Global Existence and Stability of Constant State

For dissipative one-dimensional systems (2.30) satisfying the K-condition, it is possible to prove the following global existence theorem by Hanouzet and Natalini [23]:

Theorem 2.6 (Global Existence) *Assume that the system (2.30) is strictly dissipative (2.28) and the K-condition is satisfied. Then there exists $\delta > 0$, such that, if $\|\mathbf{u}'(x,0)\|_2 \leqslant \delta$, there is a unique global smooth solution, which verifies*

$$\mathbf{u}' \in C^0([0,\infty); H^2(R) \cap C^1([0,\infty); H^1(R)).$$

This global existence theorem was generalized to a higher-dimensional case by Yong [24] and successively by Bianchini, Hanouzet and Natalini [25].

Moreover Ruggeri and Serre [16] proved that the constant equilibrium state is stable:

Theorem 2.7 (Stability of an Equilibrium State) *Under natural hypotheses of strongly convex entropy, strict dissipativeness, genuine coupling and "zero mass" initial for the perturbation of the equilibrium variables, the constant solution stabilizes*

$$\|\mathbf{u}(t)\|_2 = 0\left(t^{-1/2}\right).$$

In [23], the authors reported several examples of dissipative systems satisfying the K-condition: the p-system with damping, the Suliciu model for the isothermal viscoelasticity, the Kerr-Debye model in nonlinear electromagnetism, and the Jin-Xin relaxation model. A recent paper of Dafermos proved the existence and long time behavior of spatially periodic BV solutions [26].

2.6.3.1 An Example: Global Existence without the K-Condition

Zeng [21] considered a toy model of vibrational nonequilibrium gas in Lagrangian variables. She proved that, if the system is of composite type, the global existence holds. Therefore the K-condition is only a sufficient condition for the global existence of smooth solutions.

An intriguing open problem is to make clear the following questions: Does a weaker K-condition that is necessary to ensure global solutions exist? If such a condition exists, does it have a physical meaning that gives a possible new principle of extended thermodynamics in addition to the convexity condition of entropy?

Recently Lou and Ruggeri [27] observed that there indeed exists a weaker K-condition that is a necessary (but unfortunately not sufficient) condition for the global existence of smooth solutions. Instead of the condition that the right eigenvectors are not in the null space of $\nabla \mathbf{f}$, they posed this condition only on the right eigenvectors corresponding to genuine non-linear eigenvalues. It was proved that the assumptions of the previous theorems are fulfilled in both classical [28] and relativistic [29, 30] RET theories of monatomic gases, and also in the theory of mixtures of gases with multi-temperature [31].

2.7 Galilean Invariance

The compatibly of (2.2) and (2.8) with the Galilean invariance was studied by Ruggeri [11]. It was proved that the Galileanity imposes the strict velocity dependence of the density, flux, and production.

More precisely, if we split the field $\mathbf{u} \equiv (\mathbf{v}, \mathbf{w})$, where $\mathbf{v}$ is the velocity $\in R^3$ and $\mathbf{w} \in R^{N-3}$ are the other objective fields, we have the following theorem:

Theorem 2.8 (Galilean Invariance) *The system of balance laws (2.2) and the entropy balance law (2.8) are invariant under the Galilean transformation if there exists an $N \times N$ matrix $\mathbf{X}(\mathbf{v})$ such that:*

$$\begin{cases} \mathbf{F}^0\,(\mathbf{v},\mathbf{w}) \;= \mathbf{X}(\mathbf{v})\,\hat{\mathbf{F}}^0(\mathbf{w}), \\ \boldsymbol{\Phi}^i\,(\mathbf{v},\mathbf{w}) = \mathbf{X}(\mathbf{v})\,\hat{\boldsymbol{\Phi}}^i(\mathbf{w}), \\ \mathbf{f}\,(\mathbf{v},\mathbf{w}) \quad\; = \mathbf{X}(\mathbf{v})\,\hat{\mathbf{f}}(\mathbf{w}), \end{cases} \tag{2.36}$$

and

$$\begin{cases} h^0\,(\mathbf{v},\mathbf{w}) = \hat{h}^0\,(\mathbf{w})\,, \\ h^i\,(\mathbf{v},\mathbf{w}) = \hat{h}^0\,(\mathbf{w})\,v^i + \hat{h}^i(\mathbf{w}), \\ \Sigma\,(\mathbf{v},\mathbf{w}) = \hat{\Sigma}\,(\mathbf{w})\,, \end{cases} \tag{2.37}$$

where

$$\boldsymbol{\Phi}^i = \mathbf{F}^i - \mathbf{F}^0 v^i\,. \tag{2.38}$$

$\mathbf{X}(\mathbf{v})$ *is an exponential matrix:*

$$\mathbf{X}(\mathbf{v}) = e^{\mathbf{A}^r v_r} = \mathbf{I} + \mathbf{A}^r v_r + \frac{1}{2}\mathbf{A}^r\mathbf{A}^s v_r v_s + \dots \tag{2.39}$$

with $\mathbf{A}^r$ being three $(N \times N)$ constant matrices such that

$$\mathbf{A}^r\mathbf{A}^s = \mathbf{A}^s\mathbf{A}^r\,, \qquad \forall\; r,s = 1,2,3\,. \tag{2.40}$$

The hat on a quantity indicates, here and hereafter, the corresponding quantity evaluated at the zero velocity (intrinsic quantities):

$$\widehat{\mathbf{F}^0}(\mathbf{w}) = \mathbf{F}^0(0,\mathbf{w}), \quad \hat{\boldsymbol{\Phi}}^i(\mathbf{w}) = \boldsymbol{\Phi}^i(0,\mathbf{w}), \quad \hat{\mathbf{f}}(\mathbf{w}) = \mathbf{f}(0,\mathbf{w}), \tag{2.41}$$
$$\hat{h}^0(\mathbf{w}) = h^0(0,\mathbf{w}), \quad \hat{h}^i(\mathbf{w}) = h^i(0,\mathbf{w}), \quad \hat{\Sigma}(\mathbf{w}) = \Sigma(0,\mathbf{w}).$$

As the dependence on the velocity is prescribed by the Galilean invariance (2.36), the natural question is whether (2.36) are compatible with the constraints (2.15) arising from the entropy principle .

In [11], it was proved that the compatibility exists and that, as we expect, the entropy principle becomes a constraint only for the objective quantities $\mathbf{w}$. In fact, it was proved that

$$\mathbf{u}' = \hat{\mathbf{u}}'\mathbf{X}^{-1}(\mathbf{v}) = \hat{\mathbf{u}}'\mathbf{X}(-\mathbf{v}) \tag{2.42}$$

and, by taking into account of (2.36), (2.37), (2.39) and (2.42), (2.13) becomes

$$\hat{\mathbf{u}}' \cdot d\widehat{\mathbf{F}^0} = d\hat{h}, \qquad \hat{\mathbf{u}}' \cdot d\hat{\mathbf{F}}^i = d\hat{h}^i, \tag{2.43}$$

together with the constraints:

$$\hat{\mathbf{u}}' \cdot \mathbf{A}^r \widehat{\mathbf{F}^0} = 0, \qquad \hat{\mathbf{u}}' \cdot \mathbf{A}^r \hat{\mathbf{F}}^i = -\hat{h}'^0 \delta^{ir}. \tag{2.44}$$

While from (2.14) we obtain

$$\widehat{\mathbf{F}^0} \cdot d\hat{\mathbf{u}}' = d\hat{h}'^0, \qquad \hat{\mathbf{F}}^i \cdot d\hat{\mathbf{u}}' = d\hat{h}'^i, \tag{2.45}$$

$$h'^0 = \hat{h}'^0, \quad h'^i = \hat{h}'^0 v^i + \hat{h}'^i. \tag{2.46}$$

The presence of the three constraints $(2.44)_1$ is not surprising because, while $\mathbf{u}'$ is an R^N field, $\hat{\mathbf{u}}'$ are not independent variables. There are only $N-3$ independent components of $\hat{\mathbf{u}}'$, which, together with the three components of the velocity, forms a field. Therefore the entropy principle is full compatible with the Galilean invariance and becomes constraints for the proper constitutive functions [see (2.43)].

The convexity condition (2.16), in particular, implies (see for details [11]):

$$\hat{Q} = \delta\hat{\mathbf{u}}' \cdot \delta\hat{\mathbf{F}} < 0. \tag{2.47}$$

An alternative procedure to satisfy the Galilean invariance and the entropy principle avoiding the constraints (2.44) was made in [32].

2.7.1 Field Equations in Terms of Intrinsic Quantities

Since the velocity dependence of the quantities now becomes to be explicit, there is a lot of redundancy in the field equations. It is possible to prove that, for classical solutions, every solution of (2.2) is also the solution of

$$\frac{d\hat{\mathbf{F}}^0}{dt} + \hat{\mathbf{F}}^0 \frac{\partial v_i}{\partial x_i} + \frac{\partial \hat{\boldsymbol{\Phi}}^i}{\partial x_i} + \mathbf{A}^r \left\{ \hat{\mathbf{F}}^0 \frac{dv_r}{dt} + \hat{\boldsymbol{\Phi}}^i \frac{\partial v_r}{\partial x_i} \right\} = \hat{\mathbf{f}}. \tag{2.48}$$

These equations are much more useful for practical purposes. However, for the study of weak solutions, in particular, shock waves, we need the system of field equations in balance form.

2.7.2 Diagonal Structure in RET

In the case of the moment theory, the density $\mathbf{F}^0$, the (non-convective or intrinsic) flux $\boldsymbol{\Phi}^i$, and the production $\mathbf{f}$ are expressed as

$$\mathbf{F}^0 = \begin{pmatrix} F \\ F_{k_1} \\ F_{k_1 k_2} \\ \vdots \\ F_{k_1 k_2 \dots k_N} \end{pmatrix}, \quad \boldsymbol{\Phi}^i = \begin{pmatrix} \Phi_i \\ \Phi_{ik_1} \\ \Phi_{ik_1 k_2} \\ \vdots \\ \Phi_{ik_1 k_2 \dots k_N} \end{pmatrix}, \quad \mathbf{f} = \begin{pmatrix} f \\ f_{k_1} \\ f_{k_1 k_2} \\ \vdots \\ f_{k_1 k_2 \dots k_N} \end{pmatrix}, \tag{2.49}$$

where $F_{k_1 \dots k_j}$, $\Phi_{ik_1 \dots k_j}$, and $f_{k_1 \dots k_j}$ are symmetric tensors. Thus the system of balance equations (2.2) has a natural order with increasing tensorial rank. Each block of tensorial equations of rank j governs the evolution of a new quantity $F_{k_1 k_2 \dots k_j}$.

The field equations have the form:

$$\frac{\partial F_{i_1 \dots i_l}}{\partial t} + \frac{\partial (F_{i_1 \dots i_l} v_i + \Phi_{ii_1 \dots i_l})}{\partial x^i} = f_{i_1 \dots i_l} \quad (l = 0, 1, 2, \dots, N). \tag{2.50}$$

We assume that all subsystems that result from (2.49) by ignoring the tensor equation of rank N, or (N and $N-1$), or (N, $N-1$ and $N-2$), etc. have the Galilean invariance. Thus the matrix $\mathbf{X}(\mathbf{v})$ must be a sub-triangular block matrix. It is possible to prove (see [1, 11]) that the matrix $\mathbf{X}(\mathbf{v})$ is a polynomial matrix in $\mathbf{v}$ of order N whose diagonal elements are blocks of Kronecker delta. It reads

$$\mathbf{X}(\mathbf{v}) \equiv \begin{bmatrix} 1 & & & & & & \\ v_{k_1} & \delta^{h_1}_{k_1} & & & & & \\ v_{k_1}v_{k_2} & 2\delta^{h_1}_{(k_1}v_{k_2)} & \cdots & & & & \\ v_{k_1}v_{k_2}v_{k_3} & 3\delta^{h_1}_{(k_1}v_{k_2}v_{k_3)} & \cdots & \delta^{h_1}_{k_1}\delta^{h_2}_{k_2}\delta^{h_3}_{k_3} & & & \\ \vdots & \vdots & \vdots & \vdots & & & \\ v_{k_1}v_{k_2}\dots v_{k_n} & \binom{n}{1}\delta^{h_1}_{(k_1}v_{k_2}\dots v_{k_n)} & \cdots & \vdots & & & \\ \vdots & \vdots & \vdots & \vdots & & \vdots & \vdots \\ v_{k_1}v_{k_2}\dots v_{k_N} & \binom{N}{1}\delta^{h_1}_{(k_1}v_{k_2}\dots v_{k_N)} & \cdots & \binom{N}{3}\delta^{h_1}_{(k_1}\delta^{h_2}_{k_2}\delta^{h_3}_{k_3}v_{k_4}\dots v_{k_N)} & \cdots & & \delta^{h_1}_{k_1}\delta^{h_2}_{k_2}\dots\delta^{h_N}_{k_N} \end{bmatrix}, \tag{2.51}$$

while the matrices A^r are nil-potent matrices: $\mathbf{A}^{k_1}\mathbf{A}^{k_2}\dots\mathbf{A}^{k_{N+1}} = 0$ for all $k_1, k_2, \dots, k_{N+1}$ over $1, 2, 3$, and have the following expression:

$$\mathbf{A}^r = \begin{bmatrix} 0 & 0 & 0 & & 0 \\ \delta^r_{k_1} & 0 & 0 & & 0 \\ 0 & 2\delta^{(rh_1)}_{(k_1k_2)} & 0\cdots & & 0 \\ \vdots & \vdots & \vdots & & \vdots \\ 0 & 0 & 0\cdots & N\delta^{(rh_1h_2\dots h_{N-1})}_{(k_1k_2\dots k_N)} & 0 \end{bmatrix}. \tag{2.52}$$

Thus we are able to decompose tensors of arbitrary rank into the velocity-dependent and intrinsic parts. In particular, the decomposition of an arbitrary tensorial density is given by

$$\begin{aligned} F_{i_1\dots i_l} = \hat{F}_{i_1\dots i_l} + \binom{l}{1}\hat{F}_{(i_1\dots i_{l-1}}v_{i_l)} + \binom{l}{2}\hat{F}_{(i_1\dots i_{l-2}}v_{i_{l-1}}v_{i_l)} + \dots \\ + \binom{l}{l-1}\hat{F}_{(i_1}v_{i_2\dots}v_{i_l)} + \hat{F}v_{i_1\dots}v_{i_l}. \end{aligned} \tag{2.53}$$

The same decompositions can be made for $\boldsymbol{\Phi}^i$ and $\mathbf{f}$.

As an example we consider the Euler fluid. In this case from (2.17), (2.1) and (2.38),

$$\mathbf{F}^0 = \begin{pmatrix} \rho \\ \rho v_k \\ \frac{1}{2}\rho v^2 + \rho\varepsilon \end{pmatrix}, \quad \boldsymbol{\Phi}^i = \begin{pmatrix} 0 \\ p\delta_{ik} \\ pv_i \end{pmatrix}, \quad \mathbf{f} = \begin{pmatrix} 0 \\ 0_k \\ 0 \end{pmatrix}. \tag{2.54}$$

The matrices $\mathbf{A}^r$ and $\mathbf{X}(\mathbf{v})$ in the present case read

$$\mathbf{A}^r = \begin{bmatrix} 0 & 0 & 0 \\ \delta^r_{k_1} & 0 & 0 \\ 0 & \delta^r_{h_1} & 0 \end{bmatrix} \quad \text{and} \quad \mathbf{X}(\mathbf{v}) = \begin{bmatrix} 1 & 0 & 0 \\ v_{k_1} & \delta^k_{k_1} & 0 \\ \frac{1}{2}v^2 & v_{h_1} & 0 \end{bmatrix}. \tag{2.55}$$

References

1. I. Müller, T. Ruggeri, *Rational Extended Thermodynamics*, 2nd edn. (Springer, New York, 1998)
2. K.O. Friedrichs, P.D. Lax, Systems of conservation equation with a convex extension. Proc. Natl. Acad. Sci. USA **68**, 1686 (1971)
3. T. Ruggeri, A. Strumia, Main field and convex covariant density for quasi-linear hyperbolic systems. Relativistic fluid dynamics. Ann. Inst. H. Poincaré, Sect. A **34**, 65 (1981)
4. G. Boillat, Sur l'existence et la recherche d'équations de conservation supplémentaires pour les systémes hyperboliques. C. R. Acad. Sci. Paris A **278**, 909 (1974)
5. S.K. Godunov, An interesting class of quasilinear systems. Sov. Math. **2**, 947 (1961)
6. G. Boillat, in CIME Course, *Recent Mathematical Methods in Nonlinear Wave Propagation*, ed. by T. Ruggeri. Lecture Notes in Mathematics, vol. 1640 (Springer, Berlin, 1995), pp. 103–152
7. I.-S. Liu, Method of Lagrange multipliers for exploitation of the entropy principle. Arch. Ration. Mech. Anal. **46**, 131 (1972)
8. G. Boillat, Involutions des systèmes conservatifs. C. R. Acad. Sci. Paris **307**, 891 (1988)
9. C.M. Dafermos, Quasilinear hyperbolic systems with involutions. Arch. Ration. Mech. Anal. **94**, 373 (1986)
10. T. Ruggeri, Symmetric hyperbolic system of conservative equations for a viscous conducting fluid. Acta Mech. **47**, 167 (1983)
11. T. Ruggeri, Galilean invariance and entropy principle for systems of balance laws. The structure of the extended thermodynamics. Continuum Mech. Thermodyn. **1**, 3 (1989)
12. T. Ruggeri, Convexity and symmetrization in relativistic theories. Privileged time-like congruence and entropy. Continuum Mech. Thermodyn. **2**, 163 (1990)
13. G. Boillat, T. Ruggeri, Hyperbolic principal subsystems: entropy convexity and subcharacteristic conditions. Arch. Ration. Mech. Anal. **137**, 305 (1997)
14. G. Boillat, T. Ruggeri, On the shock structure problem for hyperbolic system of balance laws and convex entropy. Continuum Mech. Thermodyn. **10**, 285 (1998)
15. T. Ruggeri, Maximum of entropy density in equilibrium and minimax principle for an hyperbolic system of balance laws contributions to continuum theories, anniversary volume for Krzysztof Wilmanski, ed. by B. Albers. WIAS-Report No. 18 (2000), pp. 207–214
16. T. Ruggeri, D. Serre, Stability of constant equilibrium state for dissipative balance laws system with a convex entropy. Q. Appl. Math. **62**, 163 (2004)
17. S. Kawashima, Large-time behavior of solutions to hyperbolic-parabolic systems of conservation laws and applications. Proc. R. Soc. Edinb. **106A**, 169 (1987)
18. A.E. Fischer, J.E. Marsden, The Einstein evolution equations as a first-order quasi-linear symmetric hyperbolic system. Commun. Math. Phys. **28**, 1 (1972)
19. A. Majda, *Compressible Fluid Flow and Systems of Conservation Laws in Several Space Variables* (Springer, New York, 1984)
20. C.M. Dafermos, *Hyperbolic Conservation Laws in Continuum Physics*. Grundlehren der mathematischen Wissenschaften, vol. 325, 3rd edn. (Springer, Berlin, 2010)
21. Y. Zeng, Gas dynamics in thermal nonequilibrium and general hyperbolic systems with relaxation. Arch. Ration. Mech. Anal. **150**, 255 (1999)
22. Y. Shizuta, S. Kawashima, Systems of equations of hyperbolic-parabolic type with applications to the discrete Boltzmann equation. Hokkaido Math. J. **14**, 249 (1985)
23. B. Hanouzet, R. Natalini, Global existence of smooth solutions for partially dissipative hyperbolic systems with a convex entropy. Arch. Ration. Mech. Anal. **169**, 89 (2003)
24. W.-A. Yong, Entropy and global existence for hyperbolic balance laws. Arch. Ration. Mech. Anal. **172**, 247 (2004)
25. S. Bianchini, B. Hanouzet, R. Natalini, Asymptotic behavior of smooth solutions for partially dissipative hyperbolic systems with a convex entropy. Commun. Pure Appl. Math. **60**, 1559 (2007)

26. C.M. Dafermos, Periodic BV solutions of hyperbolic balance laws with dissipative source. J. Math. Anal. Appl. **428**, 405 (2015)
27. J. Lou, T. Ruggeri, Acceleration waves and weak Shizuta-Kawashima condition. Suppl. Rend. Circ. Mat. Palermo "Non Linear Hyperbolic Fields and Waves. A Tribute to Guy Boillat" **78**, 187 (2006)
28. T. Ruggeri, Global existence of smooth solutions and stability of the constant state for dissipative hyperbolic systems with applications to extended thermodynamics, in *Trends and Applications of Mathematics to Mechanics*, STAMM 2002 (Springer, Berlin, 2005)
29. T. Ruggeri, Entropy principle and relativistic extended thermodynamics: global existence of smooth solutions and stability of equilibrium state. Il Nuovo Cimento B **119**, 809 (2004); Lecture Notes of the International Conference in honour of Y. Choquet-Bruhat: Analysis, Manifolds and Geometric Structures in Physics, ed. by G. Ferrarese, T. Ruggeri (2004)
30. T. Ruggeri, Extended relativistic thermodynamics, in Y. Choquet Bruhat, *General Relativity and the Einstein equations*, (Oxford University Press, Oxford, 2009), pp. 334–340
31. T. Ruggeri, S. Simić, On the hyperbolic system of a mixture of eulerian fluids: a comparison between single and multi-temperature models. Math. Methods Appl. Sci. **30**, 827 (2007)
32. S. Pennisi, T. Ruggeri, A new method to exploit the entropy principle and galilean invariance in the macroscopic approach of extended thermodynamics. Ricerche Mat. **55**, 159 (2006)

Chapter 3
Waves in Hyperbolic Systems

Abstract Wave propagation phenomena give us an important mean to check the validation of the nonequilibrium thermodynamics theory. In this chapter, we present a short review on the modern theory of wave propagation for hyperbolic systems.

Firstly, we present the theory of linear waves emphasizing the role of the dispersion relation. The high frequency limit in the dispersion relation is also studied. Secondly, nonlinear acceleration waves are discussed together with the transport equation and the critical time. Thirdly we present the main results concerning shock waves as a particular class of weak solutions and the admissibility criterion to select physical shocks (Lax condition, entropy growth condition, and Liu condition). The chapter finishes with the discussion of traveling waves, in particular, shock waves with structure. The sub-shock formation is particularly interesting. The Riemann problem and the large time asymptotic behavior are also discussed.

3.1 Linear Wave

A typical method to test a theory of nonequilibrium thermodynamics is to study plane harmonic waves and to compare the dispersion relation with the experimental data. Let us consider here some fundamental properties of the solutions for a general linear hyperbolic system. We therefore limit our analysis within the one-dimensional problem.

3.1.1 Plane Harmonic Waves and the Dispersion Relation

Let us consider the quasi-linear system (2.3). Because of the hyperbolicity, we may write it in the normal form:

$$\frac{\partial \mathbf{u}}{\partial t} + \mathbf{A}(\mathbf{u})\frac{\partial \mathbf{u}}{\partial x} = \mathbf{f}(\mathbf{u}). \tag{3.1}$$

We linearize this equation by setting

$$\mathbf{u} = \tilde{\mathbf{u}} + \bar{\mathbf{u}}, \tag{3.2}$$

T. Ruggeri, M. Sugiyama, *Rational Extended Thermodynamics beyond the Monatomic Gas*, DOI 10.1007/978-3-319-13341-6_3

where $\bar{\mathbf{u}}$ is a small perturbation of an equilibrium state $\tilde{\mathbf{u}}$ for which $\mathbf{f}(\tilde{\mathbf{u}}) = 0$. Thus we obtain the linearized equation:

$$\frac{\partial \bar{\mathbf{u}}}{\partial t} + \tilde{\mathbf{A}} \frac{\partial \bar{\mathbf{u}}}{\partial x} = \tilde{\mathbf{B}} \bar{\mathbf{u}}, \quad \text{where} \qquad \tilde{\mathbf{A}} = \mathbf{A}(\tilde{\mathbf{u}}), \quad \tilde{\mathbf{B}} = (\nabla \mathbf{f})_{\tilde{\mathbf{u}}}, \quad \nabla \equiv \frac{\partial}{\partial \mathbf{u}}. \tag{3.3}$$

We look for the solution of the form:

$$\bar{\mathbf{u}} = \mathbf{w} e^{i(\omega t - kx)}, \tag{3.4}$$

which represents a plane harmonic wave with real (angular) frequency ω, complex wave number $k = k_r + i k_i$, and complex amplitude $\mathbf{w}$ travelling in the x-direction.

Substitution of (3.4) into (3.3) provides a homogeneous algebraic linear system of the form:

$$\left(\mathbf{I} - z\tilde{\mathbf{A}} + \frac{i}{\omega} \tilde{\mathbf{B}} \right) \mathbf{w} = 0, \tag{3.5}$$

where $\mathbf{I}$ is the unit matrix, and z stands for k/ω.

For non-trivial solutions, the *dispersion relation*

$$\det \left(\mathbf{I} - z\tilde{\mathbf{A}} + \frac{i}{\omega} \tilde{\mathbf{B}} \right) = 0 \tag{3.6}$$

must be satisfied. The dispersion relation permits the calculation of the phase velocity v_{ph} and of the attenuation factor α in terms of the frequency ω:

$$v_{ph} = \frac{\omega}{\mathscr{R}e(k)} = \frac{1}{\mathscr{R}e(z)}, \qquad \alpha = -\mathscr{I}m(k) = -\omega \mathscr{I}m(z). \tag{3.7}$$

For linear stability, $\alpha(\omega)$ must be positive (negative) for waves travelling to the right (left). When waves start at time $t = 0$ at $x = 0$, we have the linear stability condition: $\alpha(\omega) x > 0$. Since the path of the wave is given by $x = \tilde{\lambda} t$, under these circumstances, we may write the condition of linear stability as

$$\alpha(\omega) \tilde{\lambda} > 0. \tag{3.8}$$

3.1.2 High Frequency Limit

Muracchini, Ruggeri and Seccia, in the case of simple characteristic eigenvalue λ [1], and Banach, Larecki and Ruggeri, in the case of multiple eigenvalue [2], gave rigorous expressions of v_{ph} and α in the limit of high frequency, i.e., for $\omega \to \infty$. For this purpose these authors considered the formal power-series expansions of z

and $\mathbf{w}$ with respect to $1/\omega$, viz.

$$z = \sum_{\alpha \geqslant 0} \frac{z_\alpha}{\omega^\alpha}, \qquad \mathbf{w} = \sum_{\beta \geqslant 0} \frac{\mathbf{w}_\beta}{\omega^\beta}. \tag{3.9}$$

Insertion of (3.9) into (3.6) provides a recurrence formula. It was proved, in the case of multiplicity 1, that

$$\lim_{\omega \to \infty} v_{ph}(\omega) = \lambda \qquad \text{and} \qquad \lim_{\omega \to \infty} \alpha(\omega)\lambda = -\mathbf{l} \cdot \mathbf{B} \cdot \mathbf{d}. \tag{3.10}$$

The first equation says that the phase velocity coincides—in the limit of high frequency—with the characteristic velocity. The second result in (3.10) furnishes the condition of linear stability because, from (3.8), we have the condition:

$$\mathbf{l} \cdot \mathbf{B} \cdot \mathbf{d} < 0. \tag{3.11}$$

Later we shall see that the condition (3.11) guarantees non-linear stability as well.

3.2 Acceleration Wave

For a generic quasi-linear hyperbolic system, it is possible to consider a particular class of solutions that characterizes the so-called *weak discontinuity waves* or, in the language of continuum mechanics, *acceleration waves*. Let us study a moving surface (wave front) Γ prescribed by the Cartesian equation $\phi(x, t) = 0$ that separates the space into two subspaces (see Fig. 3.1). Ahead of the wave front we have a known unperturbed field $\mathbf{u}_0(x, t)$, and behind an unknown perturbed field $\mathbf{u}(x, t)$. Both the fields $\mathbf{u}_0$ and $\mathbf{u}$ are supposed to be regular solutions of (2.1) and to be continuous across the surface Γ, but to be discontinuous in the normal derivative, i.e.,

$$[\![\mathbf{u}]\!] = 0, \quad \left[\!\!\left[\frac{\partial \mathbf{u}}{\partial \phi}\right]\!\!\right] = \mathcal{A} \neq 0, \tag{3.12}$$

where the square brackets indicate the jump at the wave front[1]:

$$[\![\cdot]\!] = (\cdot)_{\phi=0^-} - (\cdot)_{\phi=0^+}.$$

[1]For simplicity, we use the symbols g and g_0 for the values of a generic quantity g evaluated at Γ with the condition that $\phi \to 0^-$ and $\phi \to 0^+$, respectively.

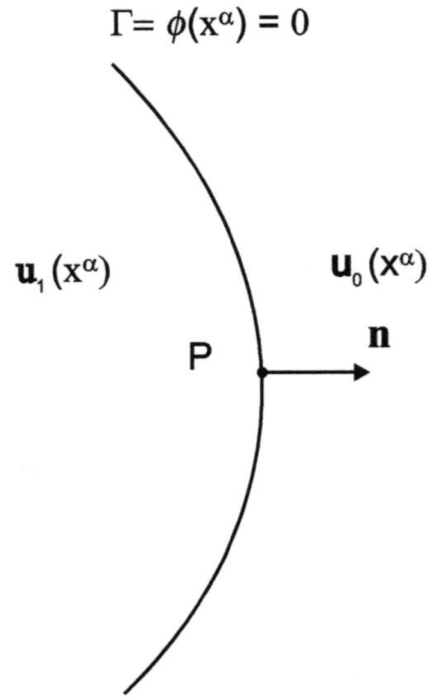

Fig. 3.1 Acceleration wave

We have the following well-known results [3–5]:

1. The normal speed $U = -\dfrac{\phi_t}{|\nabla\phi|}$ is equal to a characteristic speed evaluated in $\mathbf{u}_0$:
$$U = \lambda(\mathbf{u}_0).$$
2. The jump vector $\boldsymbol{\mathcal{A}}$ is proportional to the right eigenvector $\mathbf{d}$ of the eigenvalue λ evaluated in $\mathbf{u}_0$:
$$\boldsymbol{\mathcal{A}} = \mathscr{A}\,\mathbf{d}(\mathbf{u}_0). \tag{3.13}$$
3. The *amplitude* $\mathscr{A}$ satisfies the *Bernoulli equation* along the characteristic line:
$$\frac{d\mathscr{A}}{dt} + a(t)\mathscr{A}^2 + b(t)\mathscr{A} = 0, \tag{3.14}$$
where d/dt indicates the time derivative along the bicharacteristics lines, and $a(t)$ and $b(t)$ are known functions of the time through $\mathbf{u}_0$.

For an example in the case of one space-dimension, we have [5]:

$$\begin{aligned}
&\frac{d}{dt} = \partial_t + \lambda_0 \partial_x, \quad \frac{dx}{dt} = \lambda_0 \ (characteristic), \quad \lambda_0 = \lambda(\mathbf{u}_0) \\
&a(t) = \phi_x (\nabla \lambda \cdot \mathbf{d})_0, \qquad \nabla = \frac{\partial}{\partial \mathbf{u}}, \\
&b(t) = \left\{ d_j \left(\frac{dl_i}{du_j} - \frac{dl_j}{du_i} \right) \frac{du_i}{dt} + (\mathbf{l} \cdot \mathbf{u}_x)(\nabla \lambda \cdot \mathbf{d}) - \nabla (\mathbf{l} \cdot \mathbf{f}) \cdot \mathbf{d} \right\}_0, \\
&\frac{d\phi_x}{dt} + (\nabla \lambda \cdot \mathbf{u}_x)_0 \, \phi_x = 0, \qquad \phi_x(0) = 1.
\end{aligned} \tag{3.15}$$

The solution of (3.14) is expressed as

$$\mathscr{A}(t) = \frac{\mathscr{A}(0) exp\left(-\int_0^t b(\xi) d\xi\right)}{1 + \mathscr{A}(0) \int_0^t a(\zeta) exp\left(-\int_0^\zeta b(\xi) d\xi\right) d\zeta}. \tag{3.16}$$

In order to make an analysis of the evolution of the amplitude, we recall that, in the theory of hyperbolic systems, a wave associated to a characteristic velocity λ is called:

- *genuinely non-linear*, if

$$\nabla \lambda \cdot \mathbf{d} \neq 0 \quad \forall \mathbf{u}. \tag{3.17}$$

- *linearly degenerate* (or *exceptional*), if

$$\nabla \lambda \cdot \mathbf{d} \equiv 0 \quad \forall \mathbf{u}. \tag{3.18}$$

- *locally linearly degenerate* (or *locally exceptional*), if

$$\nabla \lambda \cdot \mathbf{d} = 0 \ \text{ for some } \mathbf{u}. \tag{3.19}$$

If a wave is genuinely nonlinear, there exists, in general, a critical time t_{cr} such that the denominator of (3.16) tends to zero and the discontinuity becomes unbounded. This instant usually corresponds to the emergence of a *strong discontinuity*, i.e., a shock wave, and the field itself presents a discontinuity across the wave front. If a wave satisfies (3.17), the coefficient $a(t) \neq 0$ and, without any loss of generality, it can always be chosen to be positive by an appropriate choice of the right eigenvector.

The qualitative analysis of the Bernoulli equation (3.14) was made by Ruggeri [5]. In particular, the stability of the zero solution of (3.14) (λ-*stability*) was proved under the conditions:

$$\int_0^\infty a(\xi) e^{-\int_0^\xi b(\zeta)d\zeta} d\xi = K < \infty, \tag{3.20}$$

$$\exists \text{ a constant } m \text{ such that } \int_0^t b(\xi)d\xi > m, \quad \forall \ t > 0. \tag{3.21}$$

In fact, if (3.20) and (3.21) are fulfilled and if

$$|\mathscr{A}(0)| < \mathscr{A}_{cr}, \qquad \mathscr{A}_{cr} = 1/K, \tag{3.22}$$

the solution $\mathscr{A}(t)$ exists for all time and remains to be bounded. Moreover, if

$$\int_0^\infty b(\xi)d\xi = +\infty, \tag{3.23}$$

then $\lim_{t\to\infty} |\mathscr{A}(t)| = 0$ and the zero solution is asymptotically stable.

If (3.22) is not satisfied and

$$\mathscr{A}(0) < 0, \qquad |\mathscr{A}(0)| > \mathscr{A}_{cr}, \tag{3.24}$$

then from (3.16) there exists a positive critical time t_{cr} given by

$$1 + \mathscr{A}(0) \int_0^{t_{cr}} a(\zeta) exp\left(-\int_0^\zeta b(\xi)d\xi\right) d\zeta = 0. \tag{3.25}$$

When the unperturbed state is an equilibrium constant state: $\mathbf{u}_0 = \mathbf{u}_E =$ constant, we have

$$\mathbf{f}(\mathbf{u}_E) = 0,$$

$$\phi_x = 1, \quad a = (\nabla\lambda \cdot \mathbf{d})_E = \text{const.}, \quad b = -(\mathbf{l} \cdot \nabla\mathbf{f} \cdot \mathbf{d})_E = \text{const.}, \tag{3.26}$$

and (3.16) becomes

$$\mathscr{A}(t) = \frac{\mathscr{A}(0)\, e^{-bt}}{1 - \mathscr{A}(0)\, \frac{a}{b}\, (e^{-bt} - 1)}. \tag{3.27}$$

In this case $\mathscr{A}_{cr} = b/a$ and the conditions (3.20) and (3.23) reduce to

$$b = -(\mathbf{l} \cdot \nabla\mathbf{f} \cdot \mathbf{d})_E > 0. \tag{3.28}$$

This is a necessary and sufficient condition for the zero solution of the Bernoulli equation to be asymptotically stable. From the definition of the K-condition (2.35), we see that, for genuinely nonlinear waves, the K-condition must be satisfied in order to satisfy (3.28) [6]. Moreover we see that linear and nonlinear λ-stability conditions, namely (3.11) and (3.28) coincide.

In the present case if the initial amplitude satisfies the inequalities (3.24) then we have a critical time:

$$t_{cr} = -\frac{1}{b}\log\left(1 + \frac{b}{a\mathscr{A}(0)}\right). \tag{3.29}$$

We notice that if the system is not dissipative and $b = 0$ the amplitude (3.16) becomes

$$\mathscr{A}(t) = \frac{\mathscr{A}(0)}{1 + \mathscr{A}(0)\,a\,t}, \tag{3.30}$$

and if $\mathscr{A}(0) < 0$ the critical time exists always and is given by

$$t_{cr} = -\frac{1}{a\mathscr{A}(0)}. \tag{3.31}$$

Instead, if $a = (\nabla\lambda \cdot \mathbf{d})_E \equiv 0$ (linearly degenerate wave), the Bernoulli equation becomes to be linear:

$$\frac{d\mathscr{A}}{dt} + b\,\mathscr{A} = 0,$$

and it does not have the critical time. Furthermore, if $b = 0$, we have $\mathscr{A}(t) = \mathscr{A}(0)$. Hence the K-condition is not necessary in this case. Therefore, we have the weaker Lou-Ruggeri K-condition:

Definition 3.1 (Weak K-Condition) A system (2.3) satisfies the weak K-condition if, in the equilibrium manifold, the right characteristic eigenvectors $\mathbf{d}$ corresponding to the genuinely nonlinear eigenvalues are not in the null space of $\nabla\mathbf{f}$.

This condition together with the dissipation condition ($b > 0$ if $a \neq 0$) is a necessary and sufficient condition such that the discontinuity wave solution exists for all time for a small initial perturbation.

Finally we observe that if we introduce the operator $\delta = [\![\partial/\partial\phi]\!]$, the characteristic velocities λ and the right eigenvectors (2.4) can be obtained by the system (2.3) with the operators chain rule:

$$\frac{\partial}{\partial t} \to -\lambda\delta, \qquad \frac{\partial}{\partial x_i} \to n_i\delta, \quad \mathbf{f} \to 0, \tag{3.32}$$

for which $\delta\mathbf{u} \propto \mathbf{d}$, and the weak K-condition reads

$$\delta\mathbf{f}|_E = (\nabla\mathbf{f}\cdot\delta\mathbf{u})\,|_E \propto (\nabla\mathbf{f}\cdot\mathbf{d})\,|_E \neq 0 \tag{3.33}$$

for all genuinely nonlinear waves. That is, the weak K-condition requires that, if the production vanishes in equilibrium: $\mathbf{f}|_E = 0$, *all the genuinely nonlinear discontinuity waves transport the disturbance of the normal derivative of the production*: $\delta\mathbf{f}|_E \neq 0$.

Remark 3.1 This weak K-condition is, in general, only necessary for global existence of smooth solutions. In fact, it is satisfied for completely linearly degenerate systems (all waves are exceptional) or for semi-linear systems. But we know that, in general, smooth solutions of these cases cannot exist for all time. Therefore we need to add more conditions to the weak K-condition in order to ensure the global existence. This problem is, however, still open!

Nevertheless, it is important to have the necessary condition in order to select physically admissible productions and also to know the possibility of the global existence of the solution when the K-condition is violated. For example, for a mixture of Euler fluids with single temperature, it was proved that the weak K-condition is violated, while for a mixture with multi-temperature the condition is satisfied [7]. This implies an important fact that the model with multi-temperature is physically more realistic than the model with single temperature!

Remark 3.2 There exists a huge literature concerning physical applications of acceleration waves. Interested readers can find some references in the papers quoted in this chapter. Moreover, other interesting examples can be seen in [8] and in the book of Sharma [9].

3.3 Shock Wave

3.3.1 Rankine-Hugoniot Relations

If the field $\mathbf{u}$ itself experiences a jump across the wave front instead of the jump of its first derivative (see Fig. 3.1), we say that it is a *shock wave*. Shock waves are possible only for systems of balance laws and they are a particular class of *weak solutions*. In fact, it is well known that a shock wave solution is a weak solution of (2.1) if and only if it satisfies the so-called *Rankine-Hugoniot relations* (RH relations, in short) or *Rankine-Hugoniot conditions* (RH conditions) across the shock front:

$$-\,s\,[\![\mathbf{F}^0]\!] + [\![\mathbf{F}^i]\!]\,n_i = 0, \tag{3.34}$$

where $[\![g(\mathbf{u})]\!] = g(\mathbf{u}_1) - g(\mathbf{u}_0)$ for a generic function g and $(\mathbf{u}_1, \mathbf{u}_0)$ are, respectively, the values of the rear (perturbed) and the front (unpertubed) sides of the wave surface. And s stands for the normal speed of a shock wave with the unit normal $\mathbf{n} \equiv (n_i)$.

The set of all perturbed states $\mathbf{u}_1$ satisfying (3.34) for a given unperturbed state $\mathbf{u}_0$ is called the *Hugoniot locus* for the point $\mathbf{u}_0$ and is denoted as $\mathscr{H}(\mathbf{u}_0)$.

If the unperturbed field $\mathbf{u}_0$ is known and we consider plane shocks with $\mathbf{n} =$ constant, the RH relations furnish a system of N equations for the $N+1$ unknowns $\mathbf{u}_1$ and s. Thus *any one* among the $(N+1)$−tuple $(\mathbf{u}_1, s)$ may be chosen as the *shock parameter* i.e., the quantity that characterizes the *strength* of the shock. Therefore, as the shock parameter, we may choose the speed s, or any component of $\mathbf{u}_1$, or a combination of these, which is generically denoted by ξ hereafter.

If we introduce the mapping:

$$\boldsymbol{\Psi}_s(\mathbf{u}) = -s\mathbf{F}^o(\mathbf{u}) + \mathbf{F}^i(\mathbf{u})n_i\,,$$

the RH relations can be written as $\boldsymbol{\Psi}_s(\mathbf{u}_1) = \boldsymbol{\Psi}_s(\mathbf{u}_0)$. Therefore the mapping $\boldsymbol{\Psi}_s(\mathbf{u})$ must be locally non-invertible in the neighborhood of $\mathbf{u}_0$:

$$\det\left(\frac{\partial \Psi_s}{\partial \mathbf{u}}\right)\bigg|_{\mathbf{u}=\mathbf{u}_0} = 0.$$

On the other hand, the Jacobian of the mapping:

$$\det\left(\frac{\partial \Psi_s}{\partial \mathbf{u}}\right)\bigg|_{\mathbf{u}=\mathbf{u}_0} = \det\left(\mathbf{A}^i n_i - s\mathbf{A}^0\right)\big|_{\mathbf{u}=\mathbf{u}_0}$$

becomes singular for $s = \lambda(\mathbf{u}_0)$ as seen from (2.5). Therefore the local non-invertibility occurs when s is equal to the characteristic speed.

In a schematic way, we may represent the values $\mathbf{u}_1$ for a given value $\mathbf{u}_0$ as the function of s. See Fig. 3.2, which indicates that *shock solutions may be seen as bifurcating curves from the trivial solution* $\mathbf{u}_1 = \mathbf{u}_0$ of (3.34) at the points where s is equal to the one of the unperturbed characteristic speeds λ_0.

3.3.2 Admissibility of Shock Waves

According to the theory of hyperbolic systems, not every solution of the RH relations corresponds to a physically meaningful shock wave. Thus, we need a criterion to select the perturbed states $\mathbf{u}_1$ that, together with $\mathbf{u}_0$, form *admissible* shocks. Since admissible shocks propagate with no change in shape when they evolve from the initial data of a Riemann problem, these solutions are sometimes called *stable shocks*.

The issue of shock admissibility, when genuinely nonlinear and linearly degenerate waves are involved, has been largely and deeply investigated in the past decades. For example, the hyperbolic system of an Euler ideal gas features only

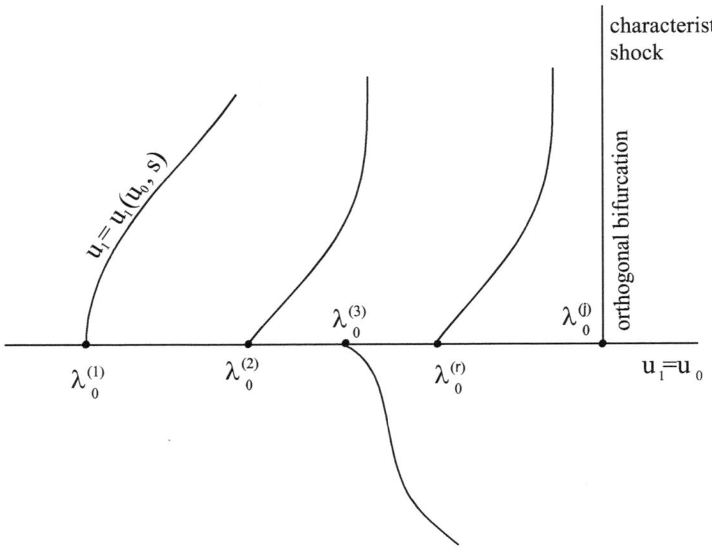

Fig. 3.2 Shocks as bifurcating branches of the trivial solution $\mathbf{u}_1 = \mathbf{u}_0$ with s being the shock parameter

waves belonging to these two types. On the contrary, the Euler system of a van der Waals fluid features linearly degenerate and locally linearly degenerate waves. A comprehensive analysis of the shock admissibility in this kind of fluids was the subject of a recent paper [10].

3.3.2.1 Lax, Entropy Growth, and Liu Admissibility Conditions

The selection rule useful to study the admissibility of shock waves depends on the type of the nonlinear waves involved. Thus, it is necessary to discuss separately the cases of genuinely non-linear, linearly degenerate, and locally linearly degenerate waves.

- When we deal with genuinely non-linear waves, the selection rule is given by the *Lax condition* [11], according to which a shock wave is admissible if the shock velocity satisfy the inequality:

$$\lambda_0 < s < \lambda_1,$$

where $\lambda_0 \equiv \lambda(\mathbf{u}_0)$ and $\lambda_1 \equiv \lambda(\mathbf{u}_1)$ are the unperturbed and perturbed characteristic velocities, respectively. Such a shock wave is called *k-shock* (being λ is the *kth* eigenvalue of the system). The Lax condition turns out to be equivalent (at least for *weak shock waves*) to the condition of entropy growth

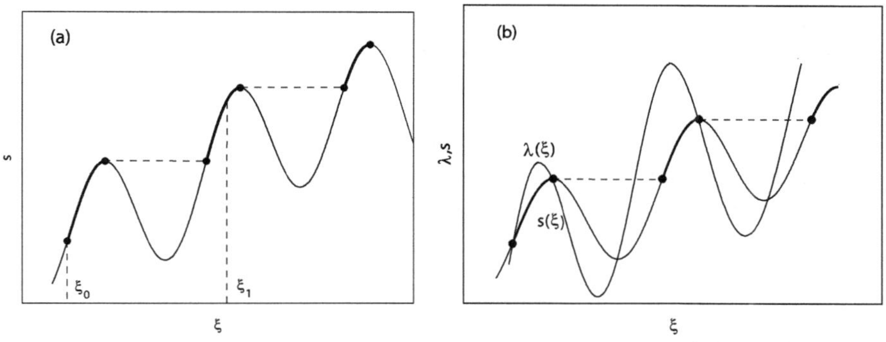

Fig. 3.3 Shock speed s as a function of the shock parameter ξ and the range of the admissible shocks (*bold curve*). λ is the characteristic speed

across the shock:

$$\eta = s[\![h^0]\!] - [\![h_n]\!] > 0, \tag{3.35}$$

where $h_n = h^i n_i$.

- When we deal with linearly degenerate waves, admissible *k-shocks* are called *characteristic shocks* and they propagate with the speed $s = \lambda_0 = \lambda_1$. In this case, there is no entropy growth across the shock, i.e., $\eta = 0$. The admissibility of a characteristic shock may thus be studied by means of the so-called *generalized Lax condition*:

$$\lambda_0 \leqslant s \leqslant \lambda_1.$$

  A characteristic shock depends on as many parameters as the multiplicity of the eigenvalue λ [12]; the system of equations of an Euler fluid in three space dimensions, for example, exhibits an eigenvalue of multiplicity three, and the shock wave associated to this eigenvalue is thus a characteristic shock depending on three parameters.
- When the system features locally linearly degenerate waves, the selection rule is given by the *Liu condition* [13, 14], stating that a shock wave is admissible if

$$s \geqslant s_*, \ \forall s_* \in \{s_* : s_* (\mathbf{u}_* - \mathbf{u}_0) = \mathbf{F}(\mathbf{u}_*) - \mathbf{F}(\mathbf{u}_0),\ \mathbf{u}_* \in \mathscr{H}(\mathbf{u}_0) \ \text{between}\, \mathbf{u}_0 \ \text{and}\, \mathbf{u}_1\}.$$

  This means that a shock is admissible if its speed, s, is not smaller than the speed of any other shock with the same unperturbed state $\mathbf{u}_0$ and with perturbed state $\mathbf{u}_*$ lying on the Hugoniot locus for $\mathbf{u}_0$ between $\mathbf{u}_0$ and $\mathbf{u}_1$ (see Fig. 3.3).

The entropy growth in this case is not sufficient for the admissibility and we need a superposition principle (see [15]).

Concerning shock waves, we want to remark interesting phenomena, that is, shock-induced phase transitions in real gases. For details, see [10, 16] for a van der Waals gas, and [17–21] for a hard-sphere gas.

Moreover there exists also a general theory concerning the interaction between shocks and discontinuity waves given in the work of Boillat and Ruggeri [22] and interesting application to fluids due to Ruggeri [23], Pandey and Sharma [24] and a complete analysis also with numerical simulation due to Mentrelli, Ruggeri, Sugiyama and Zhao [25].

3.4 Shock Structure

A shock wave is, in reality, not a discontinuous surface but has a structure with sharp but continuous transition from an unperturbed state to a perturbed state. Typically both states are two different equilibrium states. As the shock profile depends on the shock parameter, say, the unperturbed Mach number M_0, the thickness of a shock changes with this parameter. The experiments of Alsemeyer [26] in monatomic gases show that the thickness decreases until $M_0 \sim 3.1$ and then increases with the Mach number. Several authors tried to explain this behavior of the shock thickness [27–31]. The satisfactory results obtained by molecular dynamics simulations were first presented by Bird [32] in 1970. But until now no complete satisfactory phenomenological approach exists.

New interest on this subject was aroused by Ruggeri in 1993 [33]. The general mathematical structure of RET [34] was recognized, and the problem of the shock wave structure was analyzed in the context of hyperbolic systems of balance laws. With this analysis, as mentioned above, Ruggeri noticed that a singularity appears each time when the shock speed meets a characteristic eigenvalue. As a consequence, the singularity, which occurred at $M_0 = 1.65$ in the 13-moment theory, moves closer and closer to $M_0 = 1$ as the number of moments increases. Therefore the sub-shock seems to appear sooner, and no smooth shock structure could be expected or, at best, it exists only very close to $M_0 = 1$. This interpretation is, however, not correct. In fact Weiss [35], through a numerical approach, showed that at least up to 35 moments all but one of these singularities are regular ones. The singular one corresponds to the highest characteristic speed evaluated in equilibrium in front of the shock $\lambda_0^{(\max)}$. Thus he calculated smooth shock structures up to $s = \lambda_0^{(\max)}$. Beyond this value of s, the sub-shock appears. This numerical evidence of the sub-shock formation when $s > \lambda_0^{(\max)}$ was confirmed by a theorem due to Boillat and Ruggeri [36]. Therefore, as the maximum characteristic speed increases with the number of moments [37, 38], according to the Weiss conjecture, the quantitative features of the shock structure improve with more and more moments. In conclusion, RET with many moments is the natural theory that is able to explain the shock wave structure. For a comprehensive survey on this topic, see, for example, the book of RET by Müller and Ruggeri [34].

As was explained in Chap. 1, the shock wave structure in polyatomic gases are quite different from that in monatomic gases. This is one of the main topics to be studied in the present book.

In this section, we summarize mathematical background knowledge on shock structure solutions for a generic hyperbolic dissipative system of balance laws to which the models of RET belong.

3.4.1 Shock Wave Structure and Sub-shock Formation

The shock wave structure is a regular solution depending on one variable:

$$\mathbf{u} \equiv \mathbf{u}(\varphi), \qquad \varphi = x^i n_i - st, \qquad s = \text{const.}, \quad \mathbf{n} \equiv (n_i) = \text{const.}, \tag{3.36}$$

and

$$\lim_{\varphi \to \pm\infty} \mathbf{u}(\varphi) = \begin{cases} \mathbf{u}_0 \\ \mathbf{u}_1 \end{cases}; \qquad \lim_{\varphi \to \pm\infty} \frac{d\mathbf{u}}{d\varphi} = 0, \tag{3.37}$$

i.e., a plane wave solution connecting the two constant states.[2]

Substituting (3.36) into (2.1), we obtain the ordinary differential system:

$$\frac{d}{d\varphi}\left\{-s\mathbf{F}^0(\mathbf{u}) + \mathbf{F}_n(\mathbf{u})\right\} = \mathbf{f}(\mathbf{u}) \tag{3.38}$$

with the boundary conditions (3.37), where $\mathbf{F}_n = \mathbf{F}^i n_i$. This system is equivalent to the following ODE system:

$$\left(-s\mathbf{A}^0 + \mathbf{A}_n\right)\frac{d\mathbf{u}}{d\varphi} = \mathbf{f}(\mathbf{u}), \tag{3.39}$$

($\mathbf{A}_n = \mathbf{A}^i n_i$). As was noticed by Ruggeri [33], when s approaches a characteristic eigenvalue λ [see (2.4)], the solution may have a breakdown.

In physics, typically in RET, M field equations are conservation laws and $N - M$ are balance laws, i.e., these have the structure given in (2.26) and (2.27). We rewrite these by separating space and time as follows:

$$\begin{cases} \dfrac{\partial \mathbf{V}(\mathbf{u})}{\partial t} + \dfrac{\partial \mathbf{P}^i(\mathbf{u})}{\partial x^i} = 0, \\[2ex] \dfrac{\partial \mathbf{W}(\mathbf{u})}{\partial t} + \dfrac{\partial \mathbf{R}^i(\mathbf{u})}{\partial x^i} = \mathbf{g}(\mathbf{u}), \end{cases} \tag{3.40}$$

---

[2] As the balance laws satisfy the Galilean invariance, it is possible to adopt the frame moving with the shock velocity. For such an observer the wave appears stationary: $\varphi = x$.

where $\mathbf{V}, \mathbf{P}^i \in R^M$, while $\mathbf{W}, \mathbf{R}^i$ and $\mathbf{g}$ are vectors of R^{N-M}:

$$\mathbf{F}^0 \equiv \begin{bmatrix} \mathbf{V} \\ \mathbf{W} \end{bmatrix}, \qquad \mathbf{F}^i \equiv \begin{bmatrix} \mathbf{P}^i \\ \mathbf{R}^i \end{bmatrix}, \qquad \mathbf{f} \equiv \begin{bmatrix} 0 \\ \mathbf{g} \end{bmatrix}, \tag{3.41}$$

and the field $\mathbf{u}$ is expressed by a pair: $\mathbf{u} \equiv (\mathbf{v}, \mathbf{w})$ with $\mathbf{v} \in R^M$ and $\mathbf{w} \in R^{N-M}$.

On the other hand it was proved in [38] that the associated system

$$\frac{\partial \mathbf{V}(\mathbf{v}, 0)}{\partial t} + \frac{\partial \mathbf{P}^i(\mathbf{v}, 0)}{\partial x^i} = 0, \tag{3.42}$$

obtained from $(3.40)_1$ putting $\mathbf{w} \equiv 0$, is a *principal sub-system* of (3.40) (see Sect. 2.4) and then satisfies the entropy principle together with the following *sub-characteristic conditions* (see also [38])

$$\max_{k=1,2,\dots,N} \lambda^{(k)}(\mathbf{v}, 0) \geqslant \max_{J=1,2,\dots,M} \mu^{(J)}(\mathbf{v}) \tag{3.43}$$

and similarly for the minimum. The λ's are the characteristic speeds of the total system (2.1) while the μ's are the ones of the equilibrium subsystem (3.42) (in ET this system corresponds to the Euler fluid system).

Therefore (3.38) can be rewritten as:

$$\begin{cases} \frac{d}{d\varphi}\{-s\mathbf{V}(\mathbf{v}, \mathbf{w}) + \mathbf{P}_n(\mathbf{v}, \mathbf{w})\} \quad = 0, \\ \\ -s\frac{d}{d\varphi}\mathbf{W}(\mathbf{v}, \mathbf{w}) + \frac{d}{d\varphi}\mathbf{R}_n(\mathbf{v}, \mathbf{w}) = \mathbf{g}(\mathbf{v}, \mathbf{w}). \end{cases} \tag{3.44}$$

From $(3.44)_2$ evaluated at $\varphi \to \pm\infty$ with the condition $(3.37)_2$, we obtain

$$\mathbf{g}(\mathbf{v}_0, \mathbf{w}_0) = \mathbf{g}(\mathbf{v}_1, \mathbf{w}_1) = 0, \tag{3.45}$$

which implies that the solutions at infinity on both sides are equilibrium solutions (see [36, 38]):

$$\mathbf{w}_0 = \mathbf{w}_1 = 0. \tag{3.46}$$

By $(3.44)_1$, we have

$$-s\mathbf{V}(\mathbf{v}, \mathbf{w}) + \mathbf{P}_n(\mathbf{v}, \mathbf{w}) = \mathbf{c} = \text{const.} \tag{3.47}$$

That is, the quantity on the left-hand side is conserved along the process. In particular, for $\varphi \to \pm\infty$ [see (2.29)], we have

$$\mathbf{c} = -s\mathbf{V}(\mathbf{v}_0, 0) + \mathbf{P}_n(\mathbf{v}_0, 0) = -s\mathbf{V}(\mathbf{v}_1, 0) + \mathbf{P}_n(\mathbf{v}_1, 0). \tag{3.48}$$

Equation (3.48) represents the Rankine-Hugoniot conditions for shocks of the *equilibrium sub-system* (3.42), and therefore

$$\mathbf{v}_1 \equiv \mathbf{v}_1(\mathbf{v}_0, s). \tag{3.49}$$

The problem now consists in finding a C^1 solution of (3.38) connecting the two equilibrium state $\mathbf{u}_0 \equiv (\mathbf{v}_0, 0)$ and $\mathbf{u}_1 \equiv (\mathbf{v}_1(\mathbf{v}_0, s), 0)$ for increasing value of the shock velocity s and the prescribed values $\mathbf{v}_0$ and $\mathbf{n}$.

3.4.2 Non-existence of Smooth Shocks When $s > \lambda^{\max}(\mathbf{u}_o)$

As any smooth solutions of (3.38) are also the solutions of (1.22), we must have the condition for shock structure solutions:

$$\frac{d}{d\varphi}\left\{-s\,h^0(\mathbf{u}) + h_n(\mathbf{u})\right\} = \Sigma(\mathbf{u}) \geqslant 0, \tag{3.50}$$

where $h_n = h^i n_i$. This implies that the bracketed term is a increasing function of φ. Therefore we have

$$-s\,h^0(\mathbf{u}_1) + h_n(\mathbf{u}_1) \leqslant -s\,h^0(\mathbf{u}) + h_n(\mathbf{u}) \leqslant -s\,h(\mathbf{u}_0) + h_n(\mathbf{u}_0).$$

In particular, the entropy grows across the shock (3.35). The entropy growth condition (3.35) is equivalent, at least for moderately strong shocks, to the *Lax condition* for admissibility of shocks. For a fixed eigenvalue μ of the equilibrium sub-system (3.42), a shock satisfies

$$\mu(\mathbf{v}_0) < s < \mu(\mathbf{v}_1); \qquad \lim_{s \to \mu(\mathbf{v}_0)} \mathbf{v}_1(\mathbf{v}_0, s) = \mathbf{v}_0. \tag{3.51}$$

For a fixed eigenvalue $\mu(\mathbf{v}_0) \in \mu^{(J)}(\mathbf{v}_0)$, $J = 1, 2, \ldots, M$ of the equilibrium sub-system, we start from the trivial shock for which $s = \mu(\mathbf{v}_0)$ and we increase s satisfying the Lax condition (3.51). Boillat and Ruggeri was able to prove the following theorem [36]:

Theorem 3.1 (Sub-shock Formation) *Consider a system of N balance laws of which $M < N$ are conservation laws. Under the assumption that the system satisfies the entropy principle with convex entropy, a C^1 shock wave structure propagating with the velocity s greater than the maximum characteristic speed evaluated in the equilibrium state in front of the shock cannot exist, i.e., smooth solutions may exist only if $s \leqslant \lambda^{\max}(\mathbf{u}_0)$.*

The proof is based on the verification that smooth shock structure solutions must satisfy the entropy inequality (3.50) and this inequality is violated when $s > \lambda^{\max}(\mathbf{u}_0)$.

Therefore after the shock speed exceeds the maximum characteristic eigenvalue in the unperturbed state, a sub-shock arises necessarily. An open problem is to see whether there exists a physical case for which a sub-shock arises at the eigenvalues less than the maximum one. Until now, in all ET theories for any number of moments, a shock is continuous until it reaches the maximum eigenvalue. If this is a general property, it is very favorable because the increase of the number of moments implies the increase of the maximum characteristic velocity and also the increase of the critical shock velocity.

Concerning the existence of shock structure we would mention among others, the classical paper by Gilbarg [39] in the context of parabolic continuum theory of fluids, and by Yong and Zumbrun [40] for hyperbolic systems with relaxation. Interesting analysis in the context of stability and bifurcation analysis of stationary points was done by Simić [41, 42].

3.5 Riemann Problem for Balance Laws

Bernhard Riemann raised the famous question about the time-evolution of a gas under the initial condition that the gas is divided into two regions by a thin diaphragm. Each region is filled with the same gas, but with different values of the thermodynamic quantities such as the pressure, the density, and the temperature. This problem is of fundamental importance and still has many applications. An experimental apparatus called shock tube has been used in a wide variety of aerodynamic or ballistic topics like supersonic aircraft flight, gun performance, asteroid impacts, shuttle atmospheric entry, etc.

From a mathematical view point, the Riemann problem consists of an initial value problem associated to hyperbolic conservation laws. This problem, at least in one-space dimension, has been completely solved (see e.g. [11, 43–47]).

3.5.1 Riemann Problem with Structure

Liu noticed that Riemann initial data [48] can be regarded as a rough approximation of initial data containing sharp and rapid variations. Therefore this problem takes into account the presence of shock thickness, oscillations, noises, and continuous (although very steep) changes, since initial data coincide with the Riemann problem only for large $|x|$ (that is to say, they are constant equilibrium states for large $|x|$). This problem is called Riemann problem with structure (see Fig. 3.4).

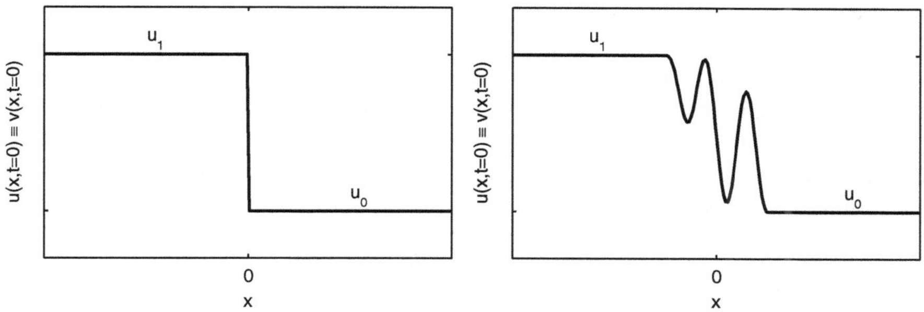

Fig. 3.4 Riemann problem (*left*) and the Riemann problem with structure (*right*)

In the case of conservation laws of hyperbolic type, Liu [48] proved, roughly speaking, that the solution of this problem converges, for large time, to the solution of the corresponding Riemann problem.

3.5.2 Conjecture Concerning Large-Time Asymptotic Behavior of Shock Structure

While, for balance laws, any mathematical theory of Riemann problem does not exist because a system of balance laws cannot admit rarefaction wave solutions depending on x/t. Recently Ruggeri and coworkers [49–51]—following Liu [52]—proposed a conjecture about the large-time behavior of the Riemann problem and the Riemann problem with structure for a system of balance laws. According to this conjecture, solutions of both Riemann problems with and without structure converge, for large time, to a solution that represents the combination of shock structures (with and without sub-shocks) and rarefactions of the equilibrium subsystem.

More precisely, let us consider Riemann data (or Riemann data with structure) for the system of balance laws (3.40) between two equilibrium constant state $(\mathbf{u}_0, \mathbf{u}_1)$ with $\mathbf{g}(\mathbf{u}_1) = \mathbf{g}(\mathbf{u}_0) = 0$. We first study the corresponding Riemann problem for the equilibrium subsystem (3.42). According with the general theory, we have a combination of rarefactions $\mathscr{R}$ and shocks $\mathscr{S}$ plus contact discontinuities and constant states. Now the conjecture is that the Riemann problem for the full system converges, for large time, to the same rarefactions $\mathscr{R}$ of the equilibrium subsystem and, instead of the shock $\mathscr{S}$, we have a corresponding sub-shock of the full system $\mathscr{S}_{struc}$ (including sub-shock) corresponding to the same state of $\mathscr{S}$. In particular, if the Riemann initial data correspond to a particular shock family $\mathscr{S}$, then for large time the Riemann problem of the full system converges to the corresponding shock structure. This means that the numerical study of the shock structure, instead of using a complex mathematical solver of ODE, can be used as a Riemann solvers [47] if we wait enough time after the initial time. This strategy seems to be useful in ET as will be discussed in the following chapters.

The conjecture was tested numerically for a Grad 13-moment system and a mixture of fluids [49, 50]. In the followings, we explain the conjecture by studying a simple 2×2 dissipative model considered by Mentrelli and Ruggeri [51] for which we can calculate analytically the shock structures and the rarefactions of the equilibrium subsystem. This toy model has all the properties of the ET systems: convexity of the entropy, dissipative character, and satisfaction of the K-condition. Its system of equations is given by

$$\begin{aligned} (u+v)_t + \left(v^2 + 2uv\right)_x &= 0, \\ u_t + \left(v^2\right)_x &= \frac{1}{\tau}(v-u). \end{aligned} \tag{3.52}$$

This system has the characteristics eigenvalues:

$$\lambda^{(1)} = u - \sqrt{u^2 + 4v^2}, \qquad \lambda^{(2)} = u + \sqrt{u^2 + 4v^2}. \tag{3.53}$$

The equilibrium subsystem of (3.52) and its eigenvalue μ are given by

$$u_t + 3uu_x = 0, \qquad \mu = 3u. \tag{3.54}$$

It is not difficult to find analytically the shock structure of the full system and the rarefaction of the equilibrium subsystem. Then, choosing the initial data corresponding to the shock $\mathscr{S}$ of (3.54), we integrate numerically the full system (3.52) by means of a numerical code based on the algorithms presented in [53]. We can see that the solution converges, in accordance with the conjecture, to the shock structure $\mathscr{S}_{struc}$. See Figs. 3.5, 3.6, 3.7 and 3.8.

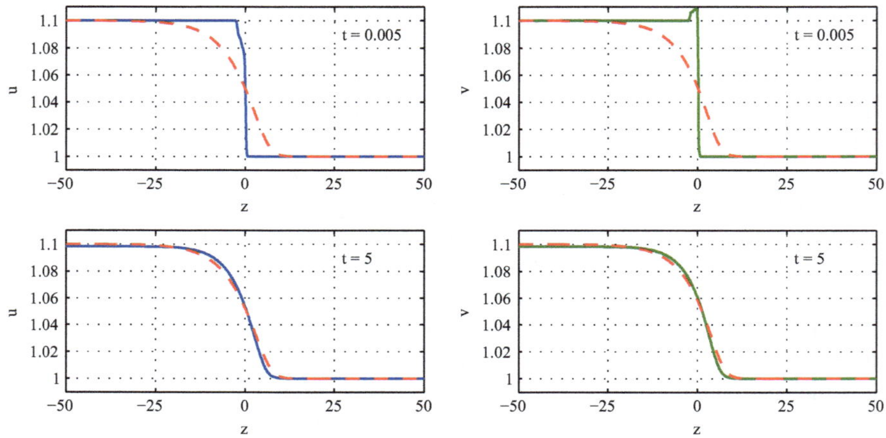

Fig. 3.5 Numerical solution of the Riemann problem with initial data (*solid line*) and analytical continuous shock structure solution (*dashed line*) at different times t

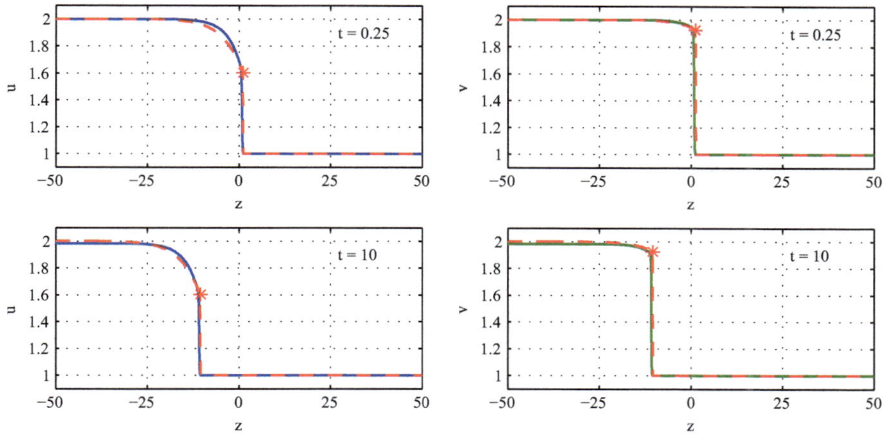

Fig. 3.6 Numerical solution of the Riemann problem (*solid line*) and analytical shock structure solution with sub-shock (*dashed line*) at different times t. The symbol "*asterisk*" indicates the point (U_*, V_*) where the continuity of the analytical shock structure solution is lost

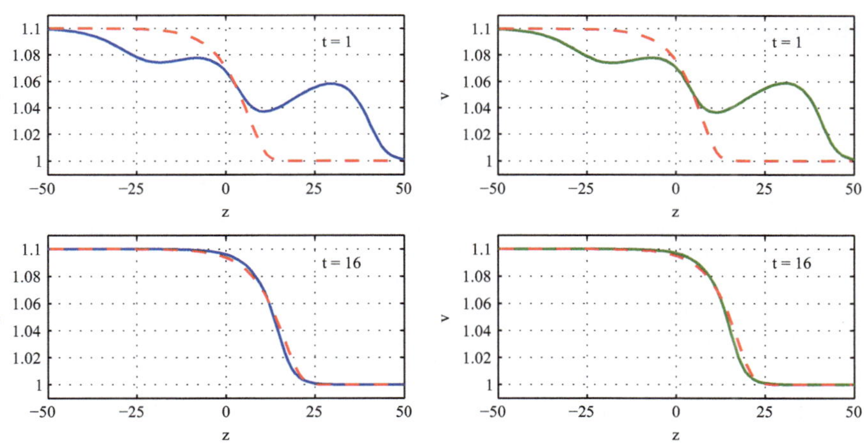

Fig. 3.7 Numerical solution of the Riemann problem (*solid line*) and analytical continuous shock structure solution (*dashed line*) at different times t

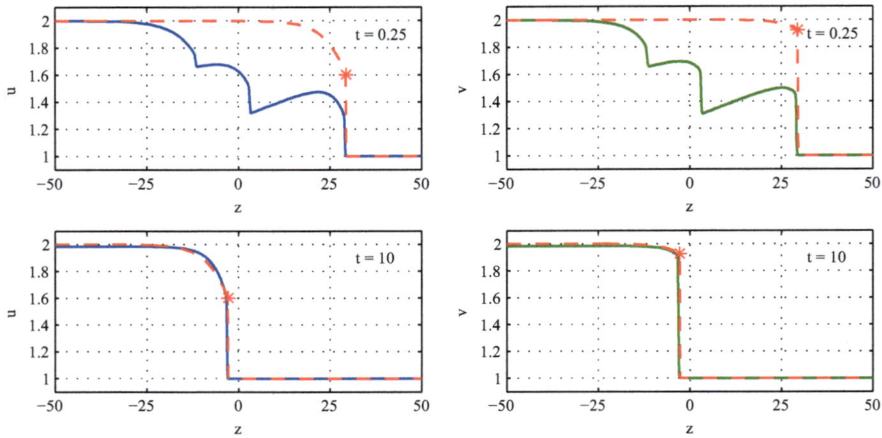

Fig. 3.8 Numerical solution of the Riemann problem (*solid line*) and analytical shock structure solution with sub-shock (*dashed line*) at different times t. The symbol "*asterisk*" indicates the point (U_*, V_*) where the continuity of the analytical shock structure solution is lost

References

1. A. Muracchini, T. Ruggeri, L. Seccia, Dispersion relation in the high frequency limit and non linear wave stability for hyperbolic dissipative systems. Wave Motion **15**(2), 143 (1992)
2. Z. Banach, W. Larecki, T. Ruggeri, Dispersion relation in the limit of high frequency for a hyperbolic system with multiple eigenvalues. Wave Motion **51**, 955 (2014)
3. G. Boillat, *La Propagation des Ondes* (Gauthier-Villars, Paris, 1965)
4. G. Boillat, T. Ruggeri, On the evolution law of the weak discontinuities for hyperbolic quasi-linear systems. Wave Motion **1**(2), 149 (1979)
5. T. Ruggeri, Stability and discontinuity waves for symmetric hyperbolic systems, in *Non-linear Wave Motion*, ed. by A. Jeffrey (Longman Press, New York, 1989), pp. 148–161
6. J. Lou, T. Ruggeri, Acceleration waves and weak Shizuta-Kawashima condition. Suppl. Rend. Circ. Mat. Palermo "Non Linear Hyperbolic Fields and Waves. A Tribute to Guy Boillat" **78**, 187 (2006)
7. T. Ruggeri, S. Simić, On the hyperbolic system of a mixture of eulerian fluids: a comparison between single and multi-temperature models. Math. Methods Appl. Sci. **30**, 827 (2007)
8. G. Boillat, in CIME Course, *Recent Mathematical Methods in Nonlinear Wave Propagation*, ed. by T. Ruggeri. Lecture Notes in Mathematics, vol. 1640 (Springer, Berlin, 1995), pp. 103–152
9. V.D. Sharma, *Quasilinear Hyperbolic Systems, Compressible Flows, and Waves*. Pure and Applied Mathematics, vol. 142 (CRC Press, Boca Raton, 2010). ISBN 978-1-4398-3690-3
10. N. Zhao, A. Mentrelli, T. Ruggeri, M. Sugiyama, Admissible shock waves and shock-induced phase transitions in a van der waals fluid. Phys. Fluids **23**, 086101 (2011)
11. P.D. Lax, Hyperbolic systems of conservation laws and the mathematical theory of shock waves, in *CBMS-NSF, Regional Conference Series in Applied Mathematics*, vol. 11 (SIAM, Providence, 1973)
12. G. Boillat, Chocs caractéeristiques C. R. Acad Sci. Paris A **274**, 1018 (1972)
13. T.-P. Liu, The entropy condition and the admissibility of shocks. J. Math. Anal. Appl. **53**, 78 (1976)

14. T.-P. Liu, Admissible solutions of hyperbolic conservation laws. Mem. Am. Math. Soc. **240**, 12 (1981)
15. T.-P. Liu, T. Ruggeri, Entropy production and admissibility of shocks. Acta Math. Appl. Sinica, English Ser. **19**, 1 (2003)
16. S. Taniguchi, A. Mentrelli, T. Ruggeri, M. Sugiyama, N. Zhao, Prediction and simulation of compressive shocks with lower perturbed density for increasing shock strength in real gases. Phys. Rev. E **82**, 036324 (2010)
17. N. Zhao, M. Sugiyama, T. Ruggeri, Phase transition induced by a shock wave in hard-sphere and hard-disc systems. J. Chem. Phys. **129**, 054506 (2008)
18. S. Taniguchi, A. Mentrelli, N. Zhao, T. Ruggeri, M. Sugiyama, Shock-induced phase transition in systems of hard spheres with internal degrees of freedom. Phys. Rev. E **81**, 066307 (2010)
19. Y. Zheng, N. Zhao, T. Ruggeri, M. Sugiyama, S. Taniguchi, Non-polytropic effect on shock-induced phase transitions in a hard-sphere system. Phys. Lett. A **374**, 3315 (2010)
20. S. Taniguchi, N. Zhao, M. Sugiyama, Shock-induced phase transitions from gas phase to solid phase. J. Phys. Soc. Jpn. **80**, 083401 (2011)
21. S. Taniguchi, M. Sugiyama, Shock-induced phase transitions in systems of hard spheres with attractive interactions. Acta Appl. Math. **122**, 473 (2012)
22. G. Boillat, T. Ruggeri, Reflection and transmission of discontinuity waves through a shock wave, general theory including also the case of characteristic shocks. Proc. R. Soc. Edinb. **83A**, 17 (1979)
23. T. Ruggeri, Interaction between a discontinuity wave and a shock wave: critical time for the fastest transmitted wave, example of the polytropic fluid. Appl. Anal. **11**, 103 (1980)
24. M. Pandey, V.D. Sharma, Interaction of a characteristic shock with a weak discontinuity in a non-ideal gas. Wave Motion **44**, 346 (2007)
25. A. Mentrelli, T. Ruggeri, M. Sugiyama, N. Zhao, Interaction between a shock and an acceleration wave in a perfect gas for increasing shock strength. Wave Motion **45**, 498 (2008)
26. H. Alsmeyer, Density profiles in argon and nitrogen shock waves measured by the absorption of an electron beam. J. Fluid Mech. **74**, 497 (1976)
27. D. Gilbarg, D. Paolucci, The structure of shock waves in the continuum theory of fluids. J. Rat. Mech. Anal. **2**, 617 (1953)
28. R. Becker, Stoßwelle und Detonation. Z. f. Physik **8**, 321 (1922)
29. H.W. Mott-Smith, The solution of the Boltzmann equation for a shock wave. Phys. Rev. **82**, 885 (1951)
30. K. Zoller, Zur Struktur des Verdichtungsstoßes. Zeitschr. f. Physik **130**, 1 (1951)
31. J.D. Foch, On higher order hydrodynamic theories of shock structure. Acta Physica Austriaca, Suppl. **10**, 123 (1973)
32. G.A. Bird, Aspects of the structure of strong shock waves. Phys. Fluids **13**, 1172 (1970)
33. T. Ruggeri, Breakdown of shock-wave-structure solutions. Phys. Rev. E **47**, 4135 (1993)
34. I. Müller, T. Ruggeri, *Extended Thermodynamics*, 1st edn. Springer Tracts in Natural Philosophy, vol. 37 (Springer, New York, 1993)
35. W. Weiss, Continuous shock structure in extended thermodynamics. Phys. Rev. E **52**, R5760 (1995)
36. G. Boillat, T. Ruggeri, On the shock structure problem for hyperbolic system of balance laws and convex entropy. Continuum Mech. Thermodyn. **10**, 285 (1998)
37. G. Boillat, T. Ruggeri, Moment equations in the kinetic theory of gases and wave velocities. Continuum Mech. Thermodyn. **9**, 205 (1997)
38. G. Boillat, T. Ruggeri, Hyperbolic principal subsystems: entropy convexity and subcharacteristic conditions. Arch. Ration. Mech. Anal. **137**, 305 (1997)
39. D. Gilbarg, The existence and limit behavior of the one-dimensional shock layer. Am. J. Math. **73**, 256 (1951)
40. K. Zumbrun, W.-A. Yong, Existence of relaxation shock profiles for hyperbolic conservation laws. SIAM J. Appl. Math. **60**, 1565 (2000)
41. S. Simić, Shock structure in continuum models of gas dynamics: stability and bifurcation analysis. Nonlinearity **22**, 1337 (2009)

42. S. Simić, Shock structure in the mixture of gases: stability and bifurcation of equilibria, in *Thermodynamics – Kinetics of Dynamic Systems*, Chap. 8, ed. by J. Carlos, M. Pirajãn (InTech, Rijeka, 2011)
43. P. Lax, Hyperbolic systems of conservation laws II. Commun. Pure Appl. Math. **10**, 537 (1957)
44. C.M. Dafermos, *Hyperbolic Conservation Laws in Continuum Physics*. Grundlehren der mathematischen Wissenschaften, vol. 325, 3rd edn. (Springer, Berlin, 2010)
45. J. Smoller, *Shock Waves and Reaction -Diffusion Equations* (Springer, New York, 1994)
46. D. Serre, *Systems of Conservation Laws 1 Hyperbolicity, Entropies, Shock Waves* (Cambridge University Press, Cambridge, 1999)
47. E. Toro, *Riemann Solvers and Numerical Methods for Fluid Dynamics* (Springer, Berlin, 2009)
48. T.-P. Liu, Linear and nonlinear large-time behavior of solutions of general systems of hyperbolic conservation laws. Commun. Pure Appl. Math. **30**, 767 (1977); Large-time behavior of Solutions of Initial and Initial-Boundary Value Problems of a general system of hyperbolic conservation laws. Commun. Math. Phys. **55**, 163 (1977)
49. F. Brini, T. Ruggeri, On the Riemann problem in extended thermodynamics, in *Proceedings of the 10th International Conference on Hyperbolic Problems (HYP2004)*, Osaka, 13–17 Sept 2004, vol. I (Yokohama Publisher Inc., Yokohama, 2006), p. 319
50. F. Brini, T. Ruggeri, On the Riemann problem with structure in extended thermodynamics. Suppl. Rend. Circ. Mat. Palermo "Non Linear Hyperbolic Fields and Waves. A Tribute to Guy Boillat", Serie II **78**, 21 (2006)
51. A. Mentrelli, T. Ruggeri, Asymptotic behavior of Riemann and Riemann with structure problems for a 2×2 hyperbolic dissipative system. Suppl. Rend. Circ. Mat. Palermo II **78**, 201 (2006)
52. T.-P. Liu, Nonlinear hyperbolic-dissipative partial differential equations, in *Recent Mathematical Methods in Nonlinear Wave Propagation*, ed. by T. Ruggeri. Lecture Notes in Mathematics, vol. 1640 (Springer, Berlin, 1996), pp. 103–136
53. S.F. Liotta, V. Romano, G. Russo, Central scheme for balance laws of relaxation type. SIAM J. Numer. Anal. **38**, 1337 (2000)

Part II
Survey of Rational Extended Thermodynamics of Monatomic Gas

Chapter 4
RET of Rarefied Monatomic Gas

Abstract In this chapter, we give a survey of the main results of RET concerning rarefied monatomic gases, some of which are explained in the Müller–Ruggeri book of RET (Müller and Ruggeri, Rational Extended Thermodynamics, Springer, New York, 1998). We start from the phenomenological RET theory with 13 fields and prove that the closure of RET coincides with the one obtained by Grad using kinetic arguments and with the MEP procedure. The theory with N-moments is also presented with the proof of nesting theories that emerge from the concept of principal subsystem. The problematic of bounded domain in RET is also considered, and a simple example of heat conduction is explained to show a significant difference of the results between RET and NSF.

A lower bound for the maximum characteristic velocity is obtained in terms of the truncation tensor index N. This quantity increases as the number of moments grows and it is unbounded when $N \to \infty$.

The relativistic counterpart is also described briefly. In this framework, the maximum characteristic velocity is bounded for any number of moments, and converges to the light velocity from the below for $N \to \infty$.

The chapter contains also comparison between the RET theory and experiments in sound waves and light scattering.

4.1 Extended Thermodynamics with 13 Fields and Subsystems

We saw, in Sect. 1.5.1, that RET of rarefied monatomic gases is intimately related to the moment theory (1.18) associated with the Boltzmann equation. As emphasized there, the 13-field RET theory is a purely phenomenological theory [1] in which the system of field equations in balance type (1.24) is adopted and the closure of the system with the use of the universal principles of physics is accomplished. We saw also, in Sect. 1.5.2, the problem of closure of the 13-moment theory.

We have three closure methods by using the universal principles of macroscopic ET [1], the perturbative method of Grad [2] at the kinetic level, and the maximum entropy principle [3]. The vital point is that three different and apparently uncorrelated closure methods give the same system (1.25).

T. Ruggeri, M. Sugiyama, *Rational Extended Thermodynamics beyond the Monatomic Gas*, DOI 10.1007/978-3-319-13341-6_4

The differential system (1.25) has been extensively studied. For example, acceleration waves, shock waves, shock wave structure, and the hyperbolicity region were studied (for details see [4] and references therein).

In this section we summarize the 13-moment system and its principal subsystems.

The equations of ET with 13 fields are given in (1.25). The components of the main field $\mathbf{u}'$:

$$\mathbf{u}' \equiv \left(\lambda, \lambda_i, \mu, \lambda_{<ij>}, \mu_i\right)$$

have the following expressions [4, 5]:

$$\begin{aligned}
\lambda &= -\frac{1}{T}\left\{g - \frac{v^2}{2} + \frac{1}{2p}\sigma_{<ij>}\, v_i\, v_j - \frac{\rho}{5p^2} q_i\, v_i\, v^2\right\}, \\
\lambda_i &= -\frac{1}{T}\left\{v_i - \frac{1}{p}\sigma_{<ij>}\, v_j + \frac{\rho}{5p^2}\left(v^2 q_i + 2q_j\, v_j\, v_i\right) - \frac{q_i}{p}\right\}, \\
\mu &= \frac{1}{2T}\left\{1 - \frac{2\rho}{3p^2} q_k v_k\right\}, \\
\lambda_{<ij>} &= \frac{1}{T}\left\{\frac{1}{2p}\sigma_{<ij>} + \frac{\rho}{5p^2}\left(v_i\, q_j + v_j\, q_i - \frac{2}{3} v_k\, q_k\, \delta_{ij}\right)\right\}, \\
\mu_i &= -\frac{\rho}{5Tp^2} q_i.
\end{aligned} \tag{4.1}$$

The maximum characteristic velocity in equilibrium is $\lambda_{\max} = 1.65c_0$ (c_0 is the sound velocity for a monatomic gas).

Let us consider possible principal subsystems following the general theory presented in Sect. 2.4.

4.1.1 10-Field Principal Subsystem

The 10-field system is a principal subsystem of the 13-field system when

$$\mu_i = 0 \quad \rightarrow \quad q_i = 0.$$

Neglecting the last block of the corresponding equation of (1.25) and inserting $q_i = 0$ in the previous equations, we have

$$\begin{aligned}
&\frac{\partial \rho}{\partial t} + \frac{\partial}{\partial x_k}(\rho v_k) = 0, \\
&\frac{\partial \rho v_i}{\partial t} + \frac{\partial}{\partial x_k}\left\{\rho v_i v_k + p\delta_{ik} - \sigma_{\langle ik\rangle}\right\} = 0,
\end{aligned}$$

$$\frac{\partial}{\partial t}(\rho v^2 + 2\rho\varepsilon) + \frac{\partial}{\partial x_k}\{\rho v^2 v_k + 2(\rho\varepsilon + p)v_k - 2\sigma_{\langle kl\rangle}v_l\} = 0, \tag{4.2}$$

$$\frac{\partial}{\partial t}\{\rho v_i v_j + p\delta_{ij} - \sigma_{\langle ij\rangle}\} +$$

$$\frac{\partial}{\partial x_k}\left\{\rho v_i v_j v_k + p(v_i\delta_{jk} + v_j\delta_{ki} + v_k\delta_{ij}) - \sigma_{\langle ij\rangle}v_k - \sigma_{\langle jk\rangle}v_i - \sigma_{\langle ki\rangle}v_j\right\} = \frac{\sigma_{\langle ij\rangle}}{\tau_S}.$$

In this case $\lambda_{\max} = 1.34c_0$.

4.1.2 Euler 5-Field Principal Subsystem

The Euler system is the 5-field principal subsystem of the 13- and 10-field systems when

$$\mu_i = 0, \quad \lambda_{<ij>} = 0 \;\rightarrow\; q_i = 0, \quad \sigma_{<ij>} = 0.$$

Neglecting the last block of the corresponding equation of (4.2), and inserting $\sigma_{\langle ij\rangle} = 0$, in the previous equations, we have

$$\frac{\partial\rho}{\partial t} + \frac{\partial}{\partial x^k}(\rho v_k) = 0,$$

$$\frac{\partial\rho v_i}{\partial t} + \frac{\partial}{\partial x^k}(\rho v_i v_k + p\delta_{ik}) = 0, \tag{4.3}$$

$$\frac{\partial}{\partial t}(\rho v^2 + 2\rho\varepsilon) + \frac{\partial}{\partial x_k}\{\rho v^2 v_k + 2(\rho\varepsilon + p)v_k\} = 0.$$

The maximum velocity is now $\lambda_{\max} = 1\,c_0$.

4.1.3 4-Field Principal Subsystem

The so called p-system is the 4-field subsystem of the Euler system (4.3) with

$$\mu_i = 0, \;\; \lambda_{<ij>} = 0, \;\; \mu = \text{const.} \;\rightarrow\; q_i = 0, \;\; \sigma_{<ij>} = 0, \;\; T = T^* = \text{const.}$$

The system of field equations is given by

$$\frac{\partial}{\partial t}\rho + \frac{\partial}{\partial x^i}(\rho v_i) = 0,$$

$$\frac{\partial}{\partial t}(\rho v_j) + \frac{\partial}{\partial x^i}(\rho v_i v_j + p^* \delta_{ij}) = 0,$$

where

$$p^* \equiv p(\rho, T^*) \quad \text{and} \quad \lambda_{\max} = 0.7746\, c_0 \ .$$

4.1.4 1-Field Principal Subsystem

Finally, the transport equation:

$$\frac{\partial}{\partial t}\rho + \frac{\partial}{\partial x^i}(\rho v_i^*) = 0$$

is a 1-field principal subsystem of all the previous ones:

$$\mu_i = 0,\ \lambda_{<ij>} = 0,\ \mu = \text{const.},\ \lambda_j = \text{const.} \ \rightarrow$$
$$q_i = 0,\ \sigma_{<ij>} = 0,\ T = T^* = \text{const.},\ v_i = v_i^* = \text{const.}$$

and we have $\lambda_{\max} = 0$.

According with the general result, the maximum characteristic velocity increases with the number of fields [6].

4.2 Bounded Domain: Heat Conduction and Problematic Boundary Data

Up to here, we have considered phenomena evolving in an unbounded domain and we have seen that ET is successful because (1) it satisfies explicitly the universal principles of physics, (2) it has a desired mathematical structure of symmetric hyperbolic systems, (3) it is perfectly consistent with the kinetic theory, and (4) the results derived from it are in good agreement with experimental data. The analysis of phenomena in a bounded domain is, however, not quite satisfactory when the number of moments is more than 13 as we will see below.

In this section, we summarize the problems encountered in the study of nonequilibrium phenomena in a bounded domain. Let us start with the heat-conduction problem by using the 13-moment ET theory. We will see that the results in non-planar geometry are very different from the results predicted by the classical Navier-Stokes Fourier theory.

4.2.1 Heat Conduction Analyzed by the 13-Moment ET Theory

Müller and Ruggeri [7] studied one-dimensional heat conduction in a gas at rest in planar, cylindrical and spherical geometries by using the 13-moment ET theory. It turns out that, in the radially symmetric cases, the stress tensor does not reduce to a scalar pressure and that the heat flux depends on the normal components of the deviatoric stress tensor. As a result, the singularities of temperature on the axis of the cylinder and in the center of the sphere—which are characteristic for the Navier-Stokes Fourier solution—disappear. In this section, we explain this result.

For the present argument the nature of the molecular interaction is quite irrelevant. Therefore we choose the simplest model and consider the system of equations based upon the BGK model. Furthermore, we spread out the covariant derivatives by using the Christoffel symbols. Then we obtain a more specific, but more complex version of the field equations (1.25), which in a stationary case becomes [7]:

$$
\begin{aligned}
& g^{ik}\frac{\partial p}{\partial x^k} - \frac{\partial \sigma^{<ik>}}{\partial x^k} - \Gamma^i_{kl}\,\sigma^{<kl>} - \Gamma^k_{kl}\,\sigma^{<il>} = 0, \\
& \frac{\partial q^k}{\partial x^k} + \Gamma^k_{kl}\,q^l = 0, \\
& \frac{2}{5}\left(g^{ik}\frac{\partial q^j}{\partial x^k} + g^{jk}\frac{\partial q^i}{\partial x^k} + g^{ik}\Gamma^j_{kl}\,q^l + g^{jk}\Gamma^i_{kl}\,q^l \right) = \frac{1}{\tau}\sigma^{<ij>}, \\
& 5pg^{ik}\frac{k_B}{m}\frac{\partial T}{\partial x^k} - 7\sigma^{<ik>}\frac{k_B}{m}\frac{\partial T}{\partial x^k} - 2\frac{k_B}{m}T\left(\frac{\partial \sigma^{<ik>}}{\partial x^k} + \Gamma^i_{kl}\,\sigma^{<kl>} + \Gamma^k_{kl}\sigma^{<il>} \right) \\
& \quad = -\frac{2}{\tau}q^i.
\end{aligned}
\tag{4.4}
$$

Here g_{ik} is the metric tensor, Γ^i_{jk} are the Christoffel symbols appropriate to the coordinates x^k.

For the planar case with rectangular Cartesian coordinates, g^{ik} is the Kronecker tensor and all Christoffel symbols vanish. In the cylindrical case with coordinates $(x^1, x^2, x^3) = (r, \vartheta, z)$ we have

$$
g^{ik} = \begin{pmatrix} 1 & & 0 \\ & \frac{1}{r^2} & \\ 0 & & 1 \end{pmatrix} \text{ and } \Gamma^1_{22} = -r, \;\; \Gamma^2_{21} = \Gamma^2_{12} = \frac{1}{r}, \;\; \Gamma^m_{kn} = 0 \text{ else.} \tag{4.5}
$$

And in the spherical case with $(x^1, x^2, x^3) = (r, \vartheta, \varphi)$ we have

$$g^{ik} = \begin{pmatrix} 1 & & 0 \\ & \frac{1}{r^2} & \\ 0 & & \frac{1}{r^2 \sin^2 \vartheta} \end{pmatrix} \quad \text{and} \quad \begin{aligned} &\Gamma^3_{31} = \Gamma^3_{13} = \Gamma^2_{21} = \Gamma^2_{12} = \tfrac{1}{r}, \Gamma^1_{22} = -r, \\ &\Gamma^1_{33} = -r\sin^2\vartheta, \;\; \Gamma^3_{32} = \Gamma^3_{23} = ctg\,\vartheta, \\ &\Gamma^2_{33} = -\sin\vartheta\cos\vartheta, \\ &\Gamma^m_{kn} = 0 \quad \text{else.} \end{aligned} \tag{4.6}$$

Note that, in all three cases, the coordinate lines are orthogonal so that the metric tensors are diagonal.

4.2.1.1 One Dimensional Solutions and Their Comparison with the Solutions Derived from the Navier-Stokes Fourier Theory

We investigate solutions, in which all fields depend on x^1 only and in which q^1 is the only non-vanishing component of the heat flux. In such a case, the Eq. (4.4) readily imply that all shear stresses vanish and that the deviatoric normal stresses are related to q^1 by

$$\sigma^{<11>} = -\frac{4}{5}\Gamma^k_{k1}\tau q^1, \; \sigma^{<22>} = \frac{4}{5}g^{22}\Gamma^2_{21}\tau q^1, \; \sigma^{<33>} = \frac{4}{5}g^{33}\Gamma^3_{31}\tau q^1. \tag{4.7}$$

We also obtain

$$\frac{dq^1}{dx^1} = -\Gamma^k_{k1}q^1. \tag{4.8}$$

We conclude from (4.4) that $p = const.$ holds and the 1-component provides a relation between the heat flux and the temperature gradient, viz.

$$q^1 = -\kappa\left(1 - \frac{7}{5}\frac{\sigma^{<11>}}{p}\right)\frac{dT}{dx^1}, \tag{4.9}$$

where κ is the heat conductivity related to the relaxation time τ by

$$\kappa = \frac{5}{2}\frac{k_B}{m}\tau p.$$

From (4.7), we conclude that, in spherical coordinates, the physical components of the shear stress are not zero. This result is particular to extended thermodynamics, because with the Navier-Stokes constitutive equations we have $\sigma^{<ij>} = 0$ in any geometry. Thus we conclude, from (4.7), that in the planar case, where $\sigma^{<11>}$ vanishes, Fourier's law with a constant heat conductivity is recovered, because the heat flux is proportional to the gradient of the temperature. In the other cases, the cylindrical and the spherical ones, Fourier's law is not valid because of the second term on the right hand side of (4.7).

We now solve the system of equations and obtain the fields $\sigma^{<11>}$, q^1, and T. The only geometric quantity left in that system is Γ^k_{k1} which reads

$$\Gamma^k_{k1} = \frac{j}{x^1} \quad \text{with } j = \begin{bmatrix} 0 & & \text{planar} \\ 1 & \text{for the} & \text{cylindrical} \\ 2 & & \text{spherical} \end{bmatrix} \text{case.} \tag{4.10}$$

The general integral is easily obtained. In the planar case we have the simple solution

$$q^1 = c_1; \quad \sigma^{<11>} = 0, \quad T = c_2 - \frac{c_1}{\kappa} x^1 \tag{4.11}$$

with a linear temperature profile which is also predicted by the Fourier theory (c_1, c_2 are integration constants). For $j = 1, 2$ we have (we write now for evident reasons $r, q^r, \sigma^{<rr>}$ instead of $x^1, q^1, \sigma^{<11>}$)

$$q^r = \frac{c_1}{r^j} \qquad \text{and} \qquad \sigma^{<rr>} = -\frac{8j}{25} \frac{m}{k_B} \frac{\kappa}{p} \frac{c_1}{r^{j+1}},$$

and (4.9) becomes

$$\frac{dT}{dr} = -\frac{c_1}{\kappa} \frac{r}{\frac{56}{125} j \frac{m}{k_B} \frac{\kappa}{p^2} c_1 + r^{j+1}}. \tag{4.12}$$

The first term in the denominator is absent in the Navier-Stokes Fourier theory, because $\sigma^{<rr>} = 0$ in that theory.

Thus we can see a first difference between ET and the Fourier theory: The derivative of the temperature tends to zero for $r \to 0$, while in the Fourier case it diverges. The general solution of (4.12) reads

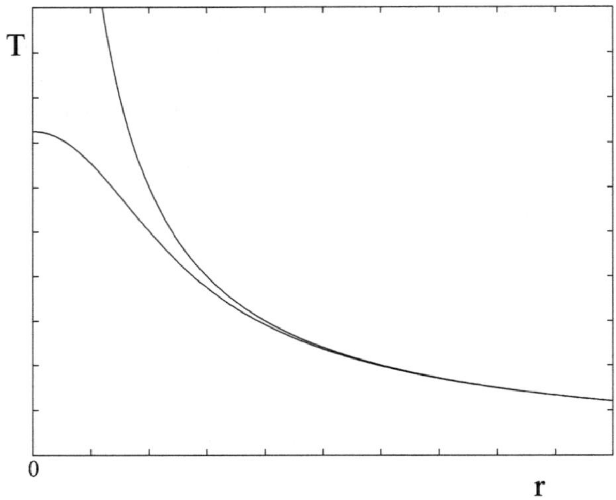

Fig. 4.1 The behavior of the general integral of the temperature in the spherical symmetry. The divergent curve represents the solution of the Fourier theory while the bounded one represents the 13-moment solution

in the cylindrical case, i.e., for $j = 1$

$$T = c_2 - \frac{c_1}{2\kappa} \log(b + r^2),$$

in the spherical case, i.e., for $j = 2$

$$T = c_2 + \frac{c_1}{3\, 2^{\frac{1}{3}}\, b^{\frac{1}{3}} \kappa} \left\{ \sqrt{3} \arctan\left(\frac{b^{\frac{1}{3}} - 2^{\frac{2}{3}} r}{\sqrt{3}\, b^{\frac{1}{3}}} \right) + \right. \tag{4.13}$$

$$\left. + \log\left(2\, b^{\frac{1}{3}} + 2^{\frac{2}{3}} r\right) - \frac{1}{2} \log\left(2\, b^{\frac{2}{3}} - 2^{\frac{2}{3}}\, b^{\frac{1}{3}} r + 2^{\frac{1}{3}} r^2\right) \right\},$$

while according to the Fourier law we have

$$T = c_2 - \frac{c_1}{\kappa} \log r \quad \text{for } j = 1 \quad \text{and} \quad T = c_2 + \frac{c_1}{\kappa} \frac{1}{r} \quad \text{for } j = 2. \tag{4.14}$$

For abbreviation we have set $b = \frac{56}{125} \frac{m}{k_B} \frac{\kappa}{p^2} c_1$.

Figure 4.1 illustrates the difference between these solutions in the spherical case for c_1 arbitrarily assigned and c_2 such that the temperature vanishes for large values of r. We see that the temperature is finite in ET, while it diverges in the Fourier theory. We also observe that the solutions coincide when the gradient is small, while they differ significantly where the curves becomes steeper. As a matter of course, ET becomes relevant when gradients are large.

4.2.1.2 Solution of a Boundary Value Problem

In the planar case, the result from the Fourier theory is identical to the one from ET. Indeed, from (4.11)$_3$, we conclude that the temperature is linear in x^1. More interesting are the radially symmetric cases, in particular, the cylindrically symmetric one.

We consider a gas between two co-axial cylinder or between two concentric spheres with inner radius r_i and outer radius r_e. We heat the gas at the inner radius with a prescribed heat flux q and the temperature at the outer radius is kept to a value T_e. Thus we solve the boundary value problem:

$$q^r(r_i) = q \quad \text{and} \qquad T(r_e) = T_e. \tag{4.15}$$

In the cylindrical case we obtain the solution:

$$q^r = \frac{qr_i}{r}, \quad \sigma^{rr} = \frac{8}{25}\frac{m}{k_B}\frac{\kappa}{p}\frac{qr_i}{r^2}, \quad T = T_e - \frac{qr_i}{2\kappa}\log\left(\frac{\frac{56}{125}\frac{m}{k_B}\frac{\kappa}{p^2}qr_i + r^2}{\frac{56}{125}\frac{m}{k_B}\frac{\kappa}{p^2}qr_i + r_e^2}\right). \tag{4.16}$$

Figure 4.2 shows this solution for T in the case of argon with the relative atomic mass $M = 40$ and for the data:

$$r_i = 10^{-3}\,\text{m}, \; r_e = 10^{-2}\,\text{m}, \quad p = 10^2\,\frac{\text{N}}{\text{m}^2},$$
$$\tau = 10^{-5}\,\text{s}, \quad q = 10^4\,\frac{\text{W}}{\text{m}^2}, \quad T_e = 300\,\text{K}. \tag{4.17}$$

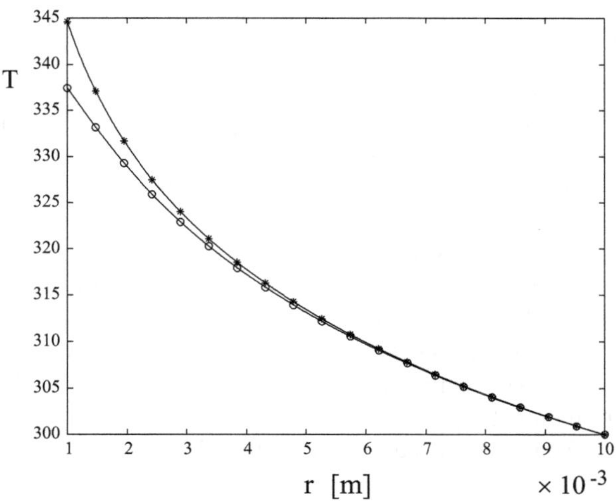

Fig. 4.2 The behavior of the temperatures in the cylindrical case with the boundary data (4.17). Fourier theory (*stars*); 13-moment theory (*circles*)

The curve marked by circles in Fig. 4.2 represents the solution of ET and the one marked by stars is the solution of the Fourier theory. At the inner cylinder, the values of the two theories differ by 7.17 K.

In the spherical case we obtain with $b = \frac{56}{125}\frac{m}{k_B}\frac{\kappa}{p^2} q r_i^2$:

$$q^r = \frac{q\, r_i^2}{r^2}, \quad \sigma^{rr} = \frac{16}{25}\frac{m}{k_B}\frac{\kappa}{p}\frac{q\, r_i^2}{r^3},$$

$$T = T_e + \frac{c_1}{32^{\frac{1}{3}} b^{\frac{1}{3}} \kappa}\left\{\sqrt{3}\left[\arctan\left(\frac{b^{\frac{1}{3}} - 2^{\frac{2}{3}} r}{\sqrt{3}\, b^{\frac{1}{3}}}\right) - \arctan\left(\frac{b^{\frac{1}{3}} - 2^{\frac{2}{3}} r_e}{\sqrt{3}\, b^{\frac{1}{3}}}\right)\right] + \right. \tag{4.18}$$

$$\left. + \log\left(\frac{2\, b^{\frac{1}{3}} + 2^{\frac{2}{3}} r}{2\, b^{\frac{1}{3}} + 2^{\frac{2}{3}} r_e}\right) - \frac{1}{2}\log\left(\frac{2\, b^{\frac{2}{3}} - 2^{\frac{2}{3}}\, b^{\frac{1}{3}} r + 2^{\frac{1}{3}} r^2}{2\, b^{\frac{2}{3}} - 2^{\frac{2}{3}}\, b^{\frac{1}{3}} r_e + 2^{\frac{1}{3}} r_e^2}\right)\right\}.$$

Both solutions (4.16) and (4.18) hold in the interval $r_i \leqslant r \leqslant r_e$.

The case of co-axial cylinders should be easy to set up experimentally. Indeed, the inner cylinder could be realized by a wire and the heating may be effected by letting an electric current run through that wire. On the other hand the corresponding case of two concentric spheres may be quite difficult to realize experimentally.

4.2.2 Difficulty in the ET Theory in a Bounded Domain When the Number of Fields is More Than 13

In the above subsection, we have seen that the 13-moment ET theory predicts the new and interesting results that show an appreciable difference from the results predicted by the Navier-Stokes Fourier theory. We understand that, when there exists a steep gradient (and/or a rapid change), the ET theory is superior to the conventional classical TIP theory. Therefore it was natural that several authors tried to understand more deeply nonequilibrium phenomena in a bounded domain by using the ET theory with more than 13 moments. However, in such studies, the authors encountered a conceptually difficult problem. This is the problem of the boundary conditions.

All quantities in the 13-moment theory have concrete physical meanings and are observable. However, the quantities expressed by higher moments have not definite physical meanings and are, in general, impossible to be measured in experiments. Therefore we cannot pose the values of these quantities at the boundary of a domain, that is, we cannot pose the boundary conditions for the ET theory with more than 13 moments appropriately.

We may have a question: Is it really necessary to know all the boundary values? Indeed let us consider, for example, heat conduction in a gas filled in between two parallel plates. In this case, we need not know all boundary values except for, say, the temperatures at the two plates in order to study the phenomena experimentally. This consideration seems to lead us to the following conjecture: if the temperature are fixed at the both sides, the other quantities adjust by themselves and have appropriate boundary values that are consistent with the experimental data. Such quantities are sometimes called uncontrollable quantities.

If this conjecture is true, we further want to know the mechanism of the self-adjustment. Several studies have been done. Struchtrup and Weiss [8] proposed the minimax principle for the entropy production, according to which the boundary values of the uncontrollable quantities are fixed. Barbera et al. also studied this problem [9]. They assume that the uncontrollable quantities fluctuate around the most probable values, and then they assume that the boundary values of the quantities are given by these most probable values.

Another strategy was proposed by Ruggeri and Lou [10]. Their approach is purely phenomenological. Let us consider the heat conduction problem again as an example with a mixture of gas. In order to obtain the temperature profile, they impose not only the boundary conditions but also some conditions inside a gas that should be given by experiments. The latter conditions are, for example, the temperatures at several points inside a gas. (See Sect. 16.6.2.)

Moreover, as pointed out by Brini and Ruggeri [11], there is another subtle problem. They proved that, if nonequilibrium variables have small values of the same order, then some derivative of these variables (critical derivatives) are not necessarily of the same order. As a consequence, the solutions violate the entropy principle and the system becomes to be inconsistent with the near-equilibrium approximation adopted.

Kinetic-theoretical study by using the concept of the accommodation factor may be helpful to solve this problem (see, for example, [12–16]). The accommodation factor is defined as the ratio between the effects of specular reflection and thermalized (diffuse) reflection of an incident molecule at the boundary wall. The introduction of this factor implies that the interaction between a gas and a boundary wall should be properly taken into consideration in order to fix the boundary conditions.

In conclusion, except for the 13-moment ET theory, the problem of the boundary conditions in ET in general still remains as a big issue to be solved even in the case of monatomic gases.

4.3 Molecular RET for Large Number of Moments

For rarefied gases, we have discovered that, in highly nonequilibrium phenomena such as sound waves with high frequencies, light scattering with large scattering angle, shock waves with large Mach number, predictions of the 13-moment theory are sill not quite satisfactory when compared with experimental data although the 13-moment theory gives us better results compared with the Navier-Stokes Fourier theory. For such phenomena, we need a theory with more moments. In these cases, it is too difficult to proceed within a purely macroscopic theory like the 13-field RET theory. Therefore let us recall that the fields F's can be regarded as the moments of a distribution function f. In order to explain this approach by using the moments F's, we first rewrite the hierarchy of balance laws in more compact notation:

$$\partial_t F_A + \partial_i F_{iA} = P_A, \tag{4.19}$$

with

$$F_A = \int_{R^3} mfc^A d\mathbf{c}, \quad F_{iA} = \int_{R^3} mfc_i\, c^A d\mathbf{c}, \tag{4.20}$$

$$P_A = \int_{R^3} Qc^A d\mathbf{c}, \tag{4.21}$$

where

$$c^A = \begin{cases} 1 & \text{for } A = 0, \\ c_{i_1} c_{i_2} \cdots c_{i_A} & \text{for } 1 \leqslant A \leqslant N, \end{cases}$$

and

$$F_A = \begin{cases} F & \text{for } A = 0, \\ F_{i_1 i_2 \cdots i_A} & \text{for } 1 \leqslant A \leqslant N, \end{cases} \qquad F_{iA} = \begin{cases} F_i & \text{for } A = 0, \\ F_{i i_1 i_2 \cdots i_A} & \text{for } 1 \leqslant A \leqslant N. \end{cases} \tag{4.22}$$

The indices i and $i_1 \leqslant i_2 \leqslant \cdots \leqslant i_A$ are defined over $1, 2, 3$. In the followings, similar notations will be adopted.

Definition 4.1 A system of moments (4.19) truncated at the tensorial index N is called (N)-system. The (4.19) is called (N^-)-system if, for the last balance equation, we consider only the trace with respect to the two indexes, $F_{k_1 k_2 \ldots k_{N-2} ll}$, instead of the full N-order tensor $F_{k_1 k_2 \ldots k_{N-2} k_{N-1} k_N}$. If, instead of two indexes, we have the contraction with respect to two couples of two indexes, we add another minus: (N^{--}).

According with this definition (that does not include all possible moment systems) the Euler fluid is a (2^-)-system and the Grad system is a (3^-)-system. Taking into account that all tensors are symmetric, the number n of moments, for an (N)-system, is given by

$$n_N = \frac{1}{6}(N+1)(N+2)(N+3), \tag{4.23}$$

and, for an (N^-)-system, is given by

$$n_{N^-} = \frac{1}{6} N(N^2 + 6N - 1). \tag{4.24}$$

4.3.1 Closure via the Entropy Principle

We require the compatibility of the truncated system (4.19) with the entropy law, i.e., all solutions of (4.19) must satisfy also the supplementary entropy balance law (1.22) where h^0, h^i and Σ are functionals of f [see (1.21)] through the moments (4.20) with $A = 0, \ldots, N$. This is a strong restriction on the distribution function f, and now the problem to be solved is as follows: Determine the distribution function f_N under the condition that any classical solution of (4.19) with (4.20) and (4.21) is also the solution of (1.22).

For a generic entropy functional, which is valid not only for classical gases but also for degenerate gases such as Bose and Fermi gases, the following theorem was proved by Boillat and Ruggeri [17]:

Theorem 4.1 *Necessary and sufficient condition such that the truncated system of moments (4.19) satisfies an entropy principle (1.22) is that the truncated distribution function f_N depends on $(\mathbf{x}, t, \mathbf{c})$ only through a single variable:*

$$f_N \equiv f_N(\chi_N),$$

where

$$\chi_N = \sum_{A=0}^{N} u'_A(\mathbf{x}, t)c^A$$

is a polynomial in $\mathbf{c}$ with the coefficients u'_A:

$$u'_A = \begin{cases} u' & \text{for } A = 0, \\ u'_{i_1 i_2 \cdots i_A} & \text{for } 1 \leqslant A \leqslant N. \end{cases}$$

The entropy density, flux and production have the following expressions:

$$h^0 = m \int_{R^3} \left(\chi_N \Omega'(\chi_N) - \Omega(\chi_N)\right) d\mathbf{c}, \tag{4.25}$$

$$h^i = m \int_{R^3} c^i \left(\chi_N \Omega'(\chi_N) - \Omega(\chi_N)\right) d\mathbf{c}, \qquad \Sigma = m \int_{R^3} Q\, \chi_N \, d\mathbf{c},$$

where the partition function $\Omega(\chi_N)$ satisfies the relation:

$$\Omega' = \frac{d\Omega}{d\chi_N} = f_N(\chi_N).$$

The system (4.19) becomes symmetric in the form (2.11) with the main field u'_A and potentials

$$h'^0 = m \int_{R^3} \Omega(\chi_N)\, d\mathbf{c}, \quad h'^i = m \int_{R^3} \Omega(\chi_N) c_i\, d\mathbf{c}, \tag{4.26}$$

provided $\Omega''(\chi_N) < 0$.

The proof can be made by the following general symmetrization Theorem 2.1. In fact (2.15) becomes in this case

$$F_A = \frac{\partial h'^{\,0}}{\partial u'_A}, \qquad F_{iA} = \frac{\partial h^{i\prime}}{\partial u'_A},$$

then

$$dh'^0 = \sum_{A=0}^{N} F_A du'_A = m \int_{R^3} f \sum_{A=0}^{N} c_A du'_A d\mathbf{c} = m \int_{R^3} f d\chi_N d\mathbf{c} = d \int_{R^3} m\Omega(\chi_N) d\mathbf{c},$$

and $(4.26)_1$ hold. From (2.14) follows the (4.25). Analogous considerations are given for h'^i and h^i.

Then the original system of the moments becomes to be closed and to be symmetric hyperbolic in terms of the main-field components (2.11) (we omit now the summation on the repeated index B):

$$J_{AB}\partial_t u'_B + J_{iAB}\partial_i u'_B = P_A(u'_C), \qquad A = 0, \ldots, N \tag{4.27}$$

where

$$J_{AB}(u'_C) = \frac{\partial^2 h'^{\,0}}{\partial u'_A \partial u'_B} = \int_{R^3} m\Omega''(\chi_N) c^A c^B d\mathbf{c},$$

$$J_{iAB}(u'_C) = \frac{\partial^2 h'_{\,i}}{\partial u'_A \partial u'_B} = \int_{R^3} m\Omega''(\chi_N) c_i\, c^A c^B\, d\mathbf{c}.$$

Here and hereafter, the summation symbol with respect to A and/or B is omitted for simplicity. Indeed, the matrix J_{AB} is negative definite provided that $\Omega''(\chi_N) < 0$ holds, since

$$J_{AB}X_A X_B = \int_{R^3} m\Omega'' \left(c^A X_A\right)^2 d\mathbf{c} < 0 \quad \forall X_A \neq 0.$$

If we require that h^0 is the usual entropy density for non-degenerate gases, viz.

$$h^0 = -k_B \int_{R^3} f_N \ln f_N\, d\mathbf{c},$$

we obtain from (4.25)

$$\left(\chi_N + \frac{k_B}{m} \ln \Omega'\right) \Omega' - \Omega = 0,$$

and by differentiation we obtain

$$f_N(\chi_N) = e^{-1-m\chi_N/k_B}. \tag{4.28}$$

In the present case we have

$$J_{AB}\left(u'_C\right) = -\frac{m^2}{k_B}\int_{R^3} f_N c_A c_B \, d\mathbf{c}, \qquad J_{iAB}\left(u'_C\right) = -\frac{m^2}{k_B}\int_{R^3} f_N c_i c_A c_B \, d\mathbf{c}. \tag{4.29}$$

In an equilibrium state, (4.28) reduces to the well-known Maxwellian distribution function. We observe that f_N is not a solution of the Boltzmann equation. But we have the conjecture (open problem) that, for $N \to \infty$, f_N tends to a solution of the Boltzmann equation.

4.3.2 Closure via the Maximum Entropy Principle

Instead of the method of the entropy principle, there is an alternative in the ET theory of moments for the determination of the phase density f_N. This is the method of maximization of the entropy under some constraints.

We have discussed the MEP in Sect. 1.5.5. In this section, we summarize the results in the case of monatomic gases (see [17] for more details). Let us treat firstly a general case where the entropy h^0 is a generic functional of f:

$$h^0 = \int_{R^3} \psi(f) d\mathbf{c}\,. \tag{4.30}$$

We ask for the phase density f_N that provides the maximum of h^0 under the constraints of fixed values F_A for the moments:

$$F_A = \int_{R^3} m f c^A d\mathbf{c}.$$

With the Lagrange multipliers λ_A, we form the expression:

$$\mathscr{L} = \int_{R^3} \psi(f) d\mathbf{c} + \lambda_A \left(F_A - m\int c^A f d\mathbf{c}\right), \tag{4.31}$$

and obtain the relation:

$$\delta\mathscr{L} = \int_{R^3}\left(\frac{d\psi}{df} - m\lambda_A c^A\right)\delta f d\mathbf{c} = 0.$$

Thus we have

$$\frac{d\psi}{df} = m\lambda_A c^A$$

as a necessary condition for an extremum. Hence it follows that f is a function of

$$\chi = \lambda_A c^A,$$

and that $\psi(f)$ has the form:

$$\psi(f) = m\left(\chi f - \int f d\chi\right). \tag{4.32}$$

Insertion of (4.32) into (4.30) gives exactly the same result as that from the entropy principle (4.25). Thus we conclude that *the maximization of the entropy leads to the same result as that from the entropy principle in molecular RET of moments* [17]. In particular, the Lagrange multipliers λ_A are identical to the main field components u'_A.

4.4 Maximum Characteristic Velocity in the Classical Theory

Characteristic velocities λ, in the propagation direction with the unit vector $\mathbf{n} \equiv (n^i)$, of the symmetric hyperbolic system (4.27) with (4.28) are eigenvalues of

$$K^{AB} = J_{iAB}n^i - \lambda J_{AB} = -\frac{m^2}{k_B}\int_{R^3} f(\chi)(\mathbf{c}\cdot\mathbf{n} - \lambda)c^A c^B \, d\mathbf{c}. \tag{4.33}$$

In particular, the wave speeds for disturbances propagating in an equilibrium state are eigenvalues of

$$\int_{R^3} f_M(\mathbf{c}\cdot\mathbf{n} - \lambda)c^A c^B \, d\mathbf{c}, \tag{4.34}$$

where f_M is the Maxwellian distribution function. As the integrals in (4.34) are known, it is easy to evaluate the maximum eigenvalues for increasing N.

Numerical results were obtained by Weiss [18] who obtained increasing value of the maximum characteristic velocity for increasing number of moments n that depends on the truncation index N through (4.23). For instance, for $n = 20$, $\lambda_{\max} = 1.8c_0$ and for $n = 15.180$, $\lambda_{\max} = 9.36c_0$, where $\lambda_{\max}$ is the maximum characteristic velocity in equilibrium in units of the sound wave velocity. Therefore an interesting problem is: what is the limit of $\lambda_{\max}$ as $N \to \infty$?

4.4.1 Lower Bound Estimate and Characteristic Velocities for Large Number of Moments

In the previous examples, we have seen the validity of the subcharacteristic conditions. Now we are able to prove the behavior of $\lambda_{\max}$ when $N \to \infty$. The $(k+1)(k+2)/2$ components of order k of the main field:

$$u'_{i_1 i_2 \dots i_k}, \qquad i_1 \leqslant i_2 \leqslant \dots \leqslant i_k$$

can be mapped in the corresponding variables

$$u'_{pqr}, \qquad p+q+r=k,$$

where p, q, r are, respectively, the number of indices over $1, 2, 3$. With this notation

$$\chi = \sum_{p,q,r} u'_{pqr} c_1^p c_2^q c_3^r, \qquad 0 \leqslant p+q+r \leqslant N.$$

Theorem 4.2 (Boillat and Ruggeri [17]) *For any N we have the lower bound condition:*

$$\frac{\lambda_{\max}}{c_0} \geqslant \sqrt{\frac{6}{5}\left(N - \frac{1}{2}\right)} \tag{4.35}$$

where c_0 is the sound velocity. Therefore, $\lambda_{\max}$ becomes unbounded when $N \to \infty$.

Sketch of the Proof By the use of the variable u'_{pqr}, the components of the matrix (4.34) are given by

$$\int_{R^3} f_M (c_i n^i - \lambda) c_1^{p+s} c_2^{q+t} c_3^{r+u} d\mathbf{c}.$$

The matrix is negative semi-definite, if λ is the largest eigenvalue $\lambda_{\max}$. As the elements a_{ij} of a semi-definite matrix satisfy the inequalities:

$$a_{ii} a_{jj} \geqslant a_{ij}^2, \tag{4.36}$$

we have

$$\int_{R^3} f_M (c_i n^i - \lambda_{\max}) c_1^{2p} c_2^{2q} c_3^{2r} d\mathbf{c} \int_{R^3} f_M (c_i n^i - \lambda_{\max}) c_1^{2s} c_2^{2t} c_3^{2u} d\mathbf{c}$$
$$\geqslant \left(\int_{R^3} f_M (c_i n^i - \lambda_{\max}) c_1^{p+s} c_2^{q+t} c_3^{r+u} d\mathbf{c} \right)^2. \tag{4.37}$$

Fig. 4.3 The behavior of the maximum characteristic velocity versus the truncation number N and the lower bound estimate (4.35)

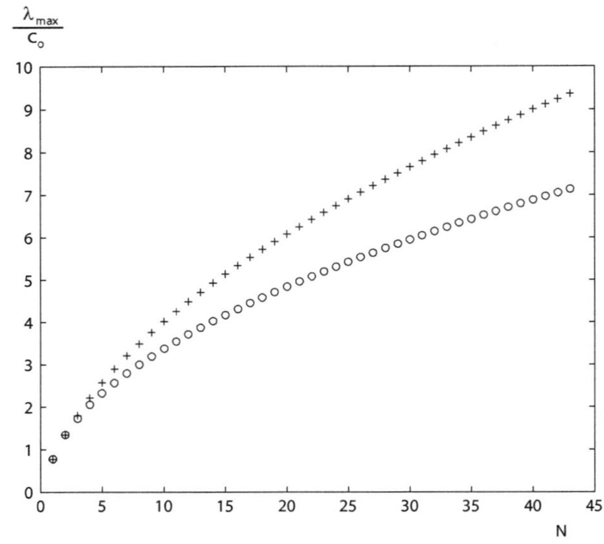

In this case (4.37) reduces to

$$\lambda_{\max}^2 \int_{R^3} f_M c_1^{2p} c_2^{2q} c_3^{2r} d\mathbf{c} \int_{R^3} f_M c_1^{2s} c_2^{2t} c_3^{2u} d\mathbf{c}$$
$$\geqslant \left(\int_{R^3} f_M (c_i n^i - \lambda_{\max}) c_1^{p+s} c_2^{q+t} c_3^{r+u} d\mathbf{c} \right)^2 . \tag{4.38}$$

With the choice $p = N, s = N-1, q = r = t = u = 0, \mathbf{n} \equiv (1,0,0)$, this inequality becomes

$$\lambda_{\max}^2 \geqslant \frac{\int_{R^3} f_M c_1^{2N} dc_1}{\int_{R^3} f_M c_1^{2(N-1)} dc_1} = \frac{1}{b} \frac{\Gamma(N+1/2)}{\Gamma(N-1/2)} = \frac{6}{5} c_0^2 \left(N - \frac{1}{2} \right)$$

and the proof is completed. Therefore

$$\lim_{N \to \infty} \lambda_{\max} = \infty.$$

In Fig. 4.3, we compare the numerical values of $\lambda_{\max}/c_0$ given by Weiss [18] with our lower bound in the right-hand side of (4.35).

This is a surprising result because the first motivation of ET was to repair the paradox of infinite velocity of the Navier-Stokes Fourier classical approach. Therefore, for any finite N, we have symmetric hyperbolic systems with finite characteristic velocities. But when we take infinite moments we have a parabolic behavior.

Instead, in the relativistic context, it was proved that the limit of the maximum characteristic velocity for $N \to \infty$ is the light velocity [19–21] (see Sect. 4.7.1.1).

4.5 Convergence Problem and a Theory Near Equilibrium State

All results explained above are valid also for a case far from equilibrium provided that the integrals in (4.20) and (4.21) are convergent. The problem of the convergence of the moments is one of the main questions in a far-from-equilibrium case. In particular, as we will see, the index of truncation N must be even. This implies, in particular, that a theory with 13 moments is not allowed when far from equilibrium!

Moreover, if the conjecture that the distribution function f_N, when $N \to \infty$, tends to the distribution function f that satisfies the Boltzmann equation is true, we need another convergence requirement.

These problems were studied by Boillat and Ruggeri in [17]. As before, χ_N is expressed as

$$\chi_N = \sum_{p,q,r} u'_{pqr} c_1^p c_2^q c_3^r, \quad 0 \leqslant p+q+r \leqslant N. \tag{4.39}$$

Since

$$\left| \sum_{p,q,r} u'_{pqr} c_1^p c_2^q c_3^r \right| \leqslant a_N c^N$$

with

$$a_N = \max_{|\mathbf{t}|=1} \nu_N(\mathbf{t}), \qquad \nu_N(\mathbf{t}) = \sum_{p,q,r} u'_{pqr} t_1^p t_2^q t_3^r,$$

when $p+q+r = N \to \infty$, the series is absolutely convergent for any $\mathbf{c}$ provided that

$$u'_{pqr} \to 0, \qquad \frac{a_{N+1}}{a_N} \to 0.$$

Hence the components of the main field become smaller and smaller when N increases. This justifies the truncation of the system. On the other hand, when N is finite, the integrals of moments must also be convergent. When c is large, $\chi_N \simeq |c|^N \nu_N$. Therefore it is easy to see, by using the spherical coordinates, that the integrals of moments converge provided that $\nu_N(\mathbf{t}) < 0$ for any unit vector $\mathbf{t}$. But, as $\nu_N(-\mathbf{t}) = (-1)^N \nu_N(\mathbf{t})$, we can conclude that N must be even and $\max_{|\mathbf{t}|=1} \nu_N(\mathbf{t}) < 0$.

Now, the distribution function (4.28) obtained as the solution of the variational problem is expanded in the neighborhood of a local equilibrium state:

$$f_N \approx f_M \left(1 - \frac{m}{k_B}\tilde{u}'_A c_A\right), \quad \tilde{u}'_A = u'_A - u'^E_A, \tag{4.40}$$

where u'^E_A are the main field components evaluated in the local equilibrium state and f_M is the Maxwellian equilibrium distribution function. Taking into account the fact that all the equilibrium components of the main field vanish except those corresponding to the first five moments [see (2.29)] and plugging (4.40) into $(4.20)_1$, we obtain a linear algebraic system that permits to evaluate the main field $\tilde{u}'_A$ in terms of the densities F_A:

$$J^{\mathscr{M}}_{AB}\,\tilde{u}'_B = F_A - F^E_A,$$

where F^E_A denotes the moments F_A evaluated in the equilibrium state and

$$J^{\mathscr{M}}_{AB} = -\frac{m^2}{k_B}\int_{R^3} f_M c_A c_B \, d\mathbf{c}. \tag{4.41}$$

And, by inserting $\tilde{u}'_A$ just obtained above into $(4.20)_2$ and (4.21), the explicit dependence of the truncated fluxes and source terms on the densities is obtained. Finally, the truncated (N)-system becomes a closed system for the densities F_A.

The generalization of this procedure to higher order expansions of the distribution function has been given in [11] and as we will see in Chap. 12 there exists an interesting model in which we can close the system also far from equilibrium. This is the case of the 6-moment theory of polyatomic gases [22].

4.6 Comparison with Experimental Data: Sound Waves and Light Scattering

The ET theory is successful when the results are compared with experimental data, in particular, the results of sound waves with high frequencies and light scattering. In Fig. 4.4, taken from the book [4], we can see that the so-called dynamic factor $S(x, y)$ obtained by the ET theory fits very well the experimental data on light scattering (represented in the figure by dots) when N is sufficiently large.

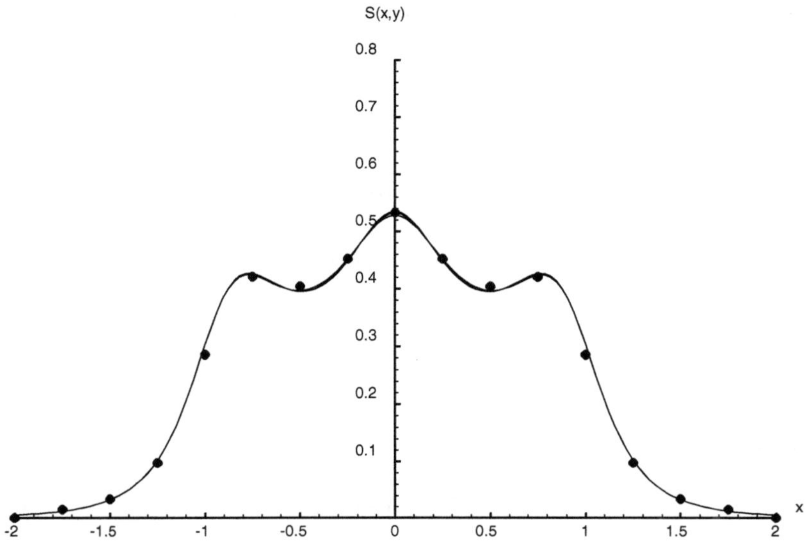

Fig. 4.4 Dynamical factor: the perfect agreement between the ET theory and the experimental data [4]

4.7 Relativistic Theory and the Limit of Maximum Characteristic Velocity

In the relativistic kinetic theory of rarefied gases, the phase density $f(x^\alpha, p^\alpha)$ $(\alpha = 0, 1, 2, 3)$ satisfies the Boltzmann-Chernikov-Lichnerowicz-Marrot equation:

$$p^\alpha \partial_\alpha f = Q, \tag{4.42}$$

where x^α and p^α are the space-time coordinates and the four-momentum of an atom, respectively. We have $p_\alpha p^\alpha = (p^0)^2 - \mathbf{p}^2 = m^2c^2$, $\mathbf{p}^2 = (p^1)^2 + (p^2)^2 + (p^3)^2$, where m is the atomic rest mass and c the speed of light. The right-hand side of (4.42) is due to collision between the atoms. Upon multiplication by p^A $(A = 0, 1, 2, \dots)$

and integration, (4.42) provides an infinite system of balance equations:

$$\begin{cases} \partial_\alpha F^\alpha = 0, \\ \partial_\alpha F^{\alpha\alpha_1} = 0, \\ \partial_\alpha F^{\alpha\alpha_1\alpha_2} = P^{\alpha_1\alpha_2}, \\ \partial_\alpha F^{\alpha\alpha_1\alpha_2\alpha_3} = P^{\alpha_1\alpha_2\alpha_3}, \\ \vdots \\ \partial_\alpha F^{\alpha\alpha_1.....\alpha_n} = P^{\alpha_1\alpha_2.....\alpha_n}, \\ \vdots \end{cases} \tag{4.43}$$

or briefly:

$$\partial_\alpha F^{\alpha A} = P_A, \quad A = 0, 1, 2, \dots \tag{4.44}$$

for the moments $F^{\alpha A}$ and productions P_A given by

$$F^{\alpha A}(x^\beta) = \int p^\alpha p^A f \, d\mathbf{p}, \tag{4.45}$$

$$P_A(x^\beta) = \int Q p^A f \, d\mathbf{p}, \tag{4.46}$$

where A is a multindex:

$$p^A = \begin{cases} 1 & \text{for } A = 0 \\ p^{\alpha_1} p^{\alpha_2} \dots p^{\alpha_A} & \text{for } A \geqslant 1 \end{cases} \tag{4.47}$$

$$F^{\alpha A} = \begin{cases} F^\alpha & \text{for } A = 0 \\ F^{\alpha\alpha_1 \dots \alpha_A} & \text{for } A \geqslant 1 \end{cases} \tag{4.48}$$

$$P_A = \begin{cases} 0 & \text{for } A = 0 \\ P^{\alpha_1 \dots \alpha_A} & \text{for } A \geqslant 1 \end{cases} \tag{4.49}$$

and $0 \leqslant \alpha_1 \leqslant \alpha_2 \leqslant \dots \leqslant \alpha_A \leqslant 3$. We recall that the first five equations of (4.44) are the conservation laws of mass, momentum and energy, according to which the first five productions vanish: $P_A = 0$ for $A = 0, 1$. The volume element of the momentum space is given by $d\mathbf{p} = \sqrt{-g}\, dp^1 dp^2 dp^3 / p^0$ and the integrals, which are supposed to be convergent, are taken over the whole $\mathbf{p}$-space.

We introduce the quantity:

$$h^\alpha = -k_B \int p^\alpha \left\{ (s^2 - 1 + \ln f) f + s(1 - sf) \ln(1 - sf) \right\} d\mathbf{p} \tag{4.50}$$

where $s = 0, -1, 1$ correspond, respectively, to the non-degenerate gas, the Fermi gas and the Bose gas. It is well known that we obtain, from (4.42), the supplementary inequality:

$$\partial_\alpha h^\alpha = \Sigma \geqslant 0. \tag{4.51}$$

This expression suggests the balance of entropy when we identify h^α and Σ with the four-entropy vector and the entropy production, respectively. Then (4.51) expresses the famous H-theorem in the relativistic framework.

4.7.1 Finite System of Moment Equations and Its Closure

We now consider a finite system of moment equations with the tensorial index $A = 0, \ldots, N$. In this case, as in the classical case, we can accomplish the closure of the system by using the entropy principle or the variational method of the maximum entropy principle (Boillat and Ruggeri [19, 20]). We obtain

$$f_N = \frac{1}{e^{-\chi/k_B} + s}, \tag{4.52}$$

where

$$\chi = u'_A p^A. \tag{4.53}$$

The system is closed in terms of the main field components u'_A:

$$H^{\alpha AB}(u'_C)\partial_\alpha u'_B = P_A(u'_C), \quad A = 0, 1 \ldots N \tag{4.54}$$

with the symmetric matrices

$$H^{\alpha AB} = \int \frac{\exp(-\chi/k_B)}{k_B\left(1 + s\exp(-\chi/k_B)\right)^2} p^\alpha p^A p^B d\mathbf{p}. \tag{4.55}$$

$H^{\alpha AB}\zeta_\alpha$ is negative definite for any timelike vector ($\zeta_\alpha \zeta^\alpha > 0$). This implies that the system (4.54) is a symmetric hyperbolic system (in the sense of Friedrichs) and is local in time. Its Cauchy problem is well-posed.

4.7.1.1 Propagation in an Equilibrium State and the Maximum Characteristics Velocity

The wave surface $\phi(x^\alpha) = 0$ is a solution of the characteristic equation:

$$\det(H^{\alpha AB}\partial_\alpha \phi) = 0.$$

As a consequence, the four gradient $\partial_\alpha \phi$ cannot be timelike and therefore the velocities of waves cannot exceed the velocity of light, i.e., λ max $\leqslant c$. When the number of equations increases, as was already shown, the maximum wave velocity cannot decrease. Now a question is: Does this velocity tend to c when N tends to infinity?

A thermodynamic equilibrium state is defined as the state for which the productions g^A vanish and the entropy production Σ reaches its minimum value, i.e., zero. According with the general result, all the main-field variables except for the first five variables are zero:

$$u' = -\frac{g}{T}, \quad u'_\alpha = \frac{u_\alpha}{T}, \tag{4.56}$$

where g, T and u_α are, respectively, the chemical potential, the absolute temperature, and the four-velocity. Therefore, for any truncation index N, the χ given by (4.53) reduces to $\chi/k_B = (-g + u_\alpha p^\alpha)/(k_B T)$ in an equilibrium state. And, in a case at rest where $u^i = 0, u^0 = c, p^0 = \sqrt{m^2c^2 + \mathbf{p}^2}$ we have

$$\frac{\chi}{k_B} = -a + \gamma\sqrt{1 + \frac{\mathbf{p}^2}{m^2c^2}},$$

where $a = g/k_B T$, $\gamma = mc^2/k_B T$. In this case, the distribution function reduces to the well-known Jüttner equilibrium distribution. We recall that for a Fermi gas a can assume all real values while for a Bose gas $a + \gamma > 0$.

The wave velocity λ in the direction of the normal $\mathbf{n}$ to the wave front is an eigenvalue of

$$\det(H^{iAB}n_i - \lambda H^{0AB}) = 0. \tag{4.57}$$

The matrix in (4.57) is negative semidefinite for the maximum eigenvalue, that is, if we take $\mathbf{n} \equiv (1, 0, 0)$, the components of the matrix:

$$H^{1AB} - \lambda_{\max} H^{0AB} = \int \frac{df}{dx} p^A p^B (p^1 - \lambda p^0) d\mathbf{p}$$

satisfy the inequalities in the form $a_{ii}a_{jj} - a_{ij}^2 \geqslant 0$ [see (4.36)]. Therefore, choosing $p^A = (p^1)^n$, $p^B = (p^1)^{n-1}$, we have

$$\frac{\lambda_{\max}^2}{c^2} \int \frac{df}{d\chi} \left(p^1\right)^{2n} d^3p \int \frac{df}{d\chi} \left(p^1\right)^{2(n-1)} d^3p \geqslant \left(\int \frac{df}{d\chi} \left(p^1\right)^{2n} \frac{d^3p}{p^0} \right)^2, \tag{4.58}$$

since the integrals of odd functions vanish.

By introducing the spherical coordinates, the above inequality, after some straightforward calculations, yields

$$\frac{\lambda_{\max}^2}{c^2} \geqslant \frac{2n-1}{2n+1} \frac{J_{n+1}^2}{I_n I_{n+1}}, \tag{4.59}$$

where

$$I_n = \int_0^\infty \phi(r) r^{2n} dr, \quad J_n = \int_0^\infty \frac{\phi(r) r^{2n}}{\sqrt{1+r^2}}\, dr,$$

$$\phi(r) = \frac{e^{\psi(r)}}{(1+se^{\psi(r)})^2},$$
$$\psi(r) = \frac{\chi_e}{k_B} = -a - \gamma\sqrt{1+r^2}, \quad s = 0, \pm 1.$$

Therefore we conclude that: *For any type of gas including the degenerate gases of fermions and bosons, the largest wave velocity has the lower and upper bounds.*

In the case of a non-degenerate gas, the previous integrals can be written in terms of the Bessel function of second kind and read

$$\frac{\lambda_{\max}^2}{c^2} \geqslant \frac{(2N-1)}{\gamma} \frac{K_{N+1}(\gamma)}{K_{N+2}(\gamma)}, \tag{4.60}$$

where $\gamma = mc^2/(k_B T)$. Using the recurrence relation for Bessel functions $K_{N+2} - K_N = 2(N+1)K_{N+1}/\gamma$, we obtain that, for $N \to \infty$, the limit value of $\lambda_{\max}$ in (4.59) is the light velocity c, since it has already been proved that it cannot be larger than c. Thus, *when the number of moments tends to infinity, the maximum velocity in equilibrium tends to the light velocity.*

This result can be proved also for degenerate gases because our proof is completely independent of the interaction term Q of (4.42).

In the *ultrarelativistic case* corresponding to small γ, taking the properties of the Bessel functions into account when $\gamma \to 0$, we obtain the simple inequality:

$$\frac{\lambda_{\max}^2}{c^2} \geqslant \frac{(2N-1)}{2(N+1)}.$$

4.7.2 The Macroscopic Relativistic 14-Field Theory

The most interesting case is the relativistic 14-moment ET theory developed by Liu et al. [23]. The system in this case coincides with the system obtained by kinetic considerations by C. Marle [24], which are similar to the ones of Grad:

$$\begin{cases} \partial_\alpha F^\alpha = 0, \\ \partial_\alpha F^{\alpha\beta} = 0, \\ \partial_\alpha F^{\alpha\beta\gamma} = P^{\beta\gamma}. \end{cases} \tag{4.61}$$

The field variables in this case are the number of particle vector F^{α} and the full energy-momentum tensor $F^{\alpha\beta} = T^{\alpha\beta}$, while $F^{\alpha\beta\gamma}$ and $Pi^{\beta\gamma}$ are constitutive functions of the field variables. By using the familiar variables (n particle number, e the energy density, $t^{<\alpha\beta>}$ the deviatoric viscous stress tensor, Π the dynamic pressure and q^{α} the heat flux four-vector), the following expressions are obtained [23]:

$$
\begin{aligned}
F^{\alpha} &= nmu^{\alpha}, \\
F^{\alpha\beta} &= t^{<\alpha\beta>} + (p+\Pi)h^{\alpha\beta} + \frac{1}{c^2}(u^{\alpha}q^{\beta} + u^{\beta}q^{\alpha}) + \frac{e}{c^2}u^{\alpha}u^{\beta}, \\
F^{\alpha\beta\gamma} &= (C_1 + C_2\Pi)u^{\alpha}u^{\beta}u^{\gamma} + \\
&\frac{c^2}{6}(nm - C_1 + C_2\Pi)\left(g^{\alpha\beta}u^{\gamma} + g^{\beta\gamma}u^{\alpha} + g^{\alpha\gamma}u^{\beta}\right) + \\
&C_3\left(g^{\alpha\beta}q^{\gamma} + g^{\beta\gamma}q^{\alpha} + g^{\alpha\gamma}q^{\beta}\right) + C_4\left(t^{<\alpha\beta>}u^{\gamma} + t^{<\beta\gamma>}u^{\gamma} + t^{<\alpha\gamma>}u^{\beta}\right).
\end{aligned}
$$

The main field becomes

$$
\begin{aligned}
u' &= -\frac{g}{T} - \tau_0\Pi, \\
u'_{\alpha} &= \frac{u_{\alpha}}{T} - \tau_1\Pi u_{\alpha} - \tau_2 q_{\alpha}, \\
u'_{\alpha\beta} &= -\tau_3 t_{<\alpha\beta>} - \tau_4\Pi h_{\alpha\beta} - \frac{1}{c^2}\tau_5(u_{\alpha}q_{\beta} + u_{\beta}q_{\alpha}) - \frac{3}{c^2}\tau_6\Pi u_{\alpha}u_{\beta},
\end{aligned}
$$

where $h^{\alpha\beta}$ is the projector and we omit the long expressions of the coefficients that appear in the previous equations.

If we put

$$
u'_{\alpha\beta} = 0 \tag{4.62}
$$

then we have:

$$
t_{<\alpha\beta>} = 0, \quad q_{\alpha} = 0, \quad \Pi = 0, \tag{4.63}
$$

and we obtain the equilibrium principal subsystem, i.e., the Euler relativistic fluid system:

$$
\begin{cases}
\partial_{\alpha}F^{\alpha} = 0, \\
\partial_{\alpha}F^{\alpha\beta} = 0
\end{cases} \tag{4.64}
$$

with

$$
F^{\alpha} = nmu^{\alpha}, \qquad F^{\alpha\beta} = ph^{\alpha\beta} + \frac{e}{c^2}u^{\alpha}u^{\beta}. \tag{4.65}
$$

4.7.2.1 Remark on the Einstein Equation

From the viewpoint of ET, we can see the Einstein equation in a different way. In fact, in the classical approach, the number of balance laws is usually five. Only one component (for example, the internal energy) in the energy-momentum tensor is considered as a field variable, and the remaining ones are prescribed by the constitutive equations. While, in ET, it is assumed that all the components of the energy-momentum tensor are field variables. Therefore, in ET, the Einstein equation:

$$R_{\mu\nu} - \frac{1}{2}g_{\mu\nu}R = -\frac{8\pi G}{c^4}T_{\mu\nu}$$

becomes a *universal* equation, independent of the constitution of the material [25].

For more details concerning RET of monatomic gases, interested readers can consult the book [4] and the survey papers [26, 27].

Finally we want to recall that many applications of ET not only to rarefied gases but also to other similar fields such as fluid-dynamic models in semiconductor physics (see for example [28–30]) have been made.

References

1. I.-S. Liu, I. Müller, Extended thermodynamics of classical and degenerate ideal gases. Arch. Ration. Mech. Anal. **83**, 285 (1983)
2. H. Grad, On the kinetic theory of rarefied gases. Commun. Pure Appl. Math. **2**(4), 331 (1949)
3. W. Dreyer, Maximization of the entropy in non-equilibrium. J. Phys. A: Math. Gen. **20**, 6505 (1987)
4. I. Müller, T. Ruggeri, *Rational Extended Thermodynamics*, 2nd edn. (Springer, New York, 1998)
5. T. Ruggeri, The entropy principle: from continuum mechanics to hyperbolic systems of balance laws. Bollettino dell'Unione Matematica Italiana **(8)8-B**, 1 (2005)
6. G. Boillat, T. Ruggeri, Hyperbolic principal subsystems: entropy convexity and subcharacteristic conditions. Arch. Ration. Mech. Anal. **137**, 305 (1997)
7. I. Müller, T. Ruggeri, Stationary heat conduction in radially, symmetric situations – an application of extended thermodynamics. J. Non-Newtonian Fluid Mech. **119**, 139 (2004)
8. H. Struchtrup, W. Weiss, Maximum of the local entropy production becomes minimal in stationary processes. Phys. Rev. Lett. **80**, 5048 (1998)
9. E. Barbera, I. Müller, D. Reitebuch, N. Zhao, Determination of boundary conditions in extended thermodynamics via fluctuation theory. Continuum Mech. Thermodyn. **16**, 411 (2004)
10. T. Ruggeri, J. Lou, Heat conduction in multi-temperature mixtures of fluids: the role of the average temperature. Phys. Lett. A **373**, 3052 (2009)
11. F. Brini, T. Ruggeri, Entropy principle for the moment systems of degree α associated to the Boltzmann equation. Critical derivatives and non controllable boundary data. Continuum Mech. Thermodyn. **14**, 165 (2002)
12. H. Struchtrup, Heat transfer in the transition regime: solution of boundary value problems for Grad's moment equations via kinetic schemes. Phys. Rev. E **65**, 041204 (2002)

13. H. Struchtrup, W. Weiss, Temperature jump and velocity slip in the moment method. Continuum Mech. Thermodyn. **12**, 1 (2000)
14. H. Struchtrup, *Macroscopic Transport Equations for Rarefied Gas Flow: Approximation Methods in Kinetic Theory* (Springer, Berlin/Heidelberg, 2005)
15. N. Zhao, M. Sugiyama, Analysis of hear conduction in a rarefied gas at rest with a temperature jump at the boundary by consistent-order extended thermodynamics. Continuum Mech. Thermodyn. **18**, 367 (2007)
16. S. Taniguchi, A. Iwasaki, M. Sugiyama, Relationship between Maxwell boundary condition and two kinds of stochastic thermal wall. J. Phys. Soc. Jpn. **77**, 124004 (2008)
17. G. Boillat, T. Ruggeri, Moment equations in the kinetic theory of gases and wave velocities. Continuum Mech. Thermodyn. **9**, 205 (1997)
18. W. Weiss, Zur Hierarchie der Erweiterten Thermodynamik. Dissertation, TU Berlin, 1990
19. G. Boillat, T. Ruggeri, Maximum wave velocity in the moments system of a relativistic gas. Continuum Mech. Thermodyn. **11**, 107 (1999)
20. G. Boillat, T. Ruggeri, Relativistic gas: moment equations and maximum wave velocity. J. Math. Phys. **40**, 6399 (1999)
21. F. Brini, T. Ruggeri, Maximum velocity for wave propagation in a relativistic rarefied gas. Continuum Mech. Thermodyn. **11**, 331 (1999)
22. T. Ruggeri, Non-linear maximum entropy principle for a polyatomic gas subject to the dynamic pressure. Bull. Inst. Math. Acad. Sin. Special Issue in honor of Tai-Ping Liu 70 birthday. **11** (2016)
23. I.-S. Liu, I. Müller, T. Ruggeri, Relativistic thermodynamics of gases. Ann. Phys. **169**, 191 (1986)
24. C. Marle, Sur l'établissement des équations de l'hydrodynamique des fluides relativistes dissipatifs I et II. Ann. Inst . Henri Poincaré A **10**, 67 (1969); A **10**, 127 (1969)
25. T. Ruggeri, *Relativistic Extended Thermodynamics: General Assumptions and Mathematical Procedure Corso CIME Noto (Giugno 1987)*, ed. by A. Anile, Y. Choquet-Bruhat. Lecture Notes in Mathematics, vol. 1385 (Springer, Berlin, 1989), pp. 269–277
26. I. Müller, Extended thermodynamics: a theory of symmetric hyperbolic field equations. Entropy **10**, 477 (2008)
27. T. Ruggeri, The entropy principle from continuum mechanics to hyperbolic systems of balance laws: the modern theory of extended thermodynamics. Entropy **10**, 319 (2008)
28. A.M. Anile, S. Pennisi, Thermodynamic derivation of the hydrodynamical model for charge transport in semiconductors. Phys. Rev. B **46**, 13186 (1992)
29. A.M. Anile, V. Romano, Non parabolic band transport in semiconductors: closure of the moment equations. Continuum Mech. Thermodyn. **11**, 307 (1999)
30. M. Trovato, L. Reggiani, Maximum entropy principle and hydrodynamic models in statistical mechanics. Riv. Nuovo Cimento Soc. Ital. Fis. **35**, 99 (2012)

Part III
RET of Polyatomic Gas and Dense Gas with 14 Fields

Chapter 5
RET 14-Field Theory of Polyatomic Gas and Dense Gas

Abstract The objective of the present chapter is to explain in detail the new ET theory of rarefied polyatomic gases and dense gases with 14 independent fields (ET14), that is, the mass density, the velocity, the temperature, the shear stress, the dynamic pressure, and the heat flux. We adopt the system of field equations with a binary hierarchy structure. We show that the constitutive equations can be determined explicitly by the caloric and thermal equations of state as in the monatomic ET13 theory. We also analyze physically important systems, that is, a rarefied polyatomic gas, a gas with the virial equations of state, a hard-sphere system, and a van der Waals gas. Lastly we show that the ET13 theory of rarefied monatomic gases is derived from the ET14 theory as a singular limit.

5.1 Previous Tentatives

The Navier-Stokes Fourier theory comes out from ET, as a limiting case, by carrying out the Maxwellian iteration [1, 2]. In this respect, the Navier-Stokes Fourier theory can be seen as an approximation of ET where the relaxation times of dissipative fluxes (viscous stress and heat flux) are negligibly small. We call this Navier-Stokes Fourier limit.

On the other hand, within its validity range, the classical Navier-Stokes Fourier theory is applicable to any fluids that are not necessarily limited to rarefied gases nor to monatomic gases. Therefore, after the successful establishment of ET for rarefied monatomic gases, there appeared many studies of ET for rarefied polyatomic gases [3–5] and also for dense gases [6–11]. In such gases in nonequilibrium, the dynamic pressure Π does not vanish identically. And, in general, no simple relationship between the pressure p and the specific internal energy ε exists.

Previous authors tried to establish ET of dense gases by postulating single hierarchy structure similar to (1.24), but with 14 densities by introducing a new *fourth*-rank tensorial density such as F_{kkll} [8, 10, 11]. However, in their theories, the other important feature of the system of field equations—a flux in an equation becomes a density in the next equation—was abandoned. Because of this generality, constitutive equations could not be fully determined from the knowledge of the equilibrium properties of gases. There remain many phenomenological constants in the constitutive equations that are impossible to be evaluated experimentally

T. Ruggeri, M. Sugiyama, *Rational Extended Thermodynamics beyond the Monatomic Gas*, DOI 10.1007/978-3-319-13341-6_5

or theoretically. Moreover, when the Navier-Stokes Fourier limit is taken, the postulation of the fourth-rank tensorial density seems to be not well justified because such a density does not have any straightforward counterpart in the Navier-Stokes Fourier theory.

5.2 Binary Hierarchy in ET of Polyatomic Gas and Dense Gas: Heuristic Viewpoint

In order to grasp the structure of the basic system appropriate for ET of polyatomic gases and dense gases, first of all, let us reconsider the structure of the classical Navier-Stokes Fourier system where, in addition to the usual conservation laws of mass, momentum, and energy (1.3), we have the constitutive equations (1.10). We observe that the Eq. (1.10) can be rewritten in the following form [12]:

$$
\begin{aligned}
&\frac{\partial}{\partial x_k}\left(v_i\delta_{jk} + v_j\delta_{ik} - \frac{2}{3}v_k\delta_{ij}\right) = \frac{\sigma_{\langle ij\rangle}}{\mu},\\
&\frac{\partial v_k}{\partial x_k} = -\frac{\Pi}{\nu},\\
&\frac{\partial T}{\partial x_k} = -\frac{q_k}{\kappa}.
\end{aligned}
\tag{5.1}
$$

The system composed of Eqs. (1.3) and (5.1) can be seen as a system of 14 equations for the 14 unknown variables: ρ, v_i , ε, q_i, $\sigma_{\langle ij\rangle}$ and Π. Its mathematical structure is in the form of balance type, but, in Eq. (5.1), we have no term with time derivative. Therefore the system is not hyperbolic but parabolic.

It is, therefore, natural to assume that the mathematical structure of balance laws in ET of dense gases is of the following type [13]:

$$
\begin{aligned}
&\frac{\partial F}{\partial t} + \frac{\partial F_k}{\partial x_k} = 0,\\
&\frac{\partial F_i}{\partial t} + \frac{\partial F_{ik}}{\partial x_k} = 0,\\
&\frac{\partial F_{ij}}{\partial t} + \frac{\partial F_{ijk}}{\partial x_k} = P_{ij}, \qquad \frac{\partial G_{ll}}{\partial t} + \frac{\partial G_{llk}}{\partial x_k} = 0,\\
&\qquad\qquad\qquad\qquad\qquad\quad\;\; \frac{\partial G_{lli}}{\partial t} + \frac{\partial G_{llik}}{\partial x_k} = Q_{lli},
\end{aligned}
\tag{5.2}
$$

where F is the mass density, F_i is the momentum density, G_{ii} is the energy density, F_{ij} is the momentum flux, and G_{lli} is the energy flux. And F_{ijk} and G_{llik} are the fluxes of F_{ij} and G_{lli}, respectively, and P_{ij} and Q_{lli} are the productions with respect to F_{ij} and G_{lli}, respectively. In order to justify this structure, we admit that Eq. (1.3) correspond to $(5.2)_{1,2,4}$ with the condition that F_{ll} is different from G_{ll} because, as mentioned before, no simple relation exists between the pressure and the internal energy in polyatomic gases or dense gases. Equation $(5.2)_3$ can be split into the deviatoric and trace parts that have the mathematical structure of $(5.1)_{1,2}$ when the terms with time derivatives are neglected. While Eq. $(5.2)_5$ in a steady case has the mathematical structure of the type of the Fourier law $(5.1)_3$.

We can see that the structure of (5.2) is much more restrictive than that adopted in the previous works, and moreover the system does not have the fourth-rank tensor in the set of densities.

To sum up, for the new ET theory, we adopt 14 independent densities:

$$\begin{aligned} &\text{mass density:} && F, \\ &\text{momentum density:} && F_i, \\ &\text{energy density:} && G_{ii}, \\ &\text{momentum flux:} && F_{ij}, \\ &\text{energy flux:} && G_{lli}. \end{aligned} \tag{5.3}$$

And we adopt also the system (5.2) that is composed of two parallel hierarchical series: The one is the series starting from the mass and momentum balance equations (F-series) and the other is from the energy balance equation (G-series). In each series, the flux in an equation becomes the density in the next equation.

This binary hierarchy will be justified also from the kinetic considerations in the next Chap. 6 at least in the case of rarefied polyatomic gases.

This chapter describes the results obtained in the paper by Arima et al. [13].

5.3 ET 14-Field Theory

In this section, we construct the ET 14-field theory (ET14) by imposing the universal physical principles explained in Sect. 2.2.

We assume that the closure quantities at one point and time depend on the independent fields at that point and time, i.e., local and instantaneous. Therefore we have

$$\psi = \psi(F, F_i, F_{ij}, G_{ll}, G_{lli}), \tag{5.4}$$

where ψ is one of the quantities $\{F_{ijk}, G_{llik}, P_{ij}, Q_i\}$. The entropy density h^0 and the entropy flux $h_k (= hv_k + \varphi_k)$ are also assumed to be in the form of (5.4). Hereafter, for simplicity, we use the notation h instead of h^0 and φ^k instead of $\hat{h}^k$ as no confusion is caused.

5.3.1 Exploitation of the Galilean Invariance

The matrices $\mathbf{X}$ and $\mathbf{A}^r$ of the Galilean invariance [see (2.36) and (2.39)] are given by

$$\mathbf{X} = \begin{pmatrix} 1 & 0 & 0 & 0 & 0 \\ v_i & \delta_i^{h_1} & 0 & 0 & 0 \\ v_i v_j & 2\delta_{(i}^{h_1} v_{j)} & \delta_i^{h_1}\delta_j^{h_2} & 0 & 0 \\ v^2 & 2v_i & 0 & 1 & 0 \\ v^2 v_i & 3v_{(l}v_l\delta_{i)}^{h_1} & 2\delta_i^{(h_1} v_{h_2)} & v_i & \delta_i^{h_1} \end{pmatrix} \tag{5.5}$$

and

$$\mathbf{A}^r = \begin{pmatrix} 0 & 0 & 0 & 0 & 0 \\ \delta_i^r & 0 & 0 & 0 & 0 \\ 0 & 2\delta_{(i}^{h_1}\delta_{j)}^r & 0 & 0 & 0 \\ 0 & 2\delta_{h_1}^r & 0 & 0 & 0 \\ 0 & 0 & 2\delta_i^{(h1}\delta_r^{h_2)} & \delta_i^r & 0 \end{pmatrix}, \qquad (r = 1, 2, 3). \tag{5.6}$$

Therefore, the velocity dependence is completely prescribed by the Galilean invariance as follows:

$$\mathbf{F}^0 = \begin{pmatrix} F \\ F_i \\ F_{ij} \\ G_{ll} \\ G_{lli} \end{pmatrix} = \begin{pmatrix} \rho \\ \rho v_i \\ \rho v_i v_j + \hat{F}_{ij} \\ \rho v^2 + \hat{G}_{ll} \\ \rho v^2 v_i + 2\hat{F}_{li} v_l + \hat{G}_{ll} v_i + \hat{G}_{lli} \end{pmatrix},$$

$$\boldsymbol{\Phi}^k = \begin{pmatrix} \Phi_i \\ \Phi_{ik} \\ \Phi_{ijk} \\ \Psi_k \\ \Psi_{ik} \end{pmatrix} = \begin{pmatrix} 0 \\ \hat{F}_{ik} \\ 2\hat{F}_{k(i}v_{j)} + \hat{F}_{ijk} \\ 2\hat{F}_{kl}v_l + \hat{G}_{llk} \\ 3\hat{F}_{k(i}v_l v_{l)} + 2\hat{F}_{ikl}v_l + \hat{G}_{llk}v_i + \hat{G}_{llik} \end{pmatrix}, \tag{5.7}$$

$$\mathbf{f} = \begin{pmatrix} 0 \\ 0 \\ P_{ij} \\ 0 \\ Q_{lli} \end{pmatrix} = \begin{pmatrix} 0 \\ 0 \\ \hat{P}_{ij} \\ 0 \\ 2\hat{P}_{li}v_l + \hat{Q}_{lli} \end{pmatrix},$$

where we recall $\hat{\boldsymbol{\Phi}}^k = \hat{\mathbf{F}}^k$.

As the balance equations of F, F_i and G_{ll} represent the conservation laws of mass, momentum and energy, the intrinsic quantities $\hat{F}_{ij}$, $\hat{G}_{ll}$ and $\hat{G}_{lli}$ have the following conventional meanings:

$$\text{stress tensor:} \qquad t_{ij} = -\hat{F}_{ij} \; (= -\,(p+\Pi)\,\delta_{ij} + \sigma_{\langle ij\rangle}), \tag{5.8}$$

$$\text{specific internal energy:} \qquad \varepsilon = \frac{1}{2\rho}\hat{G}_{ll}, \tag{5.9}$$

$$\text{heat flux:} \qquad q_i = \frac{1}{2}\hat{G}_{lli}, \tag{5.10}$$

where the pressure p depends only on ρ and ε. $\sigma_{\langle ij\rangle}$ is symmetric because we deal with non-polar materials. Therefore $\hat{P}_{ij}$ is symmetric and $\hat{F}_{ijk}$ is symmetric a priori only with respect to the first two indices.

Then the **constitutive equations of ET14** are expressed by the Galilean objective variables in the form [see the closure equations (5.4)]:

$$\hat{\psi} = \hat{\psi}(\rho,\, \varepsilon,\, \Pi,\, \sigma_{\langle ij\rangle},\, q_i). \tag{5.11}$$

The entropy density $h(= h^0)$ and the intrinsic entropy flux $\varphi_i(= \hat{h}^i)$ do not depend on the velocity [see (2.37)]. They are expressed as follows:

$$\begin{aligned} h &= h(\rho,\, \varepsilon,\, \Pi,\, \sigma_{\langle ij\rangle},\, q_i), \\ \varphi_k &= \varphi_k(\rho,\, \varepsilon,\, \Pi,\, \sigma_{\langle ij\rangle},\, q_i). \end{aligned} \tag{5.12}$$

The dependence of the Lagrange multipliers on the velocity is dictated by the relation (2.42). In the present case with $(\lambda \; \lambda_i \; \lambda_{ij} \; \mu \; \mu_i)$ as seen in (5.14) below, it is expressed as

$$\begin{aligned} \lambda &= \hat{\lambda} - \hat{\lambda}_i v_i + \hat{\lambda}_{ij} v_i v_j + \hat{\mu} v^2 - \hat{\mu}_i v^2 v_i, \\ \lambda_i &= \hat{\lambda}_i - 2\hat{\lambda}_{il} v_l - 2\hat{\mu} v_i + \hat{\mu}_i v^2 + 2\hat{\mu}_l v_i v_l, \\ \lambda_{ij} &= \hat{\lambda}_{ij} - 2\hat{\mu}_{(i} v_{j)}, \\ \mu &= \hat{\mu} - \hat{\mu}_l v_l, \\ \mu_i &= \hat{\mu}_i. \end{aligned} \tag{5.13}$$

5.3.2 Exploitation of the Entropy Principle

The entropy principle requires the following relations [see (2.13)]:

$$
\begin{aligned}
dh &= \lambda dF + \lambda_i dF_i + \lambda_{ij} dF_{ij} + \mu dG_{ll} + \mu_i dG_{lli}, \\
dh_k &= \lambda dF_k + \lambda_i dF_{ik} + \lambda_{ij} dF_{ijk} + \mu dG_{llk} + \mu_i dG_{llik}, \\
\Sigma &= \lambda_{ij} P_{ij} + \mu_i Q_{lli}.
\end{aligned} \tag{5.14}
$$

$\mathbf{u}' = \left(\lambda\ \lambda_i\ \lambda_{ij}\ \mu\ \mu_i\right)$ is the main field.

Therefore, taking (2.43) into account, we have

$$
\begin{aligned}
dh &= \hat{\lambda} d\rho + \hat{\lambda}_{ij} d\hat{F}_{ij} + \hat{\mu} d\hat{G}_{ll} + \hat{\mu}_i d\hat{G}_{lli}, \\
d\varphi_k &= \hat{\lambda}_i d\hat{F}_{ik} + \hat{\lambda}_{il} d\hat{F}_{ilk} + \hat{\mu} d\hat{G}_{llk} + \hat{\mu}_i d\hat{G}_{llik}, \\
\Sigma &= \hat{\lambda}_{ij} \hat{P}_{ij} + \hat{\mu}_i \hat{Q}_{lli}.
\end{aligned} \tag{5.15}
$$

The constraints (2.44) due to the fact that the entropy density and the intrinsic entropy flux are independent of the velocity are now expressed by

$$
\begin{aligned}
&\hat{\lambda}_i = -\frac{1}{\rho}\left(\hat{G}_{ll}\hat{\mu}_i + 2\hat{F}_{il}\hat{\mu}_l\right), \\
&\left(\hat{\lambda}\rho + \hat{\lambda}_{kj}\hat{F}_{kj} + \hat{\mu}\hat{G}_{ll} + \hat{\mu}_k\hat{G}_{llk} - h\right)\delta_{ir} + 2\hat{\lambda}_{rl}\hat{F}_{il} \\
&+ 2\hat{\mu}\hat{F}_{ir} + 2\hat{\mu}_l F_{lri} + \hat{\mu}_r\hat{G}_{lli} = 0.
\end{aligned} \tag{5.16}
$$

5.3.2.1 Equilibrium State

From Definition 2.2 of an equilibrium state, which says that the entropy production Σ becomes minimum and vanishes in equilibrium, we notice that the productions $\hat{P}_{ii}$, $\hat{P}_{\langle ij\rangle}$ and $\hat{Q}_{lli}$ must vanish together with all Lagrange multipliers corresponding to the balance laws [see (2.29)]:

$$
\hat{\lambda}^E_{ll} = 0, \qquad \hat{\lambda}^E_{\langle ij\rangle} = 0, \qquad \hat{\mu}^E_i = 0, \tag{5.17}
$$

where superscript E indicates equilibrium. Since the Lagrange multipliers $\hat{\lambda}_{ll}$, $\hat{\lambda}_{\langle ij\rangle}$ and $\hat{\mu}_i$ are non-zero only in nonequilibrium, these are called nonequilibrium variables. Taking the representation theorem for isotropic vectors and tensors into account, we can show that the requirement that the intrinsic productions and the intrinsic nonequilibrium Lagrange multipliers vanish implies that $q_i, \sigma_{\langle ij\rangle}, \Pi$ must also vanish in equilibrium.

5.3.2.2 Entropy

Let us rewrite the entropy density in the form:

$$h = \rho s + \rho k, \tag{5.18}$$

where $s(\rho, \varepsilon)$ is the specific entropy density in equilibrium and $k(\rho, \varepsilon, \Pi, \sigma_{\langle ij\rangle}, q_i)$ denotes the nonequilibrium part of the entropy and therefore satisfies the condition:

$$k(\rho, \varepsilon, 0, 0, 0) = 0. \tag{5.19}$$

The equilibrium entropy density s obeys the Gibbs equation (1.7).

From $(5.15)_1$, we obtain

$$\begin{aligned}
\hat{\lambda} &= -\frac{g}{T} + k + \rho k_\rho - \varepsilon k_\varepsilon - (\rho p_\rho - \varepsilon p_\varepsilon) k_\Pi, \\
\hat{\mu} &= \frac{1}{2T} + \frac{1}{2} k_\varepsilon - \frac{1}{2} p_\varepsilon k_\Pi, \\
\hat{\lambda}_{ll} &= \rho k_\Pi, \\
\hat{\lambda}_{\langle ij\rangle} &= -\rho k_{\sigma_{\langle ij\rangle}}, \\
\hat{\mu}_i &= \frac{\rho}{2} k_{q_i},
\end{aligned} \tag{5.20}$$

where a subscript denotes a partial differentiation with respect to the quantity, for example, $k_\Pi = \frac{\partial k}{\partial \Pi}$. In an equilibrium state we have

$$\begin{aligned}
&\hat{\lambda}^E = -\frac{g}{T}, \qquad \hat{\mu}^E = \frac{1}{2T}, \\
&\hat{\lambda}^E_{ll} = 0, \qquad \hat{\lambda}^E_{\langle ij\rangle} = 0, \qquad \hat{\mu}^E_i = 0.
\end{aligned} \tag{5.21}$$

5.3.2.3 Constitutive Equations Near Equilibrium

As usual in ET, we consider only processes not far from equilibrium, which, however, may be out of local equilibrium, in such a way that the constitutive equations are linear with respect to the nonequilibrium variables $(\Pi, \sigma_{\langle ij\rangle}, q_i)$. From (2.15), we notice that we need to represent the potentials h'^α and therefore the entropy and the entropy flux [see (2.14)] in an expansion form with respect to the nonequilibrium fields at least until second order. However, as we will see below, in order to fix all the expansion coefficients, we need to represent them up to third order. Then, as all derivatives of k with respect to the nonequilibrium variables vanish in equilibrium due to the identity (5.19), we have the expansion of k and φ_k as follows:

$$\begin{aligned} k =& k_1 \Pi^2 + k_2 \sigma_{\langle ij \rangle}\sigma_{\langle ij \rangle} + k_3 q_i q_i + k_4 \Pi^3 + k_5 \Pi \sigma_{\langle ij \rangle}\sigma_{\langle ij \rangle} + k_6 \sigma_{\langle ij \rangle}\sigma_{\langle ni \rangle}\sigma_{\langle jn \rangle} \\ &+ k_7 \Pi q_i q_i + k_8 \sigma_{\langle ij \rangle} q_i q_j + O(4), \qquad (5.22) \\ \varphi_k =& (\beta_1 + \beta_2 \Pi) q_k + \beta_3 \sigma_{\langle ki \rangle} q_i + O(3), \qquad (5.23) \end{aligned}$$

where the coefficients $k_1, \cdots, k_8$ and $\beta_1, \beta_2, \beta_3$ are functions of ρ and ε. Moreover, within this approximation, we have

$$\begin{aligned} \hat{F}_{ijk} &= f_1 q_k \delta_{ij} + f_2 q_{\langle i}\delta_{j\rangle k} + O(2), \\ \hat{G}_{llij} &= (g_1 + g_2 \Pi)\delta_{ij} - g_3 \sigma_{\langle ij \rangle} + O(2), \end{aligned} \qquad (5.24)$$

where f_1, f_2, g_1, g_2 and g_3 are functions of ρ and ε.

From (5.22), we obtain

$$\begin{aligned} dk =& \Pi^2 dk_1 + \sigma_{\langle ij \rangle}\sigma_{\langle ij \rangle} dk_2 + q_i q_i dk_3 + \Pi^3 dk_4 \\ &+ \Pi \sigma_{\langle ij \rangle}\sigma_{\langle ij \rangle} dk_5 + \sigma_{\langle ij \rangle}\sigma_{\langle in \rangle}\sigma_{\langle nj \rangle} dk_6 + \Pi q_i q_i dk_7 + q_i q_j \sigma_{\langle ij \rangle} dk_8 \\ &+ \left(2k_1 \Pi + 3k_4 \Pi^2 + k_5 \sigma_{\langle ij \rangle}\sigma_{\langle ij \rangle} + k_7 q_i q_i\right) d\Pi \\ &+ \left(2k_2 \sigma_{\langle ij \rangle} + 2k_5 \Pi \sigma_{\langle ij \rangle} + 3k_6 \sigma_{\langle n\langle i \rangle}\sigma_{\langle j \rangle n \rangle} + k_8 q_{\langle i} q_{j \rangle}\right) d\sigma_{\langle ij \rangle} \\ &+ \left(2k_3 q_i + 2k_7 \Pi q_i + 2k_8 \sigma_{\langle ij \rangle} q_j\right) dq_i + O(4). \end{aligned} \qquad (5.25)$$

Therefore, by substituting (5.25) into (5.20), we obtain the Lagrange multipliers as follows:

$$\begin{aligned} \hat{\lambda} =& -\frac{g}{T} - 2k_1 \left(\rho p_\rho - \varepsilon p_\varepsilon\right) \Pi + \left\{k_1 + \rho k_{1\rho} - \varepsilon k_{1\varepsilon} - 3k_4 \left(\rho p_\rho - \varepsilon p_\varepsilon\right)\right\} \Pi^2 \\ &+ \left\{k_2 + \rho k_{2\rho} - \varepsilon k_{2\varepsilon} - k_5 \left(\rho p_\rho - \varepsilon p_\varepsilon\right)\right\} \sigma_{\langle ij \rangle}\sigma_{\langle ij \rangle} \\ &+ \left\{k_3 + \rho k_{3\rho} - \varepsilon k_{3\varepsilon} - k_7 \left(\rho p_\rho - \varepsilon p_\varepsilon\right)\right\} q_i q_i + O(3), \\ \hat{\mu} =& \frac{1}{2T} - p_\varepsilon k_1 \Pi + \frac{1}{2}\left(k_{1\varepsilon} - 3p_\varepsilon k_4\right) \Pi^2 + \frac{1}{2}\left(k_{2\varepsilon} - p_\varepsilon k_5\right) \sigma_{\langle ij \rangle}\sigma_{\langle ij \rangle} \\ &+ \frac{1}{2}\left(k_{3\varepsilon} - p_\varepsilon k_7\right) q_i q_i + O(3), \\ \hat{\lambda}_{ll} =& 2\rho k_1 \Pi + 3\rho k_4 \Pi^2 + \rho k_5 \sigma_{\langle ij \rangle}\sigma_{\langle ij \rangle} + \rho k_7 q_i q_i + O(3), \\ \hat{\lambda}_{\langle ij \rangle} =& -2\rho k_2 \sigma_{\langle ij \rangle} - 2\rho k_5 \Pi \sigma_{\langle ij \rangle} - 3\rho k_6 \sigma_{\langle n\langle i \rangle}\sigma_{\langle j \rangle n \rangle} - \rho k_8 q_{\langle i} q_{j \rangle} + O(3), \\ \hat{\mu}_i =& \rho k_3 q_i + \rho k_7 \Pi q_i + \rho k_8 \sigma_{\langle ij \rangle} q_j + O(3). \end{aligned} \qquad (5.26)$$

Let us determine the coefficients $k_1, \cdots, k_8$ by using the constraint $(5.16)_2$ and the expressions of the Lagrange multipliers (5.26). The constraint $(5.16)_2$ can be decomposed into the trace part, traceless part and antisymmetric part as follows:

$$\begin{aligned}
&h = \rho\hat{\lambda} + \frac{5}{3}(p+\Pi)\hat{\lambda}_{ll} - \frac{5}{3}\hat{\lambda}_{\langle ij\rangle}\sigma_{\langle ij\rangle} + 2(\rho\varepsilon + p + \Pi)\hat{\mu} + \frac{8}{3}q_i\hat{\mu}_i + \frac{2}{3}\hat{\mu}_l\hat{F}_{ppl},\\
&2\hat{\lambda}_{l\langle r}\hat{F}_{i\rangle l} - 2\hat{\mu}\sigma_{\langle ir\rangle} + 2\hat{\mu}_l\hat{F}_{l\langle ri\rangle} + 2\hat{\mu}_{\langle r}q_{i\rangle} = 0,\\
&2\hat{\lambda}_{l[r}\hat{F}_{i]l} + 2\hat{\mu}_{[r}q_{i]} + 2\hat{\mu}_l\hat{F}_{l[ri]} = 0
\end{aligned} \tag{5.27}$$

where the brackets [ ] stand for the antisymmetry with respect to the suffixes inside the brackets. From $(5.27)_{1,2}$ we obtain

$$\begin{aligned}
k_1 &= -\frac{1}{2\rho T\Gamma},\\
k_2 &= -\frac{1}{4\rho T p},\\
k_4 &= -\frac{1}{3\Gamma}\left\{\left(\frac{10}{3} - 2\frac{p_\varepsilon}{\rho}\right)k_1 + \rho k_{1\rho} + \frac{p}{\rho}k_{1\varepsilon}\right\},\\
k_5 &= -\frac{1}{\Gamma}\left\{\frac{10}{3}k_2 + \rho k_{2\rho} + \frac{p}{\rho}k_{2\varepsilon}\right\},\\
k_5 &= -\frac{1}{p}\left\{\left(\frac{1}{3} - \frac{p_\varepsilon}{2\rho}\right)k_1 + k_2\right\},\\
k_6 &= -\frac{1}{6\rho T p^2},\\
k_7 &= -\frac{1}{\Gamma}\left\{\frac{2}{3}(4+3f_1)k_3 + \rho k_{3\rho} + \frac{p}{\rho}k_{3\varepsilon}\right\},\\
k_8 &= \frac{1}{p}(1+f_2)k_3,
\end{aligned} \tag{5.28}$$

where Γ is a function of ρ and ε expressed by

$$\Gamma = \frac{5}{3}p - \rho p_\rho - \frac{p}{\rho}p_\varepsilon. \tag{5.29}$$

We notice two expressions of k_5 in $(5.28)_{4,5}$ are identically equivalent.

From the antisymmetric part of the constraint (5.16), i.e., $(5.27)_3$, we obtain

$$\hat{F}_{lri} = \hat{F}_{lir}. \tag{5.30}$$

Since $\hat{F}_{lri}$ is symmetric with respect to l and r, $\hat{F}_{lri}$ is symmetric with respect to all indices. Therefore instead of our linear approximation $(5.24)_1$, it is possible to adopt

$$\hat{F}_{ijk} = \hat{F}_{\langle ijk \rangle} + \frac{3}{5}\hat{F}_{ll(i}\delta_{jk)} \tag{5.31}$$

with

$$\hat{F}_{\langle ijk \rangle} = O(2), \quad \text{and} \quad \hat{F}_{lli} = 3f_1 q_i + O(2). \tag{5.32}$$

This concludes, from $(5.24)_1$, that

$$f_1 = \frac{5}{6}f_2. \tag{5.33}$$

Next we analyze the entropy flux. From (5.23), we have

$$\begin{aligned} d\varphi_k =& q_k d\beta_1 + \Pi q_k d\beta_2 + \sigma_{\langle ki \rangle} q_i d\beta_3 + \beta_2 q_k d\Pi + \beta_3 q_i d\sigma_{\langle ki \rangle} \\ &+ \left\{(\beta_1 + \beta_2 \Pi)\delta_{ki} + \beta_3 \sigma_{\langle ki \rangle}\right\} dq_i + O(3). \end{aligned} \tag{5.34}$$

On the other hand, $d\varphi_k$ is expressed as $(5.15)_2$ with $(5.16)_1$. By comparing these expressions in terms of the derivatives of ρ, ε, Π, $\sigma_{\langle ij \rangle}$ and q_i with each other, following relations are derived:

$$\begin{aligned}
\beta_{1_\rho} &= \rho\left\{g_{1_\rho} - 2\left(\varepsilon + \frac{p}{\rho}\right)p_\rho\right\} k_3, \\
\beta_{2_\rho} &= \rho\left\{2k_1 f_{1_\rho} + k_3\left(g_{2_\rho} - 2\frac{p_\rho}{\rho}\right) + k_7\left(g_{1_\rho} - 2\left(\varepsilon + \frac{p}{\rho}\right)p_\rho\right)\right\}, \\
\beta_{3_\rho} &= -\rho\left\{2k_2 f_{2_\rho} + k_3\left(g_{3_\rho} - 2\frac{p_\rho}{\rho}\right) - k_8\left(g_{1_\rho} - 2\left(\varepsilon + \frac{p}{\rho}\right)p_\rho\right)\right\}, \\
\beta_{1_\varepsilon} &= \rho\left\{g_{1_\varepsilon} - 2\left(\varepsilon + \frac{p}{\rho}\right)p_\varepsilon\right\} k_3, \\
\beta_{2_\varepsilon} &= \rho\left\{2k_1 f_{1_\varepsilon} + k_3\left(g_{2_\varepsilon} - 2\frac{p_\varepsilon}{\rho}\right) + k_7\left(g_{1_\varepsilon} - 2\left(\varepsilon + \frac{p}{\rho}\right)p_\varepsilon\right)\right\}, \\
\beta_{3_\varepsilon} &= -\rho\left\{2k_2 f_{2_\varepsilon} + k_3\left(g_{3_\varepsilon} - 2\frac{p_\varepsilon}{\rho}\right) - k_8\left(g_{1_\varepsilon} - 2\left(\varepsilon + \frac{p}{\rho}\right)p_\varepsilon\right)\right\}, \quad (5.35) \\
\beta_1 &= \frac{1}{T}, \\
\beta_2 &= \rho k_3\left\{g_2 - 2\left(\varepsilon + \frac{p}{\rho}\right)\right\},
\end{aligned}$$

$$\beta_3 = -\rho k_3 \left\{ g_3 - 2\left(\varepsilon + \frac{p}{\rho}\right)\right\},$$

$$f_1 = \frac{\beta_2}{2\rho k_1} + \frac{p_\varepsilon}{\rho},$$

$$f_2 = -\frac{\beta_3}{2\rho k_2}.$$

In particular, from $(5.35)_{1,2,\cdots,7}$ and (5.28), we obtain

$$\begin{aligned}
dg_1 =& -\frac{1}{\rho k_3 T^2}dT + 2\left(\varepsilon + \frac{p}{\rho}\right)dp,\\
d(T\varGamma g_2) =& \varGamma \left\{2\left(\varepsilon + \frac{p}{\rho}\right) + \frac{k_7}{T\rho k_3^2}\right\}dT\\
&+ 2\left(\varepsilon + \frac{p}{\rho}\right)Td\varGamma + \frac{2T\varGamma}{\rho}dp + \frac{1}{\rho k_3}d\left(\frac{p_\varepsilon}{\rho}\right),\\
d(pg_3) =& 2\left(\varepsilon + 2\frac{p}{\rho}\right)dp + \frac{1}{T}\left\{pg_3 - \frac{1}{T\rho k_3} - 2p\left(\varepsilon + \frac{p}{\rho}\right)\right\}dT.
\end{aligned} \tag{5.36}$$

Using the integrability condition for g_1 and the relation

$$g_2 = \frac{1}{\varGamma}\left\{\frac{5}{3}pg_3 + \frac{p_\varepsilon}{\rho^2 T k_3} + 2\left(\varepsilon + \frac{p}{\rho}\right)\left(\varGamma - \frac{5}{3}p\right)\right\}, \tag{5.37}$$

which comes from (5.33), we can prove that $(5.36)_2$ and $(5.36)_3$ are equivalent.

5.3.2.4 Productions

The productions are also expanded with respect to the nonequilibrium variables $\{\varPi, \sigma_{\langle ij\rangle}, q_i\}$ around an equilibrium state. In the linear approximation, we have

$$\hat{P}_{ll} = -\frac{3}{\tau_\varPi}\varPi, \quad \hat{P}_{\langle ij\rangle} = \frac{1}{\tau_S}\sigma_{\langle ij\rangle}, \quad \hat{Q}_{lli} = -\frac{2}{\tau_q}q_i, \tag{5.38}$$

where $\tau_\varPi$, τ_S and τ_q are the expansion coefficients, meaning of which will be understood in Sect. 5.3.5. Then we obtain

$$\Sigma = \frac{1}{T\varGamma\tau_\varPi}\varPi^2 + \frac{1}{2pT\tau_S}\sigma_{\langle ij\rangle}\sigma_{\langle ij\rangle} - \frac{2\rho k_3}{\tau_q}q_i q_i \geqq 0. \tag{5.39}$$

There are three conditions for the coefficients:

$$T\varGamma\tau_\varPi > 0, \quad pT\tau_S > 0, \quad \frac{\tau_q}{k_3} < 0. \tag{5.40}$$

5.3.2.5 Linear Constitutive Equations Expressed by the Usual Fields

It is usually more convenient to take $\{\rho, T\}$ as independent variables instead of $\{\rho, \varepsilon\}$. Let us first recall the following well-known relations:

$$f_\rho = \left(\frac{\partial f}{\partial \rho}\right)_\varepsilon = \left(\frac{\partial f}{\partial \rho}\right)_T - \frac{1}{c_v}\left(\frac{\partial \varepsilon}{\partial \rho}\right)_T\left(\frac{\partial f}{\partial T}\right)_\rho, \qquad f_\varepsilon = \left(\frac{\partial f}{\partial \varepsilon}\right)_\rho = \frac{1}{c_v}\left(\frac{\partial f}{\partial T}\right)_\rho,$$

where c_v is the specific heat at constant volume defined by

$$c_v = \left(\frac{\partial \varepsilon}{\partial T}\right)_\rho. \tag{5.41}$$

Then, the coefficients are rewritten as follows:

$$\begin{aligned}
k_1 &= -\frac{1}{2\rho T\Gamma},\\
k_2 &= -\frac{1}{4\rho T p},\\
k_4 &= -\frac{1}{3\Gamma}\left\{\left(\frac{10}{3} - \frac{2}{\rho c_v}\left(\frac{\partial p}{\partial T}\right)_\rho\right)k_1 + \rho\left(\frac{\partial k_1}{\partial \rho}\right)_T + \frac{T}{\rho c_v}\left(\frac{\partial p}{\partial T}\right)_\rho\left(\frac{\partial k_1}{\partial T}\right)_\rho\right\},\\
k_5 &= \frac{1}{4\rho T p^2} + \frac{1}{4\rho T p\Gamma}\left\{\frac{2}{3} - \frac{1}{\rho c_v}\left(\frac{\partial p}{\partial T}\right)_\rho\right\},\\
k_6 &= -\frac{1}{6\rho T p^2},\\
k_7 &= -\frac{1}{\Gamma}\left\{\frac{1}{3}\left(8 + 5f_2\right)k_3 + \rho\left(\frac{\partial k_3}{\partial \rho}\right)_T + \frac{T}{\rho c_v}\left(\frac{\partial p}{\partial T}\right)_\rho\left(\frac{\partial k_3}{\partial T}\right)_\rho\right\},\\
k_8 &= \frac{1}{p}\left(1 + f_2\right)k_3,
\end{aligned} \tag{5.42}$$

with

$$\Gamma = \frac{5}{3}p - \rho\left(\frac{\partial p}{\partial \rho}\right)_T - \frac{T}{\rho c_v}\left(\frac{\partial p}{\partial T}\right)_\rho^2 \neq 0. \tag{5.43}$$

From $(5.36)_1$ we obtain

$$\left(\frac{\partial g_1}{\partial \rho}\right)_T = 2\left(\varepsilon + \frac{p}{\rho}\right)\left(\frac{\partial p}{\partial \rho}\right)_T. \tag{5.44}$$

Moreover we can obtain

$$\frac{1}{k_3} = 2\rho T^2\left(\varepsilon + \frac{p}{\rho}\right)\left(\frac{\partial p}{\partial T}\right)_\rho - \rho T^2\left(\frac{\partial g_1}{\partial T}\right)_\rho . \tag{5.45}$$

We will show in Sects. 5.4 and 5.5 that, by using the relations (5.35), (5.42), and the equations of state ($\varepsilon = \varepsilon(\rho, T)$, $p = p(\rho, T)$), we can derive uniquely the explicit expressions of k_3 except for an integration constant. Similarly g_3 are obtained from $(5.36)_3$, in particular, from the following relations

$$\left(\frac{\partial p g_3}{\partial \rho}\right)_T = 2\left(\varepsilon + 2\frac{p}{\rho}\right)\left(\frac{\partial p}{\partial \rho}\right)_T, \tag{5.46}$$

$$\left(\frac{\partial p g_3}{\partial T}\right)_\rho = \frac{p g_3}{T} - \frac{1}{T^2\rho k_3} + 2\left(\varepsilon + 2\frac{p}{\rho}\right)\left(\frac{\partial p}{\partial T}\right)_\rho - \frac{2p}{T}\left(\varepsilon + \frac{p}{\rho}\right), \tag{5.47}$$

except for an integration constant. g_2 is also obtained from $(5.36)_2$ or (5.37). Finally, the coefficient f_2 (or f_1) can be determined by the following relation:

$$f_2 = -2\rho T p\left\{g_3 - 2\left(\varepsilon + \frac{p}{\rho}\right)\right\} k_3. \tag{5.48}$$

And β_2 and β_3 are determined from $(5.35)_{8,9}$ as

$$\begin{aligned} \beta_2 &= \rho k_3\left\{g_2 - 2\left(\varepsilon + \frac{p}{\rho}\right)\right\}, \\ \beta_3 &= -\rho k_3\left\{g_3 - 2\left(\varepsilon + \frac{p}{\rho}\right)\right\}. \end{aligned} \tag{5.49}$$

5.3.3 *Convexity of the Entropy Density*

The entropy density (5.18) with (5.22) and entropy flux (5.23) with (5.28) and (5.35) are summarized as follows:

$$\begin{aligned} h &= \rho s - \frac{1}{2T\Gamma}\Pi^2 - \frac{1}{4pT}\sigma_{\langle ij\rangle}\sigma_{\langle ij\rangle} + \rho k_3 q_i q_i + O(3), \\ \varphi_k &= \frac{1}{T}q_k + \beta_2 \Pi q_k + \beta_3 q_i \sigma_{\langle ik\rangle} + O(3). \end{aligned} \tag{5.50}$$

The convexity condition (2.47) evaluated in an equilibrium state becomes

$$\hat{Q}_E = \hat{Q}_{5E} - \frac{1}{2T\Gamma}(\delta\Pi)^2 - \frac{1}{4pT}\delta\sigma_{\langle ij\rangle}\delta\sigma_{\langle ij\rangle} + \rho k_3 \delta q_i \delta q_i < 0, \tag{5.51}$$

where $\hat{Q}_{5E}$ is the corresponding quantity for Euler fluids. From the condition $\hat{Q}_{5E} < 0$, we obtain the usual thermodynamic inequalities, that is, the positivity of the heat capacity and the compressibility [see (2.20)]. Therefore the following inequalities must be fulfilled:

$$\Gamma > 0, \qquad k_3 < 0. \tag{5.52}$$

Moreover from (5.40) and (5.52), we obtain

$$\tau_\Pi > 0, \quad \tau_S > 0, \quad \tau_q > 0. \tag{5.53}$$

Remark The function Γ can be rewritten neatly in terms of $\{\rho, s\}$ as follows:

$$\Gamma = \frac{5}{3}p - \rho\left(\frac{\partial p}{\partial \rho}\right)_s > 0. \tag{5.54}$$

5.3.4 Main Field

From (5.26), the intrinsic Lagrange multipliers in the linear approximation are obtained as follows:

$$\begin{aligned}
\hat{\lambda} &= -\frac{g}{T} + \frac{1}{T\Gamma}\left\{\left(\frac{\partial p}{\partial \rho}\right)_T - \frac{1}{c_v}\left(\frac{\varepsilon}{\rho} + \left(\frac{\partial \varepsilon}{\partial \rho}\right)_T\right)\left(\frac{\partial p}{\partial T}\right)_\rho\right\}\Pi, \\
\hat{\lambda}_i &= -2\rho\left(\varepsilon + \frac{p}{\rho}\right)k_3 q_i, \\
\hat{\lambda}_{\langle ij\rangle} &= \frac{1}{2pT}\sigma_{\langle ij\rangle}, \\
\hat{\lambda}_{ll} &= -\frac{1}{T\Gamma}\Pi, \\
\hat{\mu} &= \frac{1}{2T} + \frac{1}{2\rho T c_v \Gamma}\left(\frac{\partial p}{\partial T}\right)_\rho \Pi, \\
\hat{\mu}_i &= \rho k_3 q_i.
\end{aligned} \tag{5.55}$$

Taking into account (5.13), we can evaluate the main field components for which the differential system becomes symmetric hyperbolic. The potential h':

$$h' = \rho\hat{\lambda} + 2\rho\varepsilon\hat{\mu} + (p + \Pi)\hat{\lambda}_{ll} - \sigma_{\langle ij\rangle}\hat{\lambda}_{\langle ij\rangle} + 2q_i\hat{\mu}_i - h \tag{5.56}$$

is a convex function with respect to the main field under the same inequalities (5.52).

5.3.5 Field Equations

Summarizing all the results above, we have the linear constitutive equations:

$$\begin{aligned}
&\hat{F}_{ijk} = \frac{3}{2} f_2 q_{(i}\delta_{jk)} + O(2), \\
&\left(\hat{F}_{llk} = \frac{5}{2} f_2 q_k + O(2), \; \hat{F}_{\langle ij\rangle k} = f_2 q_{\langle i}\delta_{j\rangle k} + O(2)\right) \\
&\hat{G}_{llij} = (g_1 + g_2 \Pi)\delta_{ij} - g_3 \sigma_{\langle ij\rangle} + O(2),
\end{aligned} \tag{5.57}$$

where the coefficients f_2, g_1, g_2 and g_3 are determined from the following relations:

$$\begin{aligned}
&\left(\frac{\partial g_1}{\partial \rho}\right)_T = 2\left(\varepsilon + \frac{p}{\rho}\right)\left(\frac{\partial p}{\partial \rho}\right)_T, \\
&\left(\frac{\partial\, p g_3}{\partial \rho}\right)_T = 2\left(\varepsilon + 2\frac{p}{\rho}\right)\left(\frac{\partial p}{\partial \rho}\right)_T, \\
&\frac{1}{k_3} = 2\rho T^2 \left(\varepsilon + \frac{p}{\rho}\right)\left(\frac{\partial p}{\partial T}\right)_\rho - \rho T^2 \left(\frac{\partial g_1}{\partial T}\right)_\rho, \\
&g_2 = \frac{1}{\Gamma}\left\{\frac{5}{3} p g_3 + \frac{1}{\rho^2 T c_v k_3}\left(\frac{\partial p}{\partial T}\right)_\rho + 2\left(\varepsilon + \frac{p}{\rho}\right)\left(\Gamma - \frac{5}{3}p\right)\right\}, \\
&f_2 = -2p\rho T\left\{g_3 - 2\left(\varepsilon + \frac{p}{\rho}\right)\right\} k_3, \\
&\Gamma = \frac{5}{3}p - \rho\left(\frac{\partial p}{\partial \rho}\right)_T - \frac{T}{\rho c_v}\left(\frac{\partial p}{\partial T}\right)_\rho^2.
\end{aligned} \tag{5.58}$$

The closed system of field equations is obtained by substituting these constitutive equations into the system (5.2) with (5.7):

$$\begin{aligned}
&\frac{\partial \rho}{\partial t} + \frac{\partial}{\partial x_k}(\rho v_k) = 0, \\
&\frac{\partial \rho v_i}{\partial t} + \frac{\partial}{\partial x_k}\left\{\rho v_i v_k + (p + \Pi)\delta_{ik} - \sigma_{\langle ik\rangle}\right\} = 0, \\
&\frac{\partial}{\partial t}(\rho v^2 + 2\rho\varepsilon) + \frac{\partial}{\partial x_k}\left\{\rho v^2 v_k + 2(\rho\varepsilon + p + \Pi)v_k - 2\sigma_{\langle kl\rangle} v_l + 2q_k\right\} = 0, \\
&\frac{\partial}{\partial t}\left\{\rho v_i v_j + (p + \Pi)\delta_{ij} - \sigma_{\langle ij\rangle}\right\} +
\end{aligned}$$

$$
\begin{aligned}
&+ \frac{\partial}{\partial x_k}\Bigg\{\rho v_i v_j v_k + (p+\Pi)(v_i\delta_{jk} + v_j\delta_{ki} + v_k\delta_{ij}) \\
&- \sigma_{\langle ij\rangle} v_k - \sigma_{\langle jk\rangle} v_i - \sigma_{\langle ki\rangle} v_j \\
&+ \frac{1}{2} f_2 (q_i\delta_{jk} + q_j\delta_{ki} + q_k\delta_{ij})\Bigg\} = -\frac{\Pi\delta_{ij}}{\tau_\Pi} + \frac{\sigma_{\langle ij\rangle}}{\tau_S}, \\
&\frac{\partial}{\partial t}\left\{\rho v^2 v_i + 2(\rho\varepsilon + p + \Pi)v_i - 2\sigma_{\langle li\rangle} v_l + 2q_i\right\} + \\
&+ \frac{\partial}{\partial x_k}\Bigg\{\rho v^2 v_i v_k + 2\rho\varepsilon v_i v_k + (p+\Pi)(v^2\delta_{ik} + 4v_i v_k) - \sigma_{\langle ik\rangle} v^2 - 2\sigma_{\langle li\rangle} v_l v_k \\
&- 2\sigma_{\langle lk\rangle} v_l v_i + (2+f_2) q_i v_k + (2+f_2) q_k v_i \\
&+ f_2 q_l v_l \delta_{ik} + (g_1 + g_2\Pi)\delta_{ik} - g_3\sigma_{\langle ik\rangle}\Bigg\} \\
&= -2\frac{q_i}{\tau_q} - 2\left(\frac{\Pi\delta_{ij}}{\tau_\Pi} - \frac{\sigma_{\langle ij\rangle}}{\tau_S}\right) v_j.
\end{aligned}
\tag{5.59}
$$

If we want to use the material derivative, the balance equations can be rewritten as

$$
\begin{aligned}
&\dot{\rho} + \rho\frac{\partial v_k}{\partial x_k} = 0, \\
&\rho\dot{v}_i + \frac{\partial\left\{(p+\Pi)\delta_{ij} - \sigma_{\langle ij\rangle}\right\}}{\partial x_j} = 0, \\
&\rho\left(\frac{\partial\varepsilon}{\partial T}\right)_\rho \dot{T} + \left\{p + \Pi - \rho^2\left(\frac{\partial\varepsilon}{\partial\rho}\right)_T\right\}\frac{\partial v_k}{\partial x_k} - \frac{\partial v_i}{\partial x_k}\sigma_{\langle ik\rangle} + \frac{\partial q_k}{\partial x_k} = 0, \\
&\dot{\Pi} + \Gamma\frac{\partial v_k}{\partial x_k} + \left(\frac{5}{3} - \frac{1}{\rho c_v}\left(\frac{\partial p}{\partial T}\right)_\rho\right)\Pi\frac{\partial v_k}{\partial x_k} - \left(\frac{2}{3} - \frac{1}{\rho c_v}\left(\frac{\partial p}{\partial T}\right)_\rho\right)\sigma_{\langle rk\rangle}\frac{\partial v_r}{\partial x_k} \\
&\quad + \frac{5}{6}q_k\frac{\partial f_2}{\partial x_k} + \left(\frac{5}{6}f_2 - \frac{1}{\rho c_v}\left(\frac{\partial p}{\partial T}\right)_\rho\right)\frac{\partial q_k}{\partial x_k} = -\frac{1}{\tau_\Pi}\Pi, \\
&\dot{\sigma}_{\langle ij\rangle} + \sigma_{\langle ij\rangle}\frac{\partial v_k}{\partial x_k} + 2\sigma_{\langle k\langle i\rangle}\frac{\partial v_{j\rangle}}{\partial x_k} - 2(p+\Pi)\frac{\partial v_{\langle i}}{\partial x_{j\rangle}} - q_{\langle i}\frac{\partial f_2}{\partial x_{j\rangle}} - f_2\frac{\partial q_{\langle i}}{\partial x_{j\rangle}} = -\frac{1}{\tau_S}\sigma_{\langle ij\rangle}, \\
&\dot{q}_i + \left(1 + \frac{f_2}{2}\right)q_i\frac{\partial v_k}{\partial x_k} + \left(1 + \frac{f_2}{2}\right)q_k\frac{\partial v_i}{\partial x_k} + \frac{f_2}{2}q_l\frac{\partial v_l}{\partial x_i} \\
&\quad + \left\{-\left[\frac{1}{\rho}\left(\frac{\partial p}{\partial\rho}\right)_T - \frac{1}{2}\left(\frac{\partial g_2}{\partial\rho}\right)_T\right]\Pi\delta_{ik} + \left[\frac{1}{\rho}\left(\frac{\partial p}{\partial\rho}\right)_T - \frac{1}{2}\left(\frac{\partial g_3}{\partial\rho}\right)_T\right]\sigma_{\langle ik\rangle}\right\}\frac{\partial\rho}{\partial x_k}
\end{aligned}
$$

$$
\begin{aligned}
&+\left\{-\frac{1}{2T^2\rho k_3}-\left[\frac{1}{\rho}\left(\frac{\partial p}{\partial T}\right)_\rho-\frac{1}{2}\left(\frac{\partial g_2}{\partial T}\right)_\rho\right]\Pi\delta_{ik}\right.\\
&\left.+\left[\frac{1}{\rho}\left(\frac{\partial p}{\partial T}\right)_\rho-\frac{1}{2}\left(\frac{\partial g_3}{\partial T}\right)_\rho\right]\sigma_{\langle ik\rangle}\right\}\frac{\partial T}{\partial x_k}\\
&-\left\{\left(\varepsilon+\frac{p}{\rho}-\frac{1}{2}g_2+\frac{1}{\rho}\Pi\right)\delta_{ik}-\frac{1}{\rho}\sigma_{\langle ik\rangle}\right\}\frac{\partial\Pi}{\partial x_k}\\
&+\left\{\left(\varepsilon+\frac{p}{\rho}-\frac{1}{2}g_3+\frac{1}{\rho}\Pi\right)\delta_{il}-\frac{1}{\rho}\sigma_{\langle il\rangle}\right\}\frac{\partial\sigma_{\langle lk\rangle}}{\partial x_k}=-\frac{1}{\tau_q}q_i.
\end{aligned}
\tag{5.60}
$$

We now understand that τ_Π, τ_S and τ_q can be regarded as the relaxation times of the dynamic pressure, the shear stress and the heat flux, respectively.

5.3.6 *Relationship Between ET and Navier-Stokes Fourier Theories*

We carry out the Maxwellian iteration (for more detail see Chap. 18, Sects. 18.2.1 and 18.3) in the system (5.60): The first iterates $\Pi^{(1)}$, $\sigma^{(1)}_{\langle ij\rangle}$ and $q^{(1)}_i$ are obtained by the substitution of the 0th iterates $\Pi^{(0)}=0$, $\sigma^{(0)}_{\langle ij\rangle}=0$ and $q^{(0)}_i=0$ into the left hand side of $(5.60)_{4,5,6}$. Then we obtain

$$
\Pi^{(1)}=-\tau_\Pi\Gamma\frac{\partial v_k}{\partial x_k},\qquad \sigma^{(1)}_{\langle ij\rangle}=2p\tau_S\frac{\partial v_{\langle i}}{\partial x_{j\rangle}},\qquad q^{(1)}_i=\frac{1}{2T^2\rho k_3}\tau_q\frac{\partial T}{\partial x_i}.
\tag{5.61}
$$

On the other hand, we have the laws of Navier-Stokes and Fourier expressed by (1.10). Their comparison reveals that

$$
\nu=\Gamma\tau_\Pi,\qquad \mu=p\tau_S,\qquad \kappa=-\frac{1}{2T^2\rho k_3}\tau_q.
\tag{5.62}
$$

We can therefore estimate the values of the relaxation times τ_Π, τ_S, and τ_q from the experimental data of the coefficients ν, μ, and κ.

The second iterates are obtained by substituting the first iterates into the left hand side of $(5.60)_{4,5,6}$, and higher iterates are obtained in a similar way.

In conclusion, the system can be certainly closed by the universal principles except for some nonessential constants, provided that we know the thermal and caloric equations of state and the viscosity and heat conductivity coefficients.

5.4 Rarefied-Gas Limit

It is important to study the rarefied-gas limit of the present theory in order to check its consistency with the results from the kinetic theory of gases explained in Chap. 6. The dependence of the relations in a rarefied gas on the degrees of freedom of a molecule is also made clear.

5.4.1 *Non-polytropic Gas*

From the ideal gas law, the thermal and caloric equations of state for a non-polytropic rarefied gas are expressed as

$$p = \frac{k_B}{m}\rho T, \quad \varepsilon = \varepsilon(T). \tag{5.63}$$

In this case the specific heat depends only on the temperature:

$$c_v = c_v(T). \tag{5.64}$$

The coefficients $f_2, g_1, g_2, g_3,\ \beta_2, \beta_3$ in this case are easily obtained from the relations (5.49) and (5.58). We assume that the integration constants in g_1 and pg_3 vanish. This assumption is reasonable because of the fact that our results below are consistent with those of the kinetic theory in Chap. 6. The coefficients of the linear constitutive equations are given by

$$\begin{aligned} &f_2 = \frac{2}{1+\hat{c}_v}, \quad g_1 = 2\left(\varepsilon + \frac{k_B}{m}T\right)p, \quad g_2 = g_3 = 2\left(\varepsilon + 2\frac{k_B}{m}T\right), \\ &k_3 = -\frac{1}{2Tp^2(1+\hat{c}_v)}, \end{aligned} \tag{5.65}$$

where $\hat{c}_v$ is the dimensionless specific heat:

$$\hat{c}_v = \frac{c_v}{k_B/m}. \tag{5.66}$$

The closed system of field equations with material derivative is obtained as follows:

$$\dot{\rho} + \rho\frac{\partial v_k}{\partial x_k} = 0,$$

$$\rho\dot{v}_i + \frac{\partial\left\{(p+\Pi)\delta_{ij} - \sigma_{\langle ij\rangle}\right\}}{\partial x_j} = 0,$$

$$\rho \frac{k_B}{m} \hat{c}_v \dot{T} + (p + \Pi) \frac{\partial v_k}{\partial x_k} - \frac{\partial v_i}{\partial x_k} \sigma_{\langle ik \rangle} + \frac{\partial q_k}{\partial x_k} = 0,$$

$$\dot{\Pi} + \left(\frac{2\hat{c}_v - 3}{3\hat{c}_v} p + \frac{5\hat{c}_v - 3}{3\hat{c}_v} \Pi \right) \frac{\partial v_k}{\partial x_k} - \frac{2\hat{c}_v - 3}{3\hat{c}_v} \frac{\partial v_{\langle i}}{\partial x_{k\rangle}} \sigma_{\langle ik \rangle}$$
$$- \frac{5}{3} \frac{1}{(1+\hat{c}_v)^2} \frac{d\hat{c}_v}{dT} q_k \frac{\partial T}{\partial x_k} + \frac{2\hat{c}_v - 3}{3\hat{c}_v (1+\hat{c}_v)} \frac{\partial q_k}{\partial x_k} = -\frac{1}{\tau_\Pi} \Pi,$$

$$\dot{\sigma}_{\langle ij \rangle} - 2(p+\Pi) \frac{\partial v_{\langle i}}{\partial x_{j\rangle}} + \sigma_{\langle ij \rangle} \frac{\partial v_k}{\partial x_k} + 2 \frac{\partial v_{\langle i}}{\partial x_k} \sigma_{\langle j \rangle k \rangle}$$
$$+ \frac{2}{(1+\hat{c}_v)^2} \frac{d\hat{c}_v}{dT} \frac{\partial T}{\partial x_k} q_{\langle i} \delta_{j\rangle k} - \frac{2}{1+\hat{c}_v} \frac{\partial q_{\langle i}}{\partial x_{j\rangle}} = -\frac{1}{\tau_S} \sigma_{\langle ij \rangle},$$

$$\dot{q}_i + \frac{2+\hat{c}_v}{1+\hat{c}_v} q_i \frac{\partial v_k}{\partial x_k} + \frac{1}{1+\hat{c}_v} q_k \frac{\partial v_k}{\partial x_i} + \frac{2+\hat{c}_v}{1+\hat{c}_v} q_k \frac{\partial v_i}{\partial x_k}$$
$$- \frac{k_B}{m} T \frac{\partial p}{\partial x_k} + \frac{k_B}{m} \left\{ (1+\hat{c}_v) p \delta_{ki} + (2+\hat{c}_v) \left(\Pi \delta_{ki} - \sigma_{\langle ki \rangle} \right) \right\} \frac{\partial T}{\partial x_k}$$
$$+ \frac{1}{\rho} \left\{ (p - \Pi) \delta_{ki} + \sigma_{\langle ki \rangle} \right\} \frac{\partial}{\partial x_l} \left\{ (p+\Pi) \delta_{kl} - \sigma_{\langle kl \rangle} \right\} = -\frac{1}{\tau_q} q_i. \tag{5.67}$$

The relaxation times τ_S, τ_Π and τ_q are related to the shear and bulk viscosities and the heat conductivity:

$$\mu = p\tau_S, \qquad \nu = \frac{2\hat{c}_v - 3}{3\hat{c}_v} p\tau_\Pi, \qquad \kappa = \left(\varepsilon + \frac{k_B}{m} T \right) \frac{p}{T} \tau_q. \tag{5.68}$$

The entropy density and the entropy flux are expressed as

$$h = \rho s - \frac{3\hat{c}_v}{2(2\hat{c}_v - 3)pT} \Pi^2 - \frac{1}{4pT} \sigma_{\langle ij \rangle} \sigma_{\langle ij \rangle} - \frac{\rho}{2p^2 T (1+\hat{c}_v)} q_i q_i + O(3), \tag{5.69}$$

$$\varphi_k = \frac{1}{T} q_k - \frac{1}{pT(1+\hat{c}_v)} \Pi q_k + \frac{1}{pT(1+\hat{c}_v)} q_i \sigma_{\langle ik \rangle} + O(3). \tag{5.70}$$

The convexity condition is satisfied when

$$2\hat{c}_v - 3 > 0. \tag{5.71}$$

This condition is always satisfied for a polyatomic gas $\hat{c}_v > 3/2$. However, as will be discussed in Sect. 5.6, we should be careful in the case of monatomic gases with $\hat{c}_v = 3/2$.

In the present case, the intrinsic Lagrange multipliers (5.55) have the simplified form:

$$\begin{aligned}
\hat{\lambda} &= -\frac{g}{T} - \frac{3(\varepsilon - \hat{c}_v \frac{k_B}{m}T)}{pT(2\hat{c}_v - 3)}\Pi, \\
\hat{\lambda}_i &= \frac{\rho(\varepsilon + \frac{k_B}{m}T)}{p^2T(\hat{c}_v + 1)}q_i, \\
\hat{\lambda}_{\langle ij\rangle} &= \frac{1}{2pT}\sigma_{\langle ij\rangle}, \\
\hat{\lambda}_{ll} &= -\frac{3\hat{c}_v}{pT(2\hat{c}_v - 3)}\Pi, \\
\hat{\mu} &= \frac{1}{2T} + \frac{3}{2pT(2\hat{c}_v - 3)}\Pi, \\
\hat{\mu}_i &= -\frac{\rho}{2p^2T(\hat{c}_v + 1)}q_i.
\end{aligned} \tag{5.72}$$

5.4.2 *Polytropic Gas*

For a polytropic rarefied polyatomic gas, the equations of state are given by (1.28). Therefore the dimensionless specific heat is expressed as

$$\hat{c}_v = \frac{D}{2}. \tag{5.73}$$

The coefficients of the constitutive equations are expressed as follows:

$$\begin{aligned}
&f_2 = \frac{4}{D+2}, \quad g_1 = \frac{k_B}{m}T(D+2)p, \quad g_2 = g_3 = \frac{k_B}{m}T(D+4), \\
&k_3 = -\frac{1}{Tp^2(D+2)}.
\end{aligned} \tag{5.74}$$

With these coefficients, we obtain the closed system of field equations which is already given in (1.29). If we want to use the material derivative, we have

$$\begin{aligned}
&\dot{\rho} + \rho\frac{\partial v_k}{\partial x_k} = 0, \\
&\rho\dot{v}_i + \frac{\partial\left\{(p+\Pi)\delta_{ij} - \sigma_{\langle ij\rangle}\right\}}{\partial x_j} = 0, \\
&\rho\frac{D}{2}\frac{k_B}{m}\dot{T} + (p+\Pi)\frac{\partial v_k}{\partial x_k} - \frac{\partial v_i}{\partial x_k}\sigma_{\langle ik\rangle} + \frac{\partial q_k}{\partial x_k} = 0,
\end{aligned}$$

$$
\begin{aligned}
&\dot{\Pi} + \left(\frac{2(D-3)}{3D}p + \frac{5D-6}{3D}\Pi\right)\frac{\partial v_k}{\partial x_k} - \frac{2(D-3)}{3D}\frac{\partial v_{\langle i}}{\partial x_{k\rangle}}\sigma_{\langle ik\rangle} \\
&\quad + \frac{4(D-3)}{3D(D+2)}\frac{\partial q_k}{\partial x_k} = -\frac{1}{\tau_\Pi}\Pi, \\
&\dot{\sigma}_{\langle ij\rangle} - 2p\frac{\partial v_{\langle i}}{\partial x_{j\rangle}} + \sigma_{\langle ij\rangle}\frac{\partial v_k}{\partial x_k} - 2\Pi\frac{\partial v_{\langle i}}{\partial x_{j\rangle}} + 2\frac{\partial v_{\langle i}}{\partial x_k}\sigma_{\langle j\rangle k\rangle} - \frac{4}{D+2}\frac{\partial q_{\langle i}}{\partial x_{j\rangle}} = -\frac{1}{\tau_S}\sigma_{\langle ij\rangle}, \\
&\dot{q}_i + \frac{D+4}{D+2}q_i\frac{\partial v_k}{\partial x_k} + \frac{2}{D+2}q_k\frac{\partial v_k}{\partial x_i} + \frac{D+4}{D+2}q_k\frac{\partial v_i}{\partial x_k} \\
&\quad - \frac{k_B}{m}T\frac{\partial p}{\partial x_k} + \frac{1}{2}\frac{k_B}{m}\left((D+2)p\delta_{ki} + (D+4)\left(\Pi\delta_{ki} - \sigma_{\langle ki\rangle}\right)\right)\frac{\partial T}{\partial x_k} \\
&\quad + \frac{1}{\rho}\left((p-\Pi)\delta_{ki} + \sigma_{\langle ki\rangle}\right)\frac{\partial}{\partial x_l}\left((p+\Pi)\delta_{kl} - \sigma_{\langle kl\rangle}\right) = -\frac{1}{\tau_q}q_i.
\end{aligned}
\tag{5.75}
$$

Then relaxation times are related to the phenomenological coefficients:

$$
\mu = p\tau_S, \qquad \nu = \frac{2(D-3)}{3D}p\tau_\Pi, \qquad \kappa = \frac{D+2}{2}\frac{p^2}{\rho T}\tau_q. \tag{5.76}
$$

In this case the entropy density and the entropy flux are expressed as

$$
\begin{aligned}
h &= h^E - \frac{3D}{4(D-3)pT}\Pi^2 - \frac{1}{4pT}\sigma_{\langle ij\rangle}\sigma_{\langle ij\rangle} - \frac{\rho}{(D+2)p^2T}q_iq_i + O(3), \\
\varphi_k &= \frac{1}{T}q_k - \frac{2}{(D+2)pT}\Pi q_k + \frac{2}{(D+2)pT}q_i\sigma_{\langle ik\rangle} + O(3).
\end{aligned}
\tag{5.77}
$$

The convexity condition is satisfied if $D > 3$.

The intrinsic Lagrange multipliers (5.72) have the form:

$$
\begin{aligned}
\hat{\lambda} &= -\frac{g}{T}, \\
\hat{\lambda}_i &= \frac{1}{pT}q_i, \\
\hat{\lambda}_{\langle ij\rangle} &= \frac{1}{2pT}\sigma_{\langle ij\rangle}, \\
\hat{\lambda}_{ll} &= -\frac{3D}{2pT(D-3)}\Pi, \\
\hat{\mu} &= \frac{1}{2T} + \frac{3}{2pT(D-3)}\Pi, \\
\hat{\mu}_i &= -\frac{\rho}{p^2T(D+2)}q_i.
\end{aligned}
\tag{5.78}
$$

A remarkable point is that $\hat{\lambda}$ is independent of the dynamic pressure and coincides with the equilibrium value.

We will see in the next Chap. 6 that the system (5.75) is complete equivalent to the ones obtained via MEP. Moreover for diatomic gases with $D = 5$, field equations (5.75) coincide with those derived by Mallinger using a Grad procedure [14] except for the expressions of the relaxation times.

5.5 Models of Dense Gas

As the next step, it is interesting to study explicitly the applicability of the ET14 theory to dense gases. In this section, the ET14 theory is applied to three physically important systems: (1) a gas with the virial equations of state, (2) a hard-sphere system, and (3) a van der Waals fluid.

5.5.1 *Gas with the Virial Equations of State*

The thermal and caloric equations of state are given in the form of virial expansion:

$$\begin{aligned} p &= \frac{k_B}{m}\rho T\left(1 + B_2(T)\rho + B_3(T)\rho^2 + \cdots\right), \\ \varepsilon &= \frac{D}{2}\frac{k_B}{m}T - \frac{k_B}{m}T^2\rho B_2'(T) - \frac{1}{2}\frac{k_B}{m}T^2\rho^2 B_3'(T) + \cdots, \end{aligned} \tag{5.79}$$

where $B_2(T)$, $B_3(T)$, $\cdots$ are the second and third virial coefficients, and so on. Here a prime means a derivative with respect to the temperature T. For simplicity, we have assumed that, in the rarefied-gas limit, above equations of state tend to (1.28).

Using the equations of state (5.79), we can obtain the explicit expressions of the coefficients in the constitutive equations in the following way: We obtain $k_1, k_2, k_4, \cdots, k_8$ except for k_3 from (5.42). Integrating g_1 with respect to ρ, we obtain g_1 as follows:

$$\begin{aligned} g_1 &= \int_{\rho_0}^{\rho} 2\left(\varepsilon(\bar{\rho}, T) + \frac{p(\bar{\rho}, T)}{\bar{\rho}}\right)\left(\frac{\partial p(\bar{\rho}, T)}{\partial \bar{\rho}}\right)_T d\bar{\rho} + C_1(\rho_0, T, T_0) \\ &= \left(\frac{k_B}{m}\right)^2 (D+2)T^2\rho + \left(\frac{k_B}{m}\right)^2 T^2\rho^2\left\{(D+3)B_2 - TB_2'\right\} \\ &\quad + O\left(\rho^3\right) + C_2(\rho_0, T, T_0), \end{aligned}$$

where $C_{1,2}(\rho_0, T, T_0)$ are integration functions, and ρ_0 and T_0 are, respectively, the mass density and temperature in a reference state. As g_1 at an arbitrary value of T

must asymptotically approaches $\left(\frac{k_B}{m}\right)^2 (D+2)T^2\rho$ in the rarefied-gas limit discussed in Sect. 5.4, we obtain $C_2(\rho_0, T, T_0) = 0$. As g_1 has been determined within the approximation adopted here, we get the explicit form of k_3 from $(5.58)_3$. g_2 and g_3 can be determined in a similar way.

For simplicity, we show the constitutive equations up to the first correction with respect to ρ:

$$f_2 = \frac{4}{D+2} + \frac{2(D+2)TB_2' + 4T^2B_2''}{(D+2)^2}\rho + O\left(\rho^2\right), \tag{5.80}$$

$$g_1 = \left(\frac{k_B}{m}\right)^2 \rho T^2 \left\{(D+2) + \left[(D+3)B_2 - TB_2'\right]\rho + O(\rho^2)\right\}, \tag{5.81}$$

$$g_2 = \frac{k_B}{m}T\left\{D+4+\frac{4(D-3)B_2 - 5DTB_2' - 6T^2B_2''}{2(D-3)}\rho + O\left(\rho^2\right)\right\}, \tag{5.82}$$

$$g_3 = \frac{k_B}{m}T\left\{D+4+\left(2B_2 - TB_2'\right)\rho + O\left(\rho^2\right)\right\}. \tag{5.83}$$

As can be seen above, the first correction depends on both the virial coefficients B_2 and the degrees of freedom D. The closed system of field equations can be easily obtained by using above constitutive equations. We omit its expression for simplicity.

The relaxation times are related to the shear and bulk viscosities and the heat conductivity:

$$\begin{aligned}
\mu =& \frac{k_B}{m}\rho T\left(1 + B_2(T)\rho + O\left(\rho^2\right)\right)\tau_S, \\
\nu =& \frac{1}{3D}\frac{k_B}{m}\rho T\left\{2(D-3) - \left[(D+12)B_2\right.\right. \\
&+ \left.\left.\frac{12}{D}(D+2)TB_2' + \frac{12}{D}T^2B_2''\right]\rho + O\left(\rho^2\right)\right\}\tau_\Pi, \\
\kappa =& \frac{1}{2}\left\{\frac{k_B}{m}\right)^2 \rho T\left[D+2+\left[(D+2)B_2 - T^2B_2''\right]\rho + O\left(\rho^2\right)\right\}\tau_q.
\end{aligned} \tag{5.84}$$

5.5.2 Hard-Sphere System

The thermal and caloric equations of state are given by

$$\begin{aligned}
p &= \frac{k_B}{\omega}T\eta H(\eta), \\
\varepsilon &= \frac{D}{2}\frac{k_B}{m}T,
\end{aligned} \tag{5.85}$$

where η is the packing fraction related to the mass density ρ by

$$\eta = \frac{\rho\omega}{m}, \tag{5.86}$$

and $H(\eta)$ is a function of η determined explicitly by computer experiments [15]. Here ω is the volume of a hard sphere. We use η instead of ρ.

The coefficients in the constitutive equations derived in the same way as above are given by

$$\begin{aligned}
f_2 &= \frac{4\int_0^\eta H^2(\bar{\eta})d\bar{\eta}}{D\eta H + 2\int_0^\eta H^2(\bar{\eta})d\bar{\eta}}, \\
g_1 &= \frac{k_B^2}{m\omega}\eta T^2 \left(DH + H^2 + \frac{1}{\eta}\int_0^\eta H^2(\bar{\eta})d\bar{\eta} \right), \\
g_2 &= \frac{k_B}{m} T \left\{ D + 2H - \frac{6DH^2 + \frac{2}{\eta}(-5D + 6H)\int_0^\eta H^2(\bar{\eta})d\bar{\eta}}{2DH - 6H^2 - 3D\eta H'} \right\}, \\
g_3 &= \frac{k_B}{m} T \left(D + 2H + \frac{2}{\eta H}\int_0^\eta H^2(\bar{\eta})d\bar{\eta} \right),
\end{aligned} \tag{5.87}$$

where $H' = dH(\eta)/d\eta$. The closed system of field equations can be obtained by using the constitutive equations obtained. We omit its expression for simplicity.

The relaxation times are related to the shear and bulk viscosities and the heat conductivity:

$$\begin{aligned}
\mu &= p\tau_S, \\
\nu &= \frac{1}{3D}\frac{k_B}{\omega}\eta T \left(2DH - 6H^2 - 3D\eta H' \right) \tau_\Pi, \\
\kappa &= \left(\frac{k_B}{m} \right)^2 \frac{m}{\omega}\eta \frac{T}{2} \left(DH + \frac{2}{\eta}\int_0^\eta H^2(\bar{\eta})d\bar{\eta} \right) \tau_q.
\end{aligned} \tag{5.88}$$

The convexity condition of the entropy density in this case can be expressed by only one inequality:

$$-\frac{H}{3} + \frac{H^2}{D} + \frac{\eta H'}{2} < 0. \tag{5.89}$$

Then we find that there is a critical packing fraction η_C such that the above condition is satisfied in the region $0 < \eta < \eta_C$. In the case of $D = 5$, for example, we can estimate $\eta_C = 0.0447$ by adopting the following functional form of H [16]:

$$H(\eta) = \frac{1 + \eta + \eta^2 - \eta^3}{(1 - \eta)^3}. \tag{5.90}$$

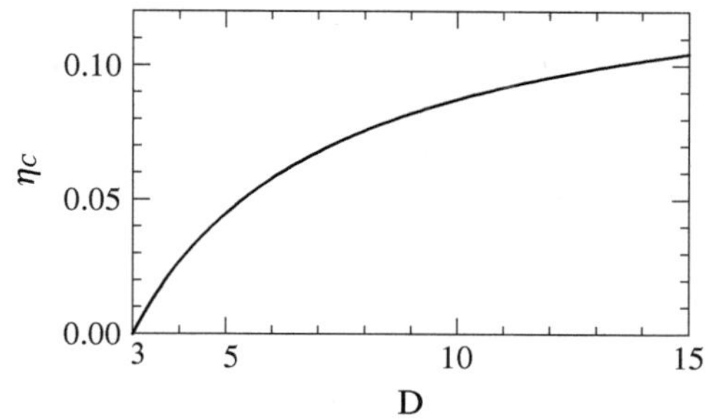

Fig. 5.1 Dependence of the critical packing fraction η_C on the degrees of freedom D of a molecule

The dependence of η_C on D is shown in Fig. 5.1. It is remarkable that $\eta_C = 0$ in the case of $D = 3$.

5.5.3 van der Waals Fluid

Let us study a gas with the thermal and caloric equations of state given by

$$\begin{aligned} p &= \frac{k_B}{m}\frac{T\rho}{1-b\rho} - a\rho^2, \\ \varepsilon &= \frac{D}{2}\frac{k_B}{m}T - a\rho, \end{aligned} \tag{5.91}$$

where the material-dependent constants a and b represent, respectively, a measure of the attraction between the constituent molecules and the effective volume (or exclusion volume) of a molecule.

The constitutive equations are obtained by using the same procedure as above:

$$f_2 = 4\frac{\frac{k_B}{m}T + \frac{a}{b^2\rho}(1-b\rho)\{b\rho + \log(1-b\rho)\}}{(D+2)\frac{k_B}{m}T - D(1-b\rho)a\rho},$$

$$g_1 = \frac{k_B}{m}T\left\{(D+2)p + \frac{k_B}{m}T\frac{b\rho^2}{(1-b\rho)^2} - 2a\rho^2\frac{1+b\rho}{1-b\rho} + \frac{8}{3}\frac{a^2\rho^3}{\frac{k_B}{m}T}\right\},$$

$$g_2 = \left\{\frac{k_B}{m}T\frac{2(D-3)(D+4) - (7D^2+18D-12)b\rho + 5D(D+2)b^2\rho^2}{1-b\rho}\right.$$

$$+\frac{a}{b}\left[D(1-b\rho)^2(10+Db\rho) + 12b\rho(2+Db\rho)\right] - 4D\frac{a^2\rho^2}{\frac{k_B}{m}T}(1-b\rho)^2$$

$$
+ 10D\frac{a}{b^2\rho}(1-b\rho)^2\log(1-b\rho)\Big\}\Big/\Big\{\frac{k_B}{m}T\,[2(D-3)-5Db\rho] + Da\rho(1-b\rho)^2\Big\},
$$

$$
g_3 = \frac{k_B}{m}T\Big\{\frac{k_B}{m}T\frac{D+4-(D+2)b\rho}{1-b\rho} + \frac{a}{b}[2-(D+8)b\rho + Db^2\rho^2]
$$

$$
+ 4\frac{a^2\rho^2}{\frac{k_B}{m}T}(1-b\rho) + 2\frac{a}{b^2\rho}(1-b\rho)\log(1-b\rho)\Big\}\Big/\left\{\frac{k_B}{m}T - a\rho(1-b\rho)\right\}. \tag{5.92}
$$

The relaxation times are related to the shear and bulk viscosities and the heat conductivity:

$$
\begin{aligned}
\mu &= p\tau_S, \\
\nu &= \left\{\frac{k_B}{m}\frac{\rho T}{3D(1-b\rho)^2}[2(D-3) - 5Db\rho] + \frac{a\rho^2}{3}\right\}\tau_\Pi, \\
\kappa &= \frac{1}{2}\left\{\left(\frac{k_B}{m}\right)^2 \frac{D+2}{1-b\rho}\rho T - \frac{k_B}{m}Da\rho^2\right\}\tau_q.
\end{aligned} \tag{5.93}
$$

We now study the convexity condition of the entropy density (5.52). For later convenience, we introduce the dimensionless variables:

$$
\hat{p} = \frac{p}{p_{cr}}, \qquad \hat{\rho} = \frac{\rho}{\rho_{cr}}, \qquad \hat{T} = \frac{T}{T_{cr}}, \tag{5.94}
$$

where $\rho_{cr} = 1/(3b)$, $p_{cr} = a/(27b^2)$ and $T_{cr} = 8a/(27\frac{k_B}{m}b)$ are, respectively, the mass density, the pressure and the temperature at the critical point. Then the thermal and caloric equations of state are rewritten in the reduced form:

$$
\begin{aligned}
\hat{p} &= \frac{8\hat{T}\hat{\rho}}{3-\hat{\rho}} - 3\hat{\rho}^2, \\
\hat{\varepsilon} &= \frac{\rho_{cr}}{p_{cr}}\varepsilon = \frac{4D\hat{T}}{3} - 3\hat{\rho}.
\end{aligned} \tag{5.95}
$$

As the inequality $(5.52)_2$ is always satisfied, the convexity condition is now expressed as

$$
\begin{aligned}
&\frac{8\hat{T}\hat{\rho}}{3-\hat{\rho}} - 3\hat{\rho}^2 > 0, \; \frac{8\hat{T}}{(3-\hat{\rho})^2} - 2\hat{\rho} > 0, \\
&\frac{18 + D(5\hat{\rho}-6)}{D(\hat{\rho}-3)^2}\hat{T}^2\hat{\rho} - \frac{3}{8}\hat{T}\hat{\rho}^2 < 0, \; \frac{8(D+2)\hat{T} + 3D(\hat{\rho}-3)\hat{\rho}}{(\hat{\rho}-3)}\hat{T}^2\hat{\rho} < 0.
\end{aligned} \tag{5.96}
$$

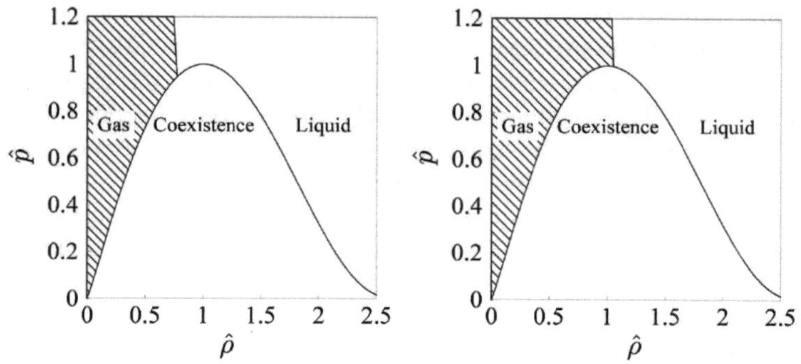

Fig. 5.2 Convexity condition is satisfied in the *shaded region*. The degrees of freedom: $D = 5$ (*left*) and 8 (*right*). The *curve* represents the coexistence curve

This convexity condition is satisfied in the shaded regions in Fig. 5.2 for $D = 5$ and 8. We notice that the condition is violated in the liquid phase.

5.5.4 *Remark*

From the results of the hard-sphere system and the van der Waals fluid, we found that the convexity condition is not always satisfied, but its region is limited up to a critical value of the mass density. Therefore, we understand that the ET14 theory developed above is applicable only to moderately dense polyatomic gases. Further developments of the ET14 theory are highly expected in order to study more dense gases.

5.6 Singular Limit from Polyatomic to Monatomic Gas

In this section, we show that a rarefied monatomic gas, where there exists no dynamic pressure, can be identified as a singular limit of a rarefied polyatomic gas. We confine our discussion within the singular limit from ET14 of a rarefied polyatomic gas to ET13 of a rarefied monatomic gas [17]. The gas is assumed to be polytropic.

Let us discuss the limiting process in the system (1.29) from polyatomic to monatomic rarefied gases when we let D approach 3 from above, where D is assumed to be a continuous variable. The limit is singular in the sense that the system for a rarefied polyatomic gas with 14 independent fields needs to converge to the system with only 13 independent fields for a rarefied monatomic gas.

The singularity can be seen also by the inequalities required for the symmetric hyperbolicity in equilibrium (5.52). This requirement is always satisfied in the ET13 theory of monatomic gases, while, in the present ET14 theory, it is expressed by the inequality $D > 3$. The condition is obviously satisfied only for polyatomic gases with $D > 3$, and the case of monatomic gases with $D = 3$ is not admissible. Therefore only the limit of D toward 3 from above is meaningful.

In the present case, the relaxation times τ_S, τ_Π and τ_q are, respectively, related to the shear viscosity μ, the bulk viscosity ν and the heat conductivity κ as seen in (5.76). We observe that the bulk viscosity vanishes when $D \to 3$ as is consistent in monatomic gases.

Let us take the limit $D \to 3$ of the system (1.29), that is, the limit from polyatomic to monatomic rarefied gases. Then we immediately notice that the limit of the system exists, but it still has 14 equations. However, we also notice the following three points (I)–(III):

(I) The limit of the equation for Π, $(1.29)_5$, is given by

$$\dot{\Pi} = -\left(\frac{1}{\tau_\Pi} + \frac{\partial v_k}{\partial x_k}\right)\Pi, \quad \longleftrightarrow \quad \left(\frac{\Pi}{\rho}\right)^\bullet = -\frac{1}{\tau_\Pi}\frac{\Pi}{\rho}. \tag{5.97}$$

This is the first-order quasi-linear partial differential equation with respect to the dynamic pressure Π.

As the limit case is the case of monatomic gases, the initial condition for (5.97) must be compatible with monatomic gases. We therefore should impose the following initial condition:

$$\Pi(0, \mathbf{x}) = 0. \tag{5.98}$$

Then, by assuming the uniqueness of the solution, the only possible solution of (5.97) under the initial condition (5.98) is given by

$$\Pi(t, \mathbf{x}) = 0 \qquad \forall\, t > 0. \tag{5.99}$$

Therefore the dynamic pressure in a monatomic gas vanishes identically for any time once we impose the initial condition (5.98).

(II) If we insert the solution (5.99) into the remaining equations in (5.75) with $D \to 3$, we confirm that the resulting equations are the same as the ones of ET13 for rarefied monatomic gases. This means that the solution of the limiting system with 14 equations is essentially equivalent to the solution of ET13 of monatomic gases.

(III) The violation of the symmetric hyperbolicity condition ($D > 3$) disappears because this inequality comes out from the non-vanishing dynamic pressure. Therefore the condition for the symmetric hyperbolicity is the same as the one in the ET13 theory, which is always satisfied in equilibrium.

To sum up, we may conclude that the ET14 theory is applicable also to rarefied monatomic gases if we impose the initial or boundary condition of zero dynamic pressure. In [17], two illustrative numerical results in the process of the singular limit, that is, the linear waves and the shock waves are shown in order to grasp the asymptotic behavior of the physical quantities, in particular, of the dynamic pressure.

References

1. I. Müller, T. Ruggeri, *Rational Extended Thermodynamics*, 2nd edn. (Springer, New York, 1998)
2. E. Ikenberry, C. Truesdell, On the pressure and the flux of energy in a gas according to Maxwell's kinetic theory. J. Ration. Mech. Anal. **5**, 1 (1956)
3. H. Engholm Jr., G.M. Kremer, Thermodynamics of a diatomic gas with rotational and vibrational degrees of freedom. Int. J. Eng. Sci. **32**(8), 1241 (1994)
4. G. M. Kremer, Extended thermodynamics and statistical mechanics of a polyatomic ideal gas. J. Non-Equilib. Thermodyn. **14**, 363 (1989)
5. G.M. Kremer, Extended thermodynamics of molecular ideal gases. Continuum Mech. Thermodyn. **1**, 21 (1989)
6. I.-S. Liu, Extended thermodynamics of fluids and virial equations of state. Arch. Ration. Mech. Anal. **88**, 1 (1985)
7. G.M. Kremer, Extended thermodynamics of non-ideal gases. Physica A **144**, 156 (1987)
8. I.-S. Liu, G.M. Kremer, Hyperbolic system of field equations for viscous fluids. Mat. Aplic. Comp. **9**(2), 123 (1990)
9. I.-S. Liu, J.A. Salvador, Hyperbolic system for viscous fluids and simulation of shock tube flows. Continuum Mech. Thermodyn. **2**, 179 (1990)
10. G.M. Kremer, On extended thermodynamics of ideal and real gases, in *Extended Thermodynamics Systems*, ed. by S. Sieniutycz, P. Salamon (Taylor and Francis, New York, 1992)
11. M.C. Carrisi, M.A. Mele, S. Pennisi, On some remarkable properties of an extended thermodynamics model for dense gases and macromolecular fluids. Proc. R. Soc. A **466**, 1645 (2010)
12. T. Ruggeri, Symmetric hyperbolic system of conservative equations for a viscous conducting fluid. Acta Mech. **47**, 167 (1983)
13. T. Arima, S. Taniguchi, T. Ruggeri, M. Sugiyama, Extended thermodynamics of dense gases. Continuum Mech. Thermodyn. **24**, 271 (2011)
14. F. Mallinger, Generalization of the Grad theory to polyatomic gases. INRIA Research Report (1998), p. 3581
15. A. Münster, *Statistical Thermodynamics*, vol. 2 (Springer, Berlin/Heidelberg/New York, 1974)
16. N.F. Carnahan, K.E. Starling, Equation of state for nonattracting rigid spheres. J. Chem. Phys. **51**, 635 (1969)
17. T. Arima, S. Taniguchi, T. Ruggeri, M. Sugiyama, Monatomic rarefied gas as a singular limit of polyatomic gas in extended thermodynamics. Phys. Lett. A **377**, 2136 (2013)

Chapter 6
Maximum Entropy Principle for Rarefied Polyatomic Gas

Abstract In this chapter, we prove, in the case of polyatomic rarefied gases, that the maximum entropy principle (MEP) gives the same closure of the system as that obtained in the phenomenological ET theory with 14 fields discussed in Chap. 5. The main idea is to consider a generalized distribution function depending not only on the velocity but also on an extra variable that connects with the internal degrees of freedom of a constituent molecule. On the basis of MEP, we again obtain the same binary hierarchy introduced in the previous chapter: the one is the usual momentum-type, F-series, and the other is the energy-type, G-series. The extra variable plays a role in the G-series. Thus we prove the perfect agreement between the ET theory and the molecular ET theory at least within 14-field theories. The agreement for any number of moments will be proved in Chap. 10.

6.1 Generalized Distribution Function for Rarefied Polyatomic Gases

We have seen that one of the most important results in the case of rarefied monatomic gases is that the closure of the system obtained by the macroscopic principles of ET is exactly same as the closures of MEP and also of the Grad procedure. Therefore, a question naturally arises: is the above statement for monatomic gases still valid for polyatomic gases? To answer to this question is the main purpose of this chapter. As seen below, the answer is affirmative. The essential contents of this chapter were firstly shown in the paper of Pavić, Ruggeri and Simić [1].

A crucial step in the developments of the kinetic theory of rarefied polyatomic gases was made by Borgnakke and Larsen [2]. It is assumed that the distribution function depends on, in addition to the velocity, a continuous variable representing the energy of the internal modes of a molecule in order to take into account the exchange of energy between translational modes and internal modes. This model was initially used for Monte Carlo simulations of polyatomic gases, and later it has been applied to the derivation of the generalized Boltzmann equation by Bourgat, Desvillettes, Le Tallec and Perthame [3].

The distribution function $f(t, \mathbf{x}, \mathbf{c}, I)$ is defined on the extended domain $[0, \infty) \times R^3 \times R^3 \times [0, \infty)$. Its rate of change is determined by the Boltzmann equation which

T. Ruggeri, M. Sugiyama, *Rational Extended Thermodynamics beyond the Monatomic Gas*, DOI 10.1007/978-3-319-13341-6_6

has the same form as the one of monatomic gases (1.16) but the collision term $Q(f)$ now takes into account the existence of the internal degrees of freedom through the collisional cross section.

6.2 Equilibrium Distribution Function for Polyatomic Gases and the Euler System

Our first aim is to recover the equilibrium distribution function and appropriate field equations for macroscopic variables at the hydrodynamic stage (i.e., transfer equations for moments) via MEP. We shall, therefore, briefly describe the kinetic model for polyatomic gases and point out the important consequences related to the internal energy density.

To this end, let us study the conservation laws of momentum and energy during a collision between two molecules:

$$\begin{aligned} \mathbf{c}' + \mathbf{c}'_* &= \mathbf{c} + \mathbf{c}_*, \\ \frac{1}{2}m|\mathbf{c}'|^2 + \frac{1}{2}m|\mathbf{c}'_*|^2 + I' + I'_* &= \frac{1}{2}mc^2 + \frac{1}{2}m|\mathbf{c}_*|^2 + I + I_*, \end{aligned} \tag{6.1}$$

where $|\mathbf{c}| = (c_i c_i)^{1/2}$. The post-collisional quantities, denoted with the prime, are described by the so-called Borgnakke-Larsen procedure, based on a repartition of the kinetic and internal energies [2–4]. We now adopt the center of mass frame and write the total energy of the incoming molecules, which is, by virtue of (6.1), the same as the one of the outgoing molecules:

$$\varepsilon = \frac{1}{4}m|\mathbf{c} - \mathbf{c}_*|^2 + I + I_* = \frac{1}{4}m|\mathbf{c}' - \mathbf{c}'_*|^2 + I' + I'_*.$$

Then, with a help of a parameter $R \in [0, 1]$, we attribute a part of total energy of the outgoing molecules to their kinetic energy, and the rest to their internal energy:

$$\begin{aligned} R\,\varepsilon &= \frac{1}{4}m|\mathbf{c}' - \mathbf{c}'_*|^2, \\ (1 - R)\,\varepsilon &= I' + I'_*. \end{aligned} \tag{6.2}$$

To distribute the internal energy itself between the two outgoing molecules, we introduce a new parameter $r \in [0, 1]$ and write

$$I' = r\,(1 - R)\,\varepsilon, \qquad I'_* = (1 - r)(1 - R)\,\varepsilon.$$

Finally, as collisions between molecules are assumed to be of specular reflection type, we parameterize (6.2) by a unitary vector $\boldsymbol{\omega} \in S^2$:

$$\mathbf{c}' - \mathbf{c}'_* = \sqrt{\frac{4R\varepsilon}{m}} T_\omega \left[\frac{\mathbf{c} - \mathbf{c}_*}{|\mathbf{c} - \mathbf{c}_*|} \right],$$

where T_ω is the symmetry operator with respect to the plane $\{\boldsymbol{\omega}\}^\perp$ defined by

$$T_\omega \mathbf{z} = \mathbf{z} - 2(\boldsymbol{\omega} \cdot \mathbf{z})\boldsymbol{\omega}, \qquad \forall \mathbf{z} \in R^3.$$

Coming back into the laboratory reference frame, we end up with expressions for post-collisional velocities:

$$\mathbf{c}' = \frac{\mathbf{c} + \mathbf{c}_*}{2} + \sqrt{\frac{R\varepsilon}{m}} T_\omega \left[\frac{\mathbf{c} - \mathbf{c}_*}{|\mathbf{c} - \mathbf{c}_*|} \right], \quad \mathbf{c}'_* = \frac{\mathbf{c} + \mathbf{c}_*}{2} - \sqrt{\frac{R\varepsilon}{m}} T_\omega \left[\frac{\mathbf{c} - \mathbf{c}_*}{|\mathbf{c} - \mathbf{c}_*|} \right].$$

The collision invariants in this model form a 5-vector:

$$m \left(1, c_i, c^2 + 2\frac{I}{m} \right)^T, \tag{6.3}$$

which leads to hydrodynamic variables in the form:

$$\begin{pmatrix} \rho \\ \rho v_i \\ \rho v^2 + 2\rho\varepsilon \end{pmatrix} = \int_{R^3} \int_0^\infty m \begin{pmatrix} 1 \\ c_i \\ c^2 + 2I/m \end{pmatrix} f(t, \mathbf{x}, \mathbf{c}, I)\varphi(I)\, dI\, d\mathbf{c}, \tag{6.4}$$

where the non-negative measure $\varphi(I)\, dI$ is introduced so as to recover the classical caloric equation of state for polyatomic gases in equilibrium. The entropy is defined by the following relation:

$$h = -k_B \int_{R^3} \int_0^\infty f \log f \varphi(I)\, dI\, d\mathbf{c}. \tag{6.5}$$

By introducing the peculiar velocity $C_i = c_i - v_i$, we rewrite (6.4) as follows:

$$\begin{pmatrix} \rho \\ 0_i \\ 2\rho\varepsilon \end{pmatrix} = \int_{R^3} \int_0^\infty m \begin{pmatrix} 1 \\ C_i \\ C^2 + 2I/m \end{pmatrix} f(t, \mathbf{x}, \mathbf{C}, I)\varphi(I)\, dI\, d\mathbf{C}. \tag{6.6}$$

Note that the internal energy density ε can be divided into the translational part ε_T and the part of the internal degrees of freedom ε_I:

$$\rho\varepsilon_T = \int_{R^3}\int_0^\infty \frac{1}{2}mC^2 f(t,\mathbf{x},\mathbf{C},I)\varphi(I)\,dI\,d\mathbf{C},$$

$$\rho\varepsilon_I = \int_{R^3}\int_0^\infty If(t,\mathbf{x},\mathbf{C},I)\varphi(I)\,dI\,d\mathbf{C}. \tag{6.7}$$

The energy ε_T is related to the kinetic temperature T:

$$\varepsilon_T = \frac{3}{2}\frac{k_B}{m}T. \tag{6.8}$$

The maximum entropy principle is expressed in terms of the following variational problem: Determine the distribution function $f(t,\mathbf{x},\mathbf{C},I)$ such that $h \to \max$, under the constraints (6.4), or equivalently, due to the Galilean invariance, under the constraints (6.6). The result is summarized as follows [1]:

Theorem 6.1 *The distribution function that maximizes the entropy (6.5) under the constraints (6.6) has the form:*

$$f_E = \frac{\rho}{mA(T)}\left(\frac{m}{2\pi k_B T}\right)^{3/2}\exp\left\{-\frac{1}{k_B T}\left(\frac{1}{2}mC^2 + I\right)\right\}, \tag{6.9}$$

where

$$A(T) = \int_0^\infty \exp\left(-\frac{I}{k_B T}\right)\varphi(I)\,dI. \tag{6.10}$$

The proof of the theorem is accomplished with the use of the Lagrange multiplier method. Introducing the vector of the multipliers $(\lambda, \lambda_i, \mu)$, we define the functional:

$$\begin{aligned}\mathscr{L} = &-\int_{R^3}\int_0^\infty k_B f\log f\varphi(I)\,dI\,d\mathbf{c} + \lambda\left(\rho - \int_{R^3}\int_0^\infty mf\varphi(I)\,dI\,d\mathbf{c}\right)\\ &+\lambda_i\left(\rho v_i - \int_{R^3}\int_0^\infty mfc_i\varphi(I)\,dI\,d\mathbf{c}\right)\\ &+\mu\left(\rho v^2 + 2\rho\varepsilon - \int_{R^3}\int_0^\infty m\left(c^2 + 2\frac{I}{m}\right)f\varphi(I)\,dI\,d\mathbf{c}\right).\end{aligned}$$

As this is a functional of the distribution function f and we want to maximize it with respect to f with the given macroscopic quantities, the previous functional can be substituted by the following one:

$$\mathscr{L} = \int_{R^3}\int_0^\infty\left\{-k_B f\log f - m\left[\lambda + \lambda_i c_i + \mu\left(c^2 + 2\frac{I}{m}\right)\right]f\right\}\varphi(I)\,dI\,d\mathbf{c}. \tag{6.11}$$

Since $\mathscr{L}$ is a scalar, it must retain the same value in the case of zero hydrodynamic velocity $\mathbf{v} = \mathbf{0}$ due to the Galilean invariance. Therefore

$$\mathscr{L}=\int_{R^3}\int_0^{\infty}\left\{-k_B f \log f - m\left[\hat{\lambda} + \hat{\lambda}_i C_i + \hat{\mu}\left(C^2 + 2\frac{I}{m}\right)\right] f\right\} \varphi(I) dI\, d\mathbf{C}. \tag{6.12}$$

Comparison between (6.11) and (6.12) yields the relations (2.42) (with matrix $\mathbf{X}$ given by (2.55)):

$$\lambda = \hat{\lambda} - \hat{\lambda}_i v_i + \hat{\mu} v^2, \quad \lambda_i = \hat{\lambda}_i - 2\hat{\mu} v_i, \quad \mu = \hat{\mu}, \tag{6.13}$$

which dictate the velocity dependence of the Lagrange multipliers in accordance with the general results of the Galilean invariance.

The Euler-Lagrange equation $\delta\mathscr{L}/\delta f = 0$ leads to the following form of the distribution function:

$$f_E = \exp\left\{-1 - \frac{m}{k_B}\left[\hat{\lambda} + \hat{\lambda}_i C_i + \hat{\mu}\left(C^2 + 2\frac{I}{m}\right)\right]\right\}.$$

Plugging this into the constraints (6.6), with the help of (6.8), one determines the zero velocity Lagrange multipliers in terms of the hydrodynamic variables:

$$\exp\left(-1 - \frac{m}{k_B}\hat{\lambda}\right) = \frac{\rho}{m A(T)}\left(\frac{m}{2\pi k_B T}\right)^{3/2}, \quad \hat{\lambda}_i = 0, \quad \hat{\mu} = \frac{1}{2T}, \tag{6.14}$$

with $A(T)$ defined by (6.10). The proof is completed.

Notice that, using (6.13) and (6.14), we can have the explicit form of the Lagrange multipliers (i.e., the main field) that symmetrize the system. These exactly coincide with the Godunov variables (2.18):

$$\lambda = \frac{1}{T}\left(-g + \frac{1}{2}v^2\right), \quad \lambda_i = -\frac{v_i}{T}, \quad \mu = \frac{1}{2T}.$$

The distribution function (6.9) is the generalization of the classical Maxwellian equilibrium distribution in the case of polyatomic gases. It was derived in [3, 4] by means of the H-theorem.

The weighting function $\varphi(I)$ is determined in such a way that it recovers the caloric equation of state for polyatomic gases. If D is the degrees of freedom of a molecule, it can be shown that $\varphi(I) = I^{\alpha}$ leads to the caloric equation of state in equilibrium provided that the relation (1.31) holds.

Statement 1 *In equilibrium, the internal energy ε_I has the following form:*

$$\varepsilon_I|_E = \frac{1}{m}\frac{B(T)}{A(T)}, \quad B(T) = \int_0^{\infty} I\varphi(I)\exp\left(-\frac{I}{k_B T}\right) dI. \tag{6.15}$$

Moreover, if weighting function is chosen to be $\varphi(I) = I^\alpha$, internal energy of a polyatomic gas reads

$$\varepsilon = \left(\frac{5}{2} + \alpha\right) \frac{k_B}{m} T, \quad \alpha > -1. \tag{6.16}$$

To prove (6.15) one just has to put the equilibrium distribution (6.9) into $(6.7)_2$. Furthermore, choosing $\varphi(I) = I^\alpha$, we have:

$$A(T) = (k_B T)^{1+\alpha} \Gamma(1+\alpha), \quad B(T) = (k_B T)^{2+\alpha} \Gamma(2+\alpha), \tag{6.17}$$

with overall restriction $\alpha > -1$. With the help of the relation for the Gamma function $\Gamma(z+1) = z\Gamma(z)$ one obtains $B(T)/A(T) = (1+\alpha)k_B T$, which in turn leads to (6.16) by combining with the translational energy (6.8).

The relation between α and D (1.31) follows directly from comparison between (6.16) and the well-known caloric equation of state for polyatomic gases:

$$\varepsilon = \frac{D}{2} \frac{k_B}{m} T. \tag{6.18}$$

Observe that the model for a monatomic gas ($D = 3$) cannot be recovered from the one with the continuous internal energy, since the value of the parameter α in monatomic case violates the overall restriction $\alpha > -1$.

Remark From (6.15) we notice that, in principle, this approach seems to be valid also for *nonpolytropic gases* for which the specific heat is not constant and ε is not linear in the temperature. A challenge is to find a measure $\varphi(I) \neq I^\alpha$ such that the specific heat have the physical meaning in a nonpolytropic gas. For the rest of this chapter, we will study only polytropic gases for which (6.18) is true with $\varphi(I) = I^\alpha$ and α is related to D through (1.31).

Now we can prove the following theorem:

Theorem 6.2 *If (6.9) is the local equilibrium distribution function with $\rho \equiv \rho(t, \mathbf{x})$, $\mathbf{v} \equiv \mathbf{v}(t, \mathbf{x})$ and $T \equiv T(t, \mathbf{x})$, then the hydrodynamic variables ρ, $\mathbf{v}$ and T satisfy the Euler system*

$$\begin{aligned}
&\frac{\partial}{\partial t}\rho + \frac{\partial}{\partial x^i}(\rho v_i) = 0, \\
&\frac{\partial}{\partial t}(\rho v_j) + \frac{\partial}{\partial x^i}\left(\rho v_i v_j + p\delta_{ij}\right) = 0, \\
&\frac{\partial}{\partial t}\left(\rho\varepsilon + \rho\frac{v^2}{2}\right) + \frac{\partial}{\partial x^k}\left\{\left(\rho\varepsilon + \rho\frac{v^2}{2} + p\right)v_k\right\} = 0
\end{aligned} \tag{6.19}$$

with $p = (k_B/m)\rho T$ and ε given by (6.18).

Proof The Euler system (6.19) is the simplest example of the binary hierarchy postulated in (5.2):

$$\frac{\partial F}{\partial t} + \frac{\partial F_k}{\partial x_k} = 0,$$

$$\frac{\partial F_i}{\partial t} + \frac{\partial F_{ik}}{\partial x_k} = 0, \tag{6.20}$$

$$\frac{\partial G_{ll}}{\partial t} + \frac{\partial G_{llk}}{\partial x_k} = 0,$$

with

$$\begin{pmatrix} F \\ F_i \end{pmatrix} = \begin{pmatrix} \rho \\ \rho v_i \end{pmatrix} = \int_{R^3}\int_0^{\infty} m \begin{pmatrix} 1 \\ c_i \end{pmatrix} f_E I^{\alpha}\, dI\, d\mathbf{c} \tag{6.21}$$

and

$$G_{ll} = \rho v^2 + 2\rho\varepsilon = \int_{R^3}\int_0^{\infty} m(c^2 + 2I/m) f I^{\alpha}\, dI\, d\mathbf{c}. \tag{6.22}$$

In the present case we obtain as closure:

$$F_{ij} = \int_{R^3}\int_0^{\infty} m c_i c_j f_E I^{\alpha}\, dI\, d\mathbf{c} = \rho v_i v_j + \frac{k_B}{m}\rho T \delta_{ij}, \tag{6.23}$$

and

$$G_{lli} = \int_{R^3}\int_0^{\infty} m(c^2 + 2I/m) c_i f_E I^{\alpha}\, dI\, d\mathbf{c} = \left(\rho v^2 + 2\rho\varepsilon + p\right) v_i, \tag{6.24}$$

with ε given by (6.18), and this completes the proof.

6.3 Justification of the Binary Hierarchy of the Moment-Equations for Polyatomic Gases

In the case of the Euler equation (6.20), we have noticed two different kinds of moments: the moment F's in (6.21), and the moment G in (6.22) where the variable I of the internal modes plays a role. Therefore it is appropriate to justify the 14-moment theory with the binary hierarchy (5.2) from the kinetic theory. Indeed we have (5.2) with

$$\begin{pmatrix} F \\ F_{i_1} \\ F_{i_1 i_2} \end{pmatrix} = \int_{\mathbb{R}^3}\int_0^{\infty} m \begin{pmatrix} 1 \\ c_{i_1} \\ c_{i_1} c_{i_2} \end{pmatrix} f(t, \mathbf{x}, \mathbf{c}, I)\, I^{\alpha}\, dI\, d\mathbf{c}, \tag{6.25}$$

$$\begin{pmatrix} G_{pp} \\ G_{ppk_1} \end{pmatrix} = \int_{\mathbb{R}^3} \int_0^\infty m \begin{pmatrix} c^2 + 2\frac{I}{m} \\ \left(c^2 + 2\frac{I}{m}\right) c_{k_1} \end{pmatrix} f(t, \mathbf{x}, \mathbf{c}, I)\, I^\alpha \, dI \, d\mathbf{c}. \tag{6.26}$$

6.3.1 MEP for the 14-Moment System for Polyatomic Gases

We now want to use the MEP in the case of the binary hierarchy with 14 fields. For the entropy defined by (6.5), the MEP poses the following variational problem: Determine the distribution function $f(t, \mathbf{x}, \mathbf{c}, I)$ such that $h \rightarrow \max$ under the constraints (6.25) and (6.26). The solution of the problem is given by the following theorem [1]:

Theorem 6.3 *The distribution function, which maximizes the entropy (6.5) under the constraints (6.25) and (6.26) and under the assumption that processes are not so far from equilibrium, has the form:*

$$f = f_E \left\{ 1 - \frac{\rho}{p^2} q_i C_i + \frac{\rho}{p^2} \left[-\sigma_{\langle ij \rangle} + \left(\frac{5}{2} + \alpha \right) (1 + \alpha)^{-1} \Pi \delta_{ij} \right] C_i C_j \right. \tag{6.27}$$
$$\left. - \frac{3}{2(1+\alpha)} \frac{\rho}{p^2} \Pi \left(\frac{1}{2} C^2 + \frac{I}{m} \right) + \left(\frac{7}{2} + \alpha \right)^{-1} \frac{\rho^2}{p^3} \left(\frac{1}{2} C^2 + \frac{I}{m} \right) q_i C_i \right\},$$

where f_E is the equilibrium distribution (6.9) with $A(T)$ being the auxiliary function (6.17)$_1$.

The functional for the constrained variational problem reads

$$\mathscr{L} = \int_{R^3} \int_0^\infty \left\{ -k_B f \log f - m \left[\lambda + \lambda_i c_i + \lambda_{ij} c_i c_j \right. \right.$$
$$\left. \left. + (\mu + \mu_i c_i) \left(c^2 + 2\frac{I}{m} \right) \right] f \right\} \varphi(I) \, dI \, d\mathbf{c}. \tag{6.28}$$

By the Galilean invariance, the functional is the same for the case with zero hydrodynamic velocity ($\mathbf{v} = \mathbf{0}$). Therefore, we have

$$\mathscr{L} = \int_{R^3} \int_0^\infty \left\{ -k_B f \log f - m \left[\hat{\lambda} + \hat{\lambda}_i C_i + \hat{\lambda}_{ij} C_i C_j \right. \right.$$
$$\left. \left. + (\hat{\mu} + \hat{\mu}_i C_i) \left(C^2 + 2\frac{I}{m} \right) \right] f \right\} \varphi(I) \, dI \, d\mathbf{C}. \tag{6.29}$$

The comparison between (6.28) and (6.29) gives the velocity dependence of the Lagrange multipliers according with the Galilean invariance (2.42). This coincides exactly with the velocity dependence for the main field in the macroscopic theory (5.13). This is in agreement with the general result given in Sect. 4.3.

From (6.29), the solution of the Euler-Lagrange equation $\delta\mathscr{L}/\delta f = 0$ is given by:

$$f = \exp(-1 - \chi/k_B), \qquad \chi = \hat{\lambda} + \hat{\lambda}_i C_i + \hat{\lambda}_{ij} C_i C_j + (\hat{\mu} + \hat{\mu}_i C_i)\left(C^2 + 2\frac{I}{m}\right).$$

As in a monatomic gas, there is the problem of the convergence of moments. We, therefore, try to obtain the approximate solution in the form of expansion around the local equilibrium:

$$f = f_E\left\{1 - \frac{m}{k_B}\left(\tilde{\lambda} + \tilde{\lambda}_i C_i + \tilde{\lambda}_{ij} C_i C_j + (\tilde{\mu} + \tilde{\mu}_i C_i)\left(C^2 + 2\frac{I}{m}\right)\right)\right\}, \tag{6.30}$$

where f_E is the equilibrium distribution (6.9) and

$$\tilde{\lambda} = \hat{\lambda} - \hat{\lambda}_E,\ \tilde{\lambda}_i = \hat{\lambda}_i - \hat{\lambda}_{iE},\ \tilde{\lambda}_{ij} = \hat{\lambda}_{ij} - \hat{\lambda}_{ijE},\ \tilde{\mu} = \hat{\mu} - \hat{\mu}_E,\ \tilde{\mu}_i = \hat{\mu}_i - \hat{\lambda}_{iE},$$

where subscript E indicates the values of the Lagrange multipliers at a local equilibrium state.

Inserting (6.30) into the densities (6.25) and (6.26), one obtains the following algebraic linear system:

$$\begin{aligned}
&\tilde{\lambda} = 0,\\
&\tilde{\lambda}_{ii} + (5 + 2\alpha)\,\tilde{\mu} = 0,\\
&\tilde{\lambda}_i + (7 + 2\alpha)\,\frac{p}{\rho}\tilde{\mu}_i = 0,\\
&2\frac{p^2}{\rho}\tilde{\lambda}_{ij} - \left(\frac{k}{m}p - \frac{2p^2}{\rho}\tilde{\mu}\right)\delta_{ij} = -\frac{k}{m}p_{ij},\\
&\left(\frac{7}{2} + \alpha\right)\tilde{\lambda}_{ii} + \left[\frac{15}{2} + 2(1+\alpha)(5+\alpha)\right]\tilde{\mu} = 0,\\
&\left(\frac{7}{2} + \alpha\right)\frac{p^2}{\rho}\tilde{\lambda}_i + \left[\frac{35}{2} + 2(1+\alpha)(7+\alpha)\right]\frac{p^3}{\rho^2}\tilde{\mu}_i = -\frac{k}{m}q_i.
\end{aligned} \tag{6.31}$$

Then the following solution is obtained:

$$\begin{aligned}
\tilde{\lambda} &= 0,\\
\tilde{\lambda}_i &= \frac{k_B}{m}\frac{\rho}{p^2}q_i,\\
\tilde{\lambda}_{ij} &= -\frac{k_B}{m}\frac{\rho}{2p^2}\left\{-\sigma_{\langle ij\rangle} + \left(\frac{5}{2} + \alpha\right)(1+\alpha)^{-1}\Pi\delta_{ij}\right\},
\end{aligned}$$

$$\tilde{\mu} = \frac{k_B}{m} \frac{3}{4(1+\alpha)} \frac{\rho}{p^2} \Pi,$$

$$\tilde{\mu}_i = -\frac{k_B}{2m} \left(\frac{7}{2} + \alpha \right)^{-1} \frac{\rho^2}{p^3} q_i,$$

which give the same intrinsic Lagrange multipliers of the phenomenological theory in the case of rarefied polyatomic gases (5.78).

Inserting these expressions into (6.30), we obtain the nonequilibrium distribution function (6.27), and the theorem is proved.

6.3.2 Non-convective Fluxes and Productions

Now, as we have the the non-equilibrium distribution function (6.27), we can evaluate the fluxes and the productions that are not in the list of density variables, and we can close the system. The non-convective fluxes $\hat{F}_{ijk}$ and $\hat{G}_{llij}$ defined as

$$\hat{F}_{ijk} = \int_{R^3} \int_0^\infty m C_i C_j C_k f(t, \mathbf{x}, \mathbf{C}, I) \, \varphi(I) \, dI \, d\mathbf{C},$$

$$\hat{G}_{llij} = \int_{R^3} \int_0^\infty \left(\frac{1}{2} m C^2 + I \right) C_i C_j f(t, \mathbf{x}, \mathbf{C}, I) \, \varphi(I) \, dI \, d\mathbf{C}$$

become, with the use of the distribution function (6.9),

$$\hat{F}_{ijk} = \left(\frac{7}{2} + \alpha \right)^{-1} \left(q_i \delta_{jk} + q_j \delta_{ki} + q_k \delta_{ij} \right), \tag{6.32}$$

$$\hat{G}_{llij} = \left(\frac{9}{2} + \alpha \right) \frac{p}{\rho} p_{ij} - \frac{p^2}{\rho} \delta_{ij}. \tag{6.33}$$

In this way we can see that the differential closed system is completely equivalent to the ones obtained by the macroscopic approach (1.29) or (5.75) as far as the principal part of the operator (left-hand side of the system) is concerned.

In reality the MEP has an advantage over the phenomenological theory. At least in principle, we can derive the explicit expressions of the production terms, while in the phenomenological approach we know only the sign of the production terms appearing in the second member (right-hand side) of (1.29). In fact we have the expressions:

$$P_{ij} = \int_{R^3} \int_0^\infty m c_i c_j Q(f) \, I^\alpha \, dI \, d\mathbf{c},$$

$$Q_{lli} = \int_{R^3} \int_0^\infty m \left(c^2 + 2\frac{I}{m} \right) c_i Q(f) \, I^\alpha \, dI \, d\mathbf{c}.$$

The main problem is, however, that, in order to have explicit expressions for the productions, we need a model for the collision term, to obtain which is, in general, quite difficult in the case of polyatomic gases

A first tentative was done in [1] in which the authors assumed the structure of the collision term given in [3, 4]:

$$Q(f)(\mathbf{c}, I) = \int_{R^3\times R_+\times[0,1]^2\times S^2} \left[f(\mathbf{c}', I')f(\mathbf{c}'_*, I'_*) - f(\mathbf{c}, I)f(\mathbf{c}_*, I_*)\right] \times$$

$$\times\, \mathscr{B}(c, c_*, I, I_*, r, R, \boldsymbol{\omega})(1-R)\,|c - c_*|\,\frac{1}{\varphi(I)} d\boldsymbol{\omega}\, dr\, dR\, dI_*\, d\mathbf{c}_*,$$

where $r \in [0,1]$ and $R \in [0,1]$ are parameters (see Sect. 6.2) describing the exchange of internal energy during molecular collision and $\boldsymbol{\omega} \in S^2$ is a unit sphere vector. As in the classical case, the model of interaction between molecules is reflected on the collision cross section $\mathscr{B}$. Here, we assume the following form of the cross section:

$$\mathscr{B} = R^{1/2}\,|c - c_*|\left|\omega \cdot \frac{c - c_*}{|c - c_*|}\right|,$$

which resembles the variable in the hard-sphere model. Furthermore, they assumed that a state of the gas during processes is not far from local equilibrium. Therefore, products of non-equilibrium distribution functions which appear in the collision term may be linearized with respect to the moments of the distribution functions, i.e., viscous stress tensor σ_{ij} and heat flux q_i.

For this particular toy model Pavić, Ruggeri and Simić [1] obtained the following expressions for the production terms:

$$P_{ij} = -\frac{4}{15}\frac{m\pi}{A^2(T)}\frac{p^2}{\rho}\left\{-\sigma_{\langle ij\rangle} + \frac{4}{7}\left(\frac{5}{2}+\alpha\right)(1+\alpha)^{-1}\Pi\delta_{ij}\right\}, \tag{6.34}$$

$$Q_{lli} = -\frac{8}{15}\frac{m\pi}{A^2(T)}\frac{p^2}{\rho}\left\{\left(-\sigma_{\langle ij\rangle} + \frac{4}{7}\left(\frac{5}{2}+\alpha\right)(1+\alpha)^{-1}\Pi\delta_{ij}\right)v_j \right.$$
$$\left. +\frac{10}{7}\left(\frac{7}{2}+\alpha\right)^{-1} q_i\right\}. \tag{6.35}$$

Comparing these expressions with the production terms of the phenomenological theory (see left side of (5.75)), we obtain the following explicit expressions for the relaxation times:

$$\frac{1}{\tau_S} = \frac{4}{15}\frac{m\pi}{A^2(T)}\frac{p^2}{\rho},$$

$$\frac{1}{\tau_\Pi} = \frac{16}{105}\left(\frac{5}{2}+\alpha\right)(1+\alpha)^{-1}\frac{m\pi}{A^2(T)}\frac{p^2}{\rho},$$

$$\frac{1}{\tau_q} = \frac{8}{21}\left(\frac{7}{2}+\alpha\right)^{-1}\frac{m\pi}{A^2(T)}\frac{p^2}{\rho}.$$

Taking into account the relationship between the relaxation times and the phenomenolgical coefficients (5.76), we can also evaluate the viscosities and the heat conductivity as follows:

$$\begin{aligned}
\mu &= \frac{15}{4}\frac{\Gamma^2(1+\alpha)}{\pi}(k_B T)^{1+2\alpha}, \\
\nu &= \frac{35}{8}(1+\alpha)^2\left(\frac{5}{2}+\alpha\right)^{-2}\frac{\Gamma^2(1+\alpha)}{\pi}(k_B T)^{1+2\alpha}, \\
\kappa &= \frac{21}{8}\left(\frac{7}{2}+\alpha\right)^{2}\frac{k_B}{m}\frac{\Gamma^2(1+\alpha)}{\pi}(k_B T)^{1+2\alpha}.
\end{aligned} \tag{6.36}$$

6.3.3 Entropy and Entropy Flux

If we insert the nonequilibrium distribution function (6.27) into the expressions of the entropy density and the intrinsic entropy flux:

$$h^0 = -k_B\int_{R^3}\int_0^\infty f\log f\,\varphi(I)\,dI\,d\mathbf{c},$$

$$\varphi_j = -k_B\int_{R^3}\int_0^\infty C_j f\log f\,\varphi(I)\,dI\,d\mathbf{c},$$

we obtain the same expressions obtained in the phenomenological approach (5.77).

In this way we have now proved the perfect matching between the phenomenological ET and the molecular ET of rarefied polyatomic gases.

6.3.4 Remark

As seen above, for polyatomic gases, the system of field equations should properly describe the nonequilibrium processes involving the internal (rotational and/or vibrational) modes of a molecule. Concerning their importance, see [5–9]. If we adopt more moments than 14, we can study more details of such processes.

References

1. M. Pavić, T. Ruggeri, S. Simić, Maximum entropy principle for rarefied polyatomic gases. Physica A **392**, 1302 (2013)
2. C. Borgnakke, P.S. Larsen, Statistical collision model for Monte Carlo simulation of polyatomic gas mixture. J. Comput. Phys. **18**, 405 (1975)
3. J.-F. Bourgat, L. Desvillettes, P. Le Tallec, B. Perthame, Microreversible collisions for polyatomic gases. Eur. J. Mech. B Fluids **13**, 237 (1994)
4. L. Desvillettes, R. Monaco, F. Salvarani, A kinetic model allowing to obtain the energy law of polytropic gases in the presence of chemical reactions. Eur. J. Mech. B Fluids **24**, 219 (2005)
5. Ya.B. Zel'dovich, Yu.P. Raizer, *Physics of Shock Waves and High-Temperature Hydrodynamic Phenomena* (Dover Publications, New York, 2002)
6. E. Nagnibeda, E. Kustova, *Non-equilibrium Reacting gas flow* (Springer, Berlin, 2009)
7. M. Groppi, G. Spiga, F. Zus, Euler closure of the Boltzmann equations for resonant bimolecular reactions. Phys. Fluids **18**, 057105 (2006)
8. M. Groppi, G. Spiga, Kinetic approach to chemical reactions and inelastic transitions in a rarefied gas. J. Math. Chem. **26**, 197 (1999)
9. M. Bisi, G. Martaló, G. Spiga, Multi-temperature hydrodynamic limit from kinetic theory in a mixture of rarefied gases. Acta Appl. Math. **122**, 37 (2012)

Part IV
Applications of the ET14 Theory

Chapter 7
Linear Wave in a Polyatomic Gas

Abstract In this chapter, we study a linear sound wave in a rarefied polyatomic gas in equilibrium with the aim of clarifying the validity and the features of the ET14 theory established in Chap. 5. We derive the dispersion relations on the basis of the ET14 theory and of the classical Navier-Stokes Fourier (NSF) theory. Comparison of these relations with experimental data reveals clearly the superiority of the ET14 theory to the NSF theory. We confine our analysis within sound waves in some rarefied diatomic gases (hydrogen, deuterium, and hydrogen deuteride gases) because suitable experimental data are scarce and are mainly restricted to rarefied gases. We also evaluate the relaxation times, and the shear and bulk viscosities and the heat conductivity of the gases.

7.1 Basic Equations

We study a linear sound wave in a rarefied polyatomic gas in equilibrium with the thermal and caloric equations of state given by (5.63). We assume that a nonequilibrium state can be characterized by the 14 independent field variables $\mathbf{u} \equiv (\rho, v_i, T, \Pi, \sigma_{\langle ij \rangle}, q_i)$. Let $\mathbf{u}_0 \equiv (\rho_0, 0, T_0, 0, 0, 0)$ be an equilibrium state, then, from (5.59) and (5.65), the linearized system in the neighborhood of $\mathbf{u}_0$ for the perturbed field $\mathbf{u}$ is given by

$$
\begin{aligned}
&\frac{\partial \rho}{\partial t} + \rho_0 \frac{\partial v_k}{\partial x_k} = 0, \\
&\rho_0 \frac{\partial v_i}{\partial t} + \frac{k_B}{m} T_0 \frac{\partial \rho}{\partial x_i} + \frac{k_B}{m} \rho_0 \frac{\partial T}{\partial x_i} - \frac{\partial \sigma_{\langle ij \rangle}}{\partial x_j} + \frac{\partial \Pi}{\partial x_i} = 0, \qquad (7.1)\\
&\frac{k_B}{m} \rho_0 \hat{c}_v \frac{\partial T}{\partial t} + \frac{k_B}{m} \rho_0 T_0 \frac{\partial v_k}{\partial x_k} + \frac{\partial q_k}{\partial x_k} = 0,
\end{aligned}
$$

T. Ruggeri, M. Sugiyama, *Rational Extended Thermodynamics beyond the Monatomic Gas*, DOI 10.1007/978-3-319-13341-6_7

$$\frac{\partial \Pi}{\partial t} + \left(\frac{2}{3} - \frac{1}{\hat{c}_v}\right)\frac{k_B}{m}\rho_0 T_0 \frac{\partial v_k}{\partial x_k} + \frac{2\hat{c}_v - 3}{3\hat{c}_v(1+\hat{c}_v)}\frac{\partial q_k}{\partial x_k} = -\frac{1}{\tau_\Pi}\Pi,$$

$$\frac{\partial \sigma_{\langle ij\rangle}}{\partial t} - 2\frac{k_B}{m}\rho_0 T_0 \frac{\partial v_{\langle i}}{\partial x_{j\rangle}} - \frac{2}{1+\hat{c}_v}\frac{\partial q_{\langle i}}{\partial x_{j\rangle}} = -\frac{1}{\tau_S}\sigma_{\langle ij\rangle},$$

$$\frac{\partial q_i}{\partial t} + (1+\hat{c}_v)\left(\frac{k_B}{m}\right)^2 \rho_0 T_0 \frac{\partial T}{\partial x_i} - \frac{k_B}{m}T_0\frac{\partial \sigma_{\langle ik\rangle}}{\partial x_k} + \frac{k_B}{m}T_0\frac{\partial \Pi}{\partial x_i} = -\frac{1}{\tau_q}q_i.$$

The dimensionless specific heat $\hat{c}_v$ (see (5.66)) and the relaxation times τ_S, τ_Π and τ_q in (7.1) are evaluated at the equilibrium state. The relations between the relaxation times and the shear viscosity μ, the bulk viscosity ν, and the heat conductivity κ are given by (5.68).

Let us confine our study within a one-dimensional problem, that is, a plane longitudinal wave. Therefore, by considering the symmetry of the wave, we assume

$$v_i \equiv \begin{pmatrix} v \\ 0 \\ 0 \end{pmatrix}, \quad \sigma_{\langle ij\rangle} \equiv \begin{pmatrix} \sigma & 0 & 0 \\ 0 & -\frac{1}{2}\sigma & 0 \\ 0 & 0 & -\frac{1}{2}\sigma \end{pmatrix}, \quad q_i \equiv \begin{pmatrix} q \\ 0 \\ 0 \end{pmatrix}. \tag{7.2}$$

Then, the linearized basic field equations (7.1) are neatly written as

$$\frac{\partial \boldsymbol{u}}{\partial t} + \boldsymbol{A}_0 \frac{\partial \boldsymbol{u}}{\partial x} = \boldsymbol{B}_0 \boldsymbol{u}, \tag{7.3}$$

where $\mathbf{u}$ is now redefined as $\mathbf{u} \equiv (\rho, v, T, \Pi, \sigma, q)$, and $\boldsymbol{A}_0$ and $\boldsymbol{B}_0$ are given by

$$\boldsymbol{A}_0 = \begin{pmatrix} 0 & \rho_0 & 0 & 0 & 0 & 0 \\ \frac{k_B}{m}\frac{T_0}{\rho_0} & 0 & \frac{k_B}{m} & \frac{1}{\rho_0} & -\frac{1}{\rho_0} & 0 \\ 0 & \frac{T_0}{\hat{c}_v} & 0 & 0 & 0 & \frac{m}{k_B \hat{c}_v \rho_0} \\ 0 & \left(\frac{2}{3} - \frac{1}{\hat{c}_v}\right)\frac{k_B}{m}\rho_0 T_0 & 0 & 0 & 0 & \frac{2\hat{c}_v - 3}{3\hat{c}_v(1+\hat{c}_v)} \\ 0 & -\frac{4}{3}\frac{k_B}{m}\rho_0 T_0 & 0 & 0 & 0 & -\frac{4}{3(1+\hat{c}_v)} \\ 0 & 0 & (1+\hat{c}_v)\left(\frac{k_B}{m}\right)^2 \rho_0 T_0 & \frac{k_B}{m}T_0 & -\frac{k_B}{m}T_0 & 0 \end{pmatrix}, \tag{7.4}$$

$$
\boldsymbol{B}_0 = \begin{pmatrix}
0\,0\,0 & 0 & 0 & 0 \\
0\,0\,0 & 0 & 0 & 0 \\
0\,0\,0 & 0 & 0 & 0 \\
0\,0\,0 & -\dfrac{1}{\tau_\Pi} & 0 & 0 \\
0\,0\,0 & 0 & -\dfrac{1}{\tau_S} & 0 \\
0\,0\,0 & 0 & 0 & -\dfrac{1}{\tau_q}
\end{pmatrix}. \tag{7.5}
$$

7.2 Dispersion Relation for Sound

In this section, using the general theory of linear wave in Sect. 3.1, we derive the dispersion relation, and then obtain the high-frequency limit of the phase velocity and the attenuation factor.

7.2.1 Dispersion Relation, Phase Velocity and Attenuation Factor

We study a plane harmonic wave propagating in the positive x-direction. From Eq. (7.3), the dispersion relation is expressed by (see (3.6))

$$
\det\left(\boldsymbol{I} - z\boldsymbol{A}_0 + \frac{i}{\omega}\boldsymbol{B}_0\right) = 0, \tag{7.6}
$$

where $z \equiv k/\omega$ and $\boldsymbol{I}$ is the unit matrix. Then the phase velocity v_{ph} and the attenuation factor α are calculated as the functions of the frequency ω by using the relation (3.7). In addition, it is useful to introduce the attenuation per wavelength:

$$
\alpha_\Lambda(\omega) = \alpha\Lambda = \frac{2\pi v_{ph}\alpha}{\omega} = -2\pi\frac{\mathcal{I}m(z)}{\mathcal{R}e(z)},
$$

where Λ is the wavelength.

By introducing the dimensionless parameters defined by

$$
\Omega = \tau_S\omega, \quad \tau_{qs} = \frac{\tau_q}{\tau_S}, \quad \tau_{ps} = \frac{\tau_\Pi}{\tau_S}, \tag{7.7}
$$

the dispersion relation (7.6) is expressed explicitly as

$$\frac{\hat{c}_v (c_0 z)^4}{3\Omega^2 (1+\hat{c}_v)^2 \tau_{ps}} \left\{-3(1+\hat{c}_v) - i\Omega\left(3+7\hat{c}_v+5\hat{c}_v\tau_{ps}\right) + 9\Omega^2\hat{c}_v\tau_{ps}\right\}$$

$$+\frac{(c_0 z)^2}{3\Omega^3(1+\hat{c}_v)^2\tau_{qs}\tau_{ps}}\Bigg\{ -3\mathrm{i}\,(1+\hat{c}_v)^2 + \Omega\,(1+\hat{c}_v)$$

$$\left[3+7\hat{c}_v+5\hat{c}_v\tau_{ps}+6\,(1+\hat{c}_v)\,\tau_{qs}\right] + \mathrm{i}\Omega^2\left[2\left(3+10\hat{c}_v+5\hat{c}_v^2\right)\tau_{qs}\right.$$

$$\left. +9\hat{c}_v\,(1+\hat{c}_v)\,\tau_{ps} + \hat{c}_v\,(13+8\hat{c}_v)\,\tau_{qs}\tau_{ps}\right] - 3\Omega^3\hat{c}_v\,(7+4\hat{c}_v)\,\tau_{ps}\tau_{qs}\Bigg\}$$

$$+\frac{(\Omega-i)(\tau_{ps}\Omega-i)(\tau_{qs}\Omega-i)}{\Omega^3\tau_{ps}\tau_{qs}} = 0 \tag{7.8}$$

with c_0 being the sound velocity in an equilibrium state:

$$c_0 = \sqrt{\left(\frac{\partial p}{\partial \rho}\right)_0 + \frac{T_0\left(\frac{\partial p}{\partial T}\right)_0^2}{\rho_0^2\left(\frac{\partial \varepsilon}{\partial T}\right)_0}} = \sqrt{\frac{k_B}{m}T_0\left(1+\frac{1}{\hat{c}_v}\right)} = \sqrt{\frac{k_B}{m}T_0\gamma(T)}, \tag{7.9}$$

where the suffix 0 indicates the values at the equilibrium state and

$$\gamma(T) = (1+\hat{c}_v(T))/\hat{c}_v(T) \tag{7.10}$$

is the ratio of the specific heats. Therefore, for given $\hat{c}_v$, τ_{qs} and τ_{ps}, the quantity $c_0 z (= c_0 k/\omega)$ is calculated from (7.8) as the function of $\Omega\ (=\tau_S\omega)$. Hereafter in this chapter, we will confine our study within the *fastest* sound wave because the experiments give us the data on this wave.

7.2.2 High Frequency Limit of the Phase Velocity and the Attenuation Factor

From (3.10), we have the relations:

$$v_{ph}^{(\infty)} \equiv \lim_{\omega\to\infty} v_{ph}(\omega) = \lambda_0, \quad \alpha^{(\infty)}\lambda_0 \equiv \lim_{\omega\to\infty}\alpha(\omega)\lambda_0 = -\boldsymbol{l}_0\cdot\boldsymbol{B}_0\cdot\boldsymbol{d}_0, \tag{7.11}$$

where the characteristic velocity λ_0 is the largest eigenvalue of $\boldsymbol{A}_0$, and $\boldsymbol{l}_0$ and $\boldsymbol{d}_0$ are the corresponding left and right eigenvectors of $\boldsymbol{A}_0$. Then we obtain the limits:

$$v_{ph}^{(\infty)} = \sqrt{\frac{\frac{k_B}{m}T_0(4\hat{c}_v+7+F)}{2(1+\hat{c}_v)}}, \tag{7.12}$$

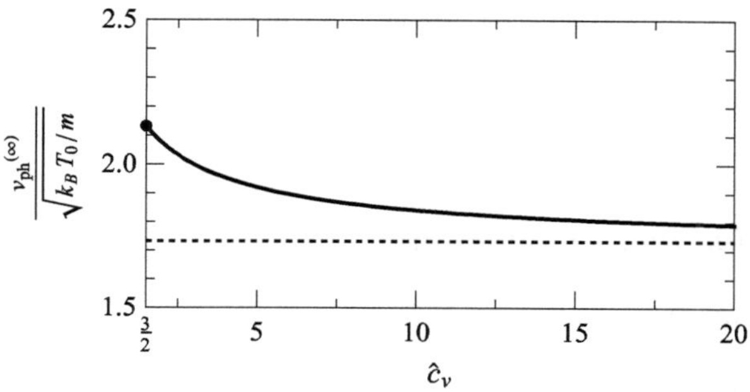

Fig. 7.1 Dependence of the phase velocity in the high frequency limit $v_{ph}^{(\infty)}$ on the dimensionless specific heat $\hat{c}_v$. Rarefied monatomic gases correspond to the case with $\hat{c}_v = 3/2$. The *dotted line* is the asymptote

$$\alpha^{(\infty)} = \frac{\sqrt{2(1+\hat{c}_v)^3}\left(F(4+\hat{c}_v) - 22 - 11\hat{c}_v + 2\hat{c}_v^2\right)}{9\hat{c}_v\tau_S\sqrt{\frac{k_B}{m}T_0}\sqrt{7+4\hat{c}_v+F}\,(7+4\hat{c}_v-F)^2 F}$$
$$\left(4\hat{c}_v + \frac{3\hat{c}_v(8+2\hat{c}_v-F)}{\tau_{qs}} + \frac{-3+2\hat{c}_v}{\tau_{ps}}\right), \tag{7.13}$$

where F is given by

$$F = \sqrt{37 + 32\hat{c}_v + 4\hat{c}_v^2}. \tag{7.14}$$

The dependence of the phase velocity $v_{ph}^{(\infty)}$ on $\hat{c}_v$ is shown in Fig. 7.1. In a rarefied monatomic gas with $\hat{c}_v = 3/2$, the phase velocity $v_{ph}^{(\infty)}$ is given by $2.13051\sqrt{k_B T_0/m}$ [1]. For large $\hat{c}_v$, it approaches $\sqrt{3k_B T_0/m}$. On the other hand, the attenuation factor $\alpha^{(\infty)}$ depends not only on $\hat{c}_v$ but also on the relaxation times. In a rarefied monatomic gas, the attenuation factor $\alpha^{(\infty)}$ is given by $\left(0.0951852 + 0.0931368/\tau_{qs}\right)/\left(\tau_S\sqrt{k_B T_0/m}\right)$ [1]. For large $\hat{c}_v$, it approaches $(1 + 2\tau_{ps})/(9\sqrt{3}\tau_{ps}\tau_S\sqrt{k_B T_0/m})$.

7.3 Comparison with Experimental Data

The dispersion relation obtained above, in particular, the phase velocity v_{ph}, the attenuation factor α and the attenuation per wavelength α_Λ as the functions of the frequency ω are compared with the experimental data on normal hydrogen (n-H_2), para hydrogen (p-H_2), normal deuterium (n-D_2), ortho deuterium (o-D_2) and hydrogen deuteride (HD) gases at temperatures from 77.3 to 1073.15 K [2–5]. The comparison is also made with the predictions by the classical NSF theory.

Before discussing the subject, we need to make preliminary calculations for determining the values of $\hat{c}_v$, τ_{qs} and τ_{ps} defined in (5.66) and $(7.7)_{2,3}$ at the equilibrium state.

7.3.1 Preliminary Calculations

7.3.1.1 Specific Heat

We calculate the specific heat $\hat{c}_v$ of hydrogen, deuterium and hydrogen deuteride gases on the basis of statistical mechanics [6, 7]. We assume that the translational mode satisfies the equipartition law of energy. Then $\hat{c}_v$ is expressed as

$$\hat{c}_v = \frac{3}{2} + \hat{c}_{v,\mathrm{rot}} + \hat{c}_{v,\mathrm{vib}},$$
$$\hat{c}_{v,\mathrm{rot}} = \beta^2 \frac{\partial^2 \log Z_{\mathrm{rot}}}{\partial \beta^2}, \quad \hat{c}_{v,\mathrm{vib}} = \beta^2 \frac{\partial^2 \log Z_{\mathrm{vib}}}{\partial \beta^2}, \quad (\beta \equiv \frac{1}{k_B T}) \tag{7.15}$$

where $\hat{c}_{v,\mathrm{rot}}$, Z_{rot} and $\hat{c}_{v,\mathrm{vib}}$, Z_{vib} are the specific heat and the partition function due to the rotational and vibrational modes, respectively.

For gases composed of heteronuclear diatomic molecules (HD), the partition function of rotational motion is given by

$$Z_{\mathrm{rot}} = \sum_{l=0}^{\infty} (2l+1) \exp\left[-\beta B l(l+1)\right], \tag{7.16}$$

where l is the quantum number of the orbital angular momentum and $B = \hbar^2/2I$ with I and $\hbar$ being the moment of inertia of a molecule and the Planck constant divided by 2π, respectively. While, for gases composed of diatomic homonuclear molecules (H_2 and D_2), the partition function of rotational motion is given by

$$Z_{\mathrm{rot}} = Z_g^{g_g} Z_u^{g_u},$$
$$Z_g = \sum_{l=even} (2l+1) \exp\left[-\beta B l(l+1)\right],$$
$$Z_u = \sum_{l=odd} (2l+1) \exp\left[-\beta B l(l+1)\right], \tag{7.17}$$

where g_g and g_u are defined by

$$H_2 \begin{cases} \text{normal} - H_2 : g_u = 3/4, \; g_g = 1/4 \\ \text{para} - H_2 : \quad g_u = 0, \quad g_g = 1 \end{cases}, \quad D_2 \begin{cases} \text{normal} - D_2 : g_u = 1/3, \; g_g = 2/3 \\ \text{ortho} - D_2 : \quad g_u = 0, \quad g_g = 1 \end{cases}. \tag{7.18}$$

For the vibrational modes, we assume the harmonic oscillator model. Then the partition function of the vibrational modes is obtained as follows:

$$Z_{\mathrm{vib}} = \frac{e^{-\frac{\hbar\omega_v\beta}{2}}}{1 - e^{-\hbar\omega_v\beta}}, \tag{7.19}$$

where ω_v is the characteristic frequency.

Numerically calculated values of $\hat{c}_v$ are shown in Table 7.1 and in Fig. 7.2. The values of B of H_2, D_2 and HD adopted are 12.09×10^{-22} [J], 6.047×10^{-22} [J] and 9.068×10^{-22} [J], respectively, and the values of ω_v are $6332k_B/\hbar$ [Hz], $4483k_B/\hbar$ [Hz] and $5486k_B/\hbar$ [Hz], respectively [8].

Table 7.1 Values of the temperature T_0, dimensionless specific heat $\hat{c}_v$, sound speed in equilibrium c_0, shear viscosity μ [4, 9–12], heat conductivity κ [4, 9–12] and the ratio of the relaxation times of the heat flux and the deviatoric part of the viscous stress τ_{qs} adopted in the present analysis. And the values of the parameter φ, bulk viscosity ν, and the ratio of the relaxation times of the bulk viscosity and the deviatoric part of the viscous stress τ_{ps} evaluated by the present analysis

Gas	T_0(K)	$\hat{c}_v$	c_0 (m/s)	μ (μPa s)	κ (mW/m K)	τ_{qs}	φ	ν (μPa s)	τ_{ps}
n-H_2	77.3	1.57	723	3.50	49.8	1.34	27.2	95.2	960
	273	2.42	1260	8.33	173	1.47	41.9	349	165
	295.15	2.45	1310	8.95	187	1.47	33.1	296	128
	873.15	2.54	2240	18.7	403	1.48	36.8	685	135
	1073.15	2.60	2480	21.0	462	1.49	40.3	846	143
p-H_2	77.3	1.76	707	3.50	52.7	1.33	75.0	263	773
	90.2	1.99	748	3.97	63.6	1.30	83.6	332	512
	170	2.96	968	6.10	113	1.14	54.8	334	166
	293	2.61	1290	8.82	192	1.46	28.8	254	101
n-D_2	77.3	2.54	472	4.82	45.6	1.30	35.7	172	131
	273.15	2.50	888	11.8	136	1.60	24.7	291	92.6
	295.15	2.50	923	12.6	141	1.55	20.9	264	78.3
	773.15	2.60	1490	24.2	260	1.45	30.9	747	109
	1073.15	2.78	1740	30.4	337	1.42	35.9	1092	117
o-D_2	77.3	2.93	463	4.82	49.4	1.26	45.4	219	140
	90.2	2.96	499	5.50	55.6	1.24	33.6	185	102
	293	2.50	920	12.3	131	1.47	22.6	278	84.7
HD	77.3	2.55	544	4.21	51.9	1.26	1.84	7.75	6.72
	293	2.50	1060	10.8	149	1.43	2.27	24.5	8.51

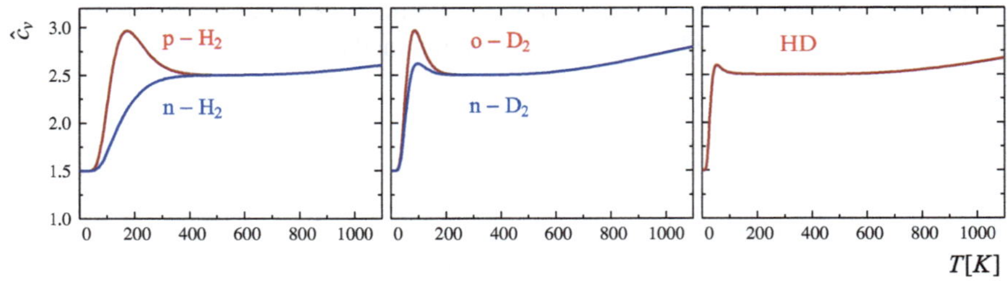

Fig. 7.2 Dependence of the dimensionless specific heat $\hat{c}_v$ for n-H_2 and p-H_2 (*left*), n-D_2 and o-D_2 (*center*), and HD (*right*) on the temperature T

7.3.2 Relaxation Times

From (5.68), we have the following relations for the ratios τ_{qs} and τ_{ps}:

$$\tau_{qs} = (1 + \hat{c}_v)^{-1} \frac{\kappa}{\frac{k_B}{m}\mu}, \tag{7.20}$$

$$\tau_{ps} = \left(\frac{2}{3} - \frac{1}{\hat{c}_v}\right)^{-1} \frac{\nu}{\mu}. \tag{7.21}$$

Therefore, in principle, with the help of the experimental data on μ, κ and ν, we can estimate the values of τ_{qs} and τ_{ps}. However, at present, as we have the reliable data only on μ and κ [4, 9–12], we adopt, in the analysis below, an adjustable parameter:

$$\varphi = \frac{\nu}{\mu}. \tag{7.22}$$

We summarize the adopted values of $\hat{c}_v$, c_0, μ, κ, τ_{qs}, and the evaluated values of φ and τ_{ps} in Table 7.1, details of which will be discussed in the next subsection.

7.3.3 Experimental Data and Theoretical Predictions for the Dispersion Relation

7.3.3.1 Hydrogen Gases: n-H_2 and p-H_2

For n-H_2, the dimensionless phase velocity, v_{ph}/c_0, the dimensionless attenuation factor, $c_0\tau_S\alpha$, and the attenuation per wavelength, α_Λ, are shown as the functions of the dimensionless frequency Ω in Fig. 7.3. We see the experimental data on the phase velocity at $T_0 = 295.15$ and 296.8 K by Winter and Hill [2] and Rhodes [3], on the attenuation factor at $T_0 = 293$ K by Sluijter et al. [4] and on the attenuation per

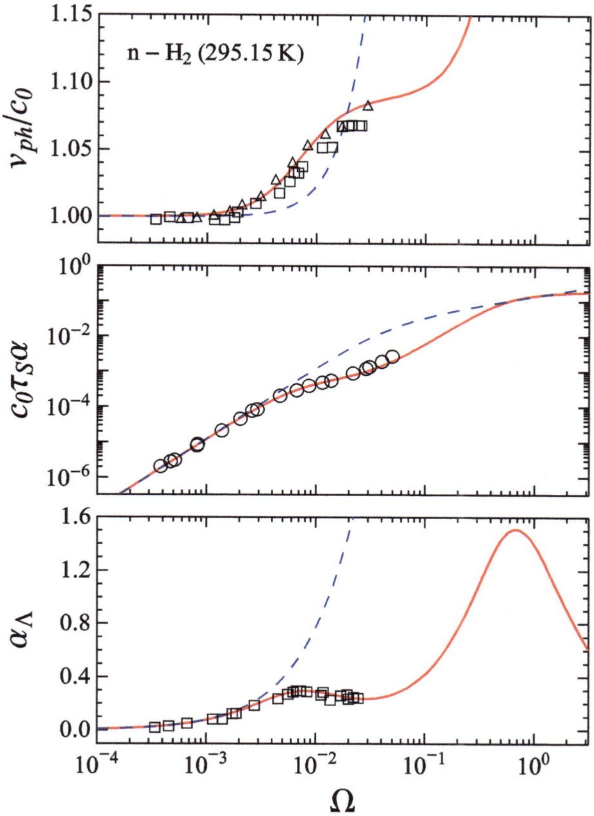

Fig. 7.3 Dependence of the dimensionless phase velocity v_{ph}/c_0, the attenuation factor $c_0\tau_S\alpha$ and the attenuation per wavelength α_Λ on the dimensionless frequency Ω for n-H_2. The *circles*, *squares* and *triangles* in the figures are, respectively, the experimental data at 293 K by Sluijter et al. [4], those at 295.15 K by Winter and Hill [2] and those at 296.8 K by Rhodes [3]. The *solid* and *dashed lines* are predictions at 295.15 K by the ET14 and NSF theories, respectively. We adopt $\varphi = 33.1$

wavelength at $T_0 = 295.15\,\mathrm{K}$ by Winter and Hill [2] accompanied by the theoretical results at $T_0 = 295.15\,\mathrm{K}$ predicted by the ET14 theory and the NSF theory.

Noticeable points in Fig. 7.3 are summarized as follows: (1) In the region with small Ω, as is expected, the predictions by the two theories coincide with each other. The value of the parameter φ is determined to be 33.1 as the best fit with the experimental data in this region. This procedure of determining φ will be adopted throughout the present chapter. (2) When we go into the ultrasonic frequency region with larger Ω, the prediction by the ET14 theory is evidently superior to that by the NSF theory. The difference between the two theories emerges around $\Omega = \omega\tau_S = 10^{-3}$. We will evaluate τ_S, which depends on T_0 and p_0, later. (3) The ET14 theory seems to be valid at least up to the experimental data with the maximum dimensionless frequency $\Omega = 10^{-1}$. (4) The large value of φ means that $\nu >> \mu$. We will discuss its physical meaning below.

At other temperatures, there exists the experimental data of α at $T_0 = 77.3\,\mathrm{K}$ by Sluijter et al. [4], those of v_{ph} and α_Λ at around 273 K, respectively, by Rhodes [3] (at $T_0 = 273.5\,\mathrm{K}$) and Stewart and Stewart [5] (at $T_0 = 273.15\,\mathrm{K}$) and those of v_{ph} and α_Λ at $T_0 = 873.15, 1073.15\,\mathrm{K}$ by Winter and Hill [2]. The comparison of the

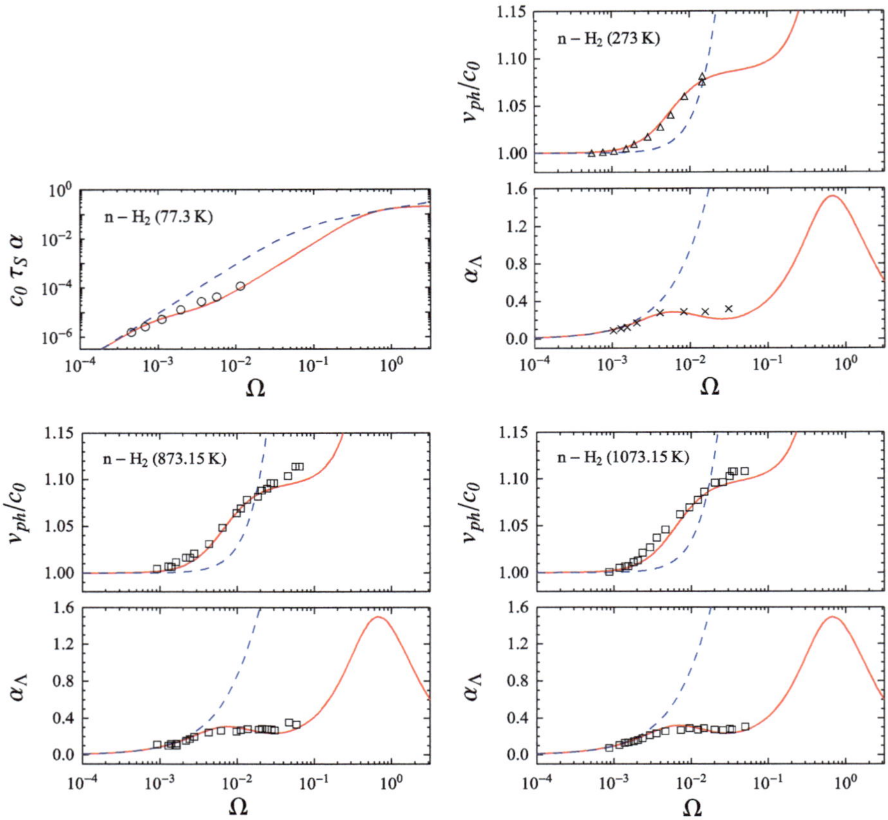

Fig. 7.4 Dependence of the dimensionless phase velocity v_{ph}/c_0, the attenuation factor $c_0\tau_S\alpha$ and the attenuation per wavelength α_Λ on the dimensionless frequency Ω for n-H_2. The *circles*, *squares*, *triangles* and *crosses* in the figures are, respectively, the experimental data by Sluijter et al. [4] at 77.3 K, Winter and Hill [2] at 873.15, 1073.15 K, Rhodes [3] at 273.5 K and Stewart and Stewart [5] at 273.15 K. The *solid* and *dashed lines* are predictions at $T_0 = 77.3, 273, 873.15, 1073.15$ K by the ET14 and NSF theories, respectively. We adopt $\varphi = 27.2, 41.9, 36.8, 40.3$, respectively, for $T_0 = 77.3, 273, 873.15, 1073.15$ K

theoretical predictions with these experimental data are shown in Fig. 7.4. We see again that the ET14 theory can describe the experimental data very well. The values of the parameter φ are selected to be 27.2, 41.9, 36.8 and 40.3, respectively.

For p-H_2, we compare the theoretical predictions with the experimental data on the phase velocity at $T_0 = 273.8, 298.4$ K by Rhodes [3] and on the attenuation factor at $T_0 = 77.3, 90.2, 170, 293$ K by Sluijter et al. [4, 9].

We have a similar result as shown in Fig. 7.5, where the selected value of the parameter φ is 75.0, 83.6, 54.8, 28.8, respectively, for $T_0 = 77.3, 90.2, 170, 293$ K. Remarkable points in this case are qualitatively the same as in the case of n-H_2 above.

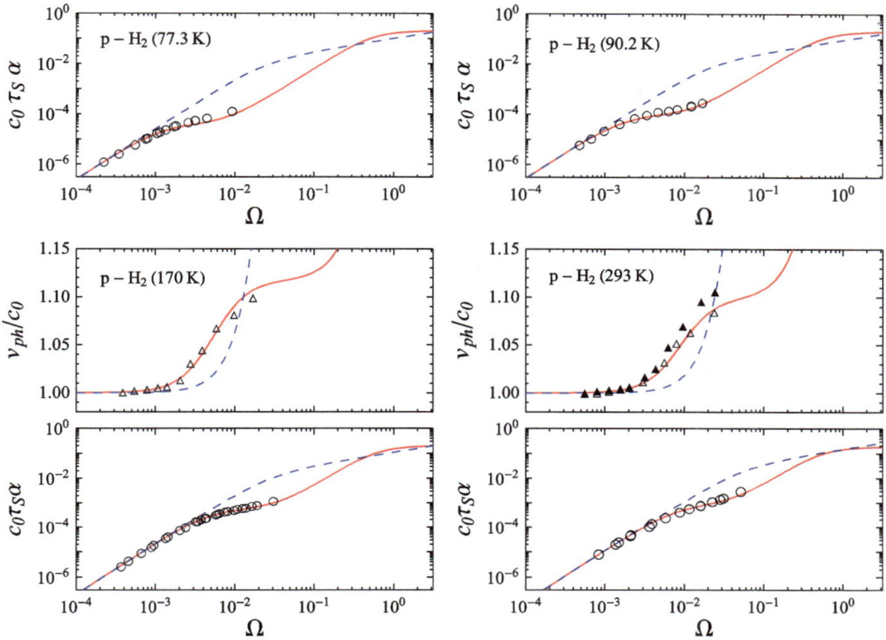

Fig. 7.5 Dependence of the dimensionless phase velocity v_{ph}/c_0 and the attenuation factor $c_0\tau_S\alpha$ on the dimensionless frequency Ω for p-H_2. The *circles* and *empty triangles* in the figures are, respectively, the experimental data by Sluijter et al. [4, 9] at $T_0 = 77.3, 90.2, 170, 293$ K and by Rhodes [3] at $T_0 = 298.4$ K. As a reference the experimental data by Rhodes [3] at $T_0 = 273.8$ K are on the figure as the *filled triangles*. The *solid* and *dashed lines* are predictions at $T_0 = 77.3, 90.2, 170, 293$ K by the ET14 and NSF theories, respectively. We adopt φ=75.0, 83.6, 54.8, 28.8, respectively, for $T_0 = 77.3, 90.2, 170, 293$ K

7.3.3.2 Deuterium Gases: n-D_2 and o-D_2

Comparisons are also made for n-D_2 at T_0 = 295.15 K with φ=20.9 in Fig. 7.6, and it shows the superiority of the prediction by the ET14 theory to that by the NSF theory in high frequency region. Also comparisons at other temperatures $T_0 = 77.3, 273.15, 773.15$ and 1073.15 K with φ= 35.7, 24.7, 30.9 and 35.9, respectively, are shown in Fig. 7.7 and for o-D_2 at T_0 = 77.3, 90.2 and 293 K with φ=45.4, 33.6 and 22.6 are shown in Fig. 7.8. From these figures, we have qualitatively the same observations as those in the case of hydrogen gases.

7.3.3.3 Hydrogen Deuteride Gases: HD

Lastly we show the results of HD gases at T_0 = 77.3 and 293 K in Fig. 7.9.We notice the following points: (1) The difference between the two theories is small and the theoretical predictions are consistent with the experimental data in the range:

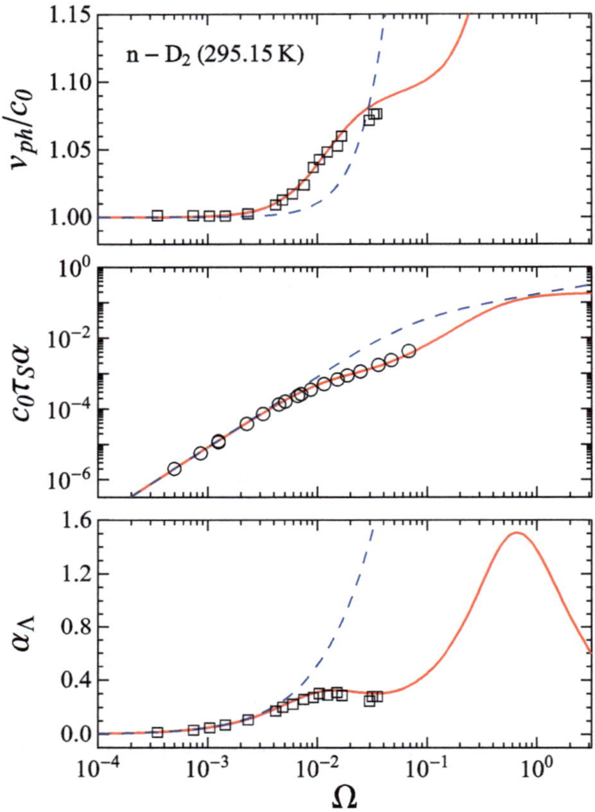

Fig. 7.6 Dependence of the dimensionless phase velocity v_{ph}/c_0, the attenuation factor $c_0\tau_S\alpha$ and the attenuation per wavelength α_Λ on the dimensionless frequency Ω for n-D_2. The *circles* and *squares* in the figures are, respectively, the experimental data by Sluijter et al. [4] at 293 K and Winter and Hill [2] at 295.15 K. The *solid* and *dashed lines* are predictions at 295.15 K by the ET14 and NSF theories, respectively. We adopt $\varphi = 20.9$

$\Omega \leqslant 10^{-1}$.This means that the local equilibrium assumption holds well up to $\Omega \fallingdotseq 10^{-1}$, while, for the other gases analyzed above, the assumption holds until $\Omega \fallingdotseq 10^{-3}$. (2) The values of φ adopted here are 1.84 and 2.27.These values are $O(1)$, that is, $\nu \sim \mu$, and are very small compared with those obtained for the other gases discussed above.We will discuss this interesting fact below.

7.3.4 Remarks

Remarks (A)–(C) are made.

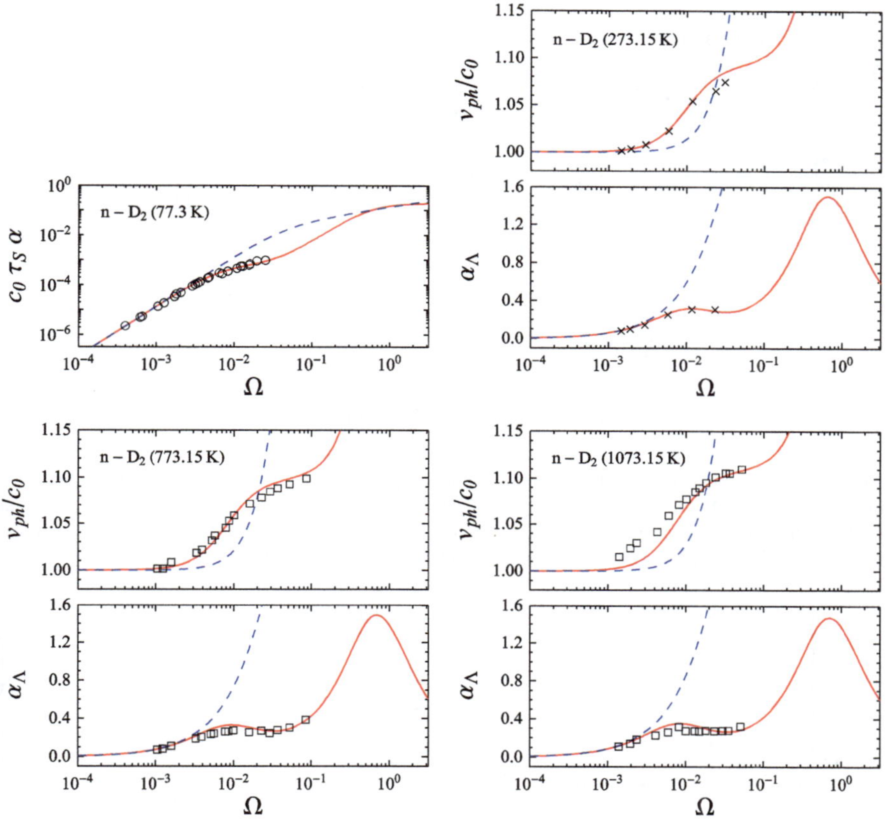

Fig. 7.7 Dependence of the dimensionless phase velocity v_{ph}/c_0, the attenuation factor $c_0\tau_S\alpha$ and the attenuation per wavelength α_Λ on the dimensionless frequency Ω for n-D_2. The *circles*, *squares* and *crosses* in the figures are, respectively, the experimental data by Sluijter et al. [4] at 77.3 K, Winter and Hill [2] at 773.15, 1073.15 K and Stewart and Stewart [5] at 273.15 K. The *solid* and *dashed lines* are predictions by the ET14 and NSF theories, respectively. We adopt φ=35.7, 24.7, 30.9 and 35.9, respectively for $T_0 = 77.3$, 273.15, 773.15 and 1073.15 K

(A) We have seen clearly that the ET14 theory is consistent with the experimental data even in the high frequency range where the local equilibrium assumption is no longer valid. There are potentially many research fields where the ET14 theory may play a crucial role, for example, fields of acoustics [13] and gas dynamics [14].

(B) From the values of the ratios τ_{qs} and τ_{ps} in Table 7.1, we have noticed an interesting fact that, except for HD gases, τ_Π is much larger than τ_S, while τ_S and τ_q are comparable with each other. This fact was reported also in some kinetic theoretical studies [15, 16]. By using the result summarized in Table 7.2, the relaxation times for given T_0 and p_0 can be estimated. For example, the relaxation times in a n-H_2 gas at $p_0 = 10^3$ [Pa] and $T_0 = 77.3$ [K] can be calculated: $\tau_S = 3.50 \times 10^{-9}$ [s], $\tau_\Pi = 3.36 \times 10^{-6}$ [s] and $\tau_q = 4.70 \times 10^{-9}$ [s].

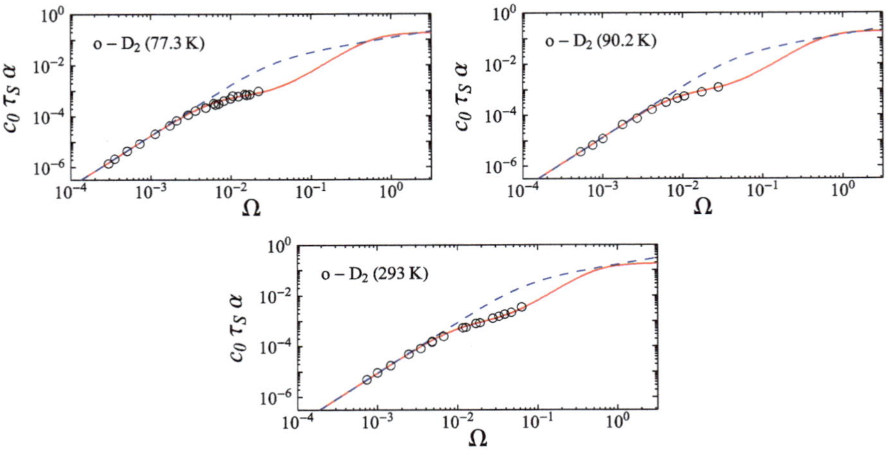

Fig. 7.8 Dependence of the dimensionless attenuation factor $c_0\tau_S\alpha$ on the dimensionless frequency Ω for o-D_2 at $T_0 = 77.3$ K, 90.2 K and 293 K. The *circles* are the experimental data by Sluijter et al. [4]. The *solid* and *dashed lines* are predictions by the ET14 and NSF theories, respectively. We adopt $\varphi = 45.4, 33.6$ and 22.6 from 77.3 K to 293 K

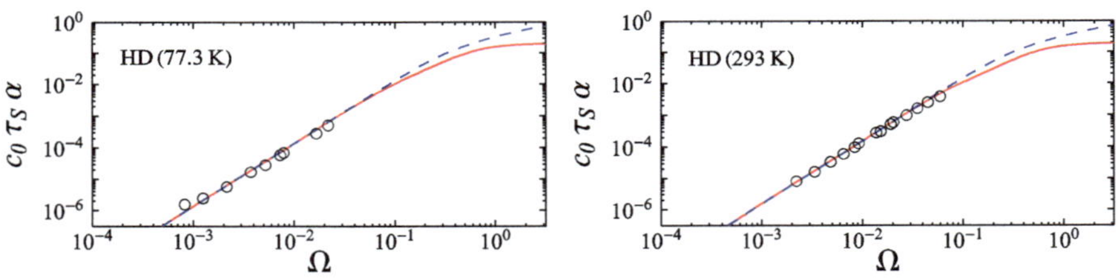

Fig. 7.9 Dependence of the dimensionless attenuation factor $c_0\tau_S\alpha$ on the dimensionless frequency Ω for HD at T_0=77.3 K and 293 K. The *circles* are the experimental data by Sluijter et al. [4]. The *solid* and *dashed lines* are predictions by the ET14 and NSF theories, respectively. We adopt φ = 1.84 at 77.3 K and 2.27 at 293 K

In the paper [17], it was pointed out that the relaxation time τ_Π is in the same order of magnitude as the relaxation time of the energy exchange between the molecular translational mode and the internal modes. The results obtained above suggest that the sharp temperature change of the specific heat due to the rotational modes $\hat{c}_{v,\mathrm{rot}}$ depicted in Fig. 7.2 is somehow related to the emergence of the large value of τ_Π. The detailed study of this subject is, however, beyond the scope of the present phenomenological study, and its statistical-mechanical or kinetic-theoretical study by taking into account the realistic collision processes between the constituent molecules is required.

(C) From the values of φ in Table 7.1, we have also noticed a similar fact that, except for HD gases, the bulk viscosity ν is much larger than the shear viscosity μ. The similarity is natural because there are relations between the viscosities and

Table 7.2 Relaxation times of the deviatoric part of the viscous stress τ_S, dynamic pressure τ_Π and heat flux τ_q multiplied by the pressure p_0 for several values of T_0 in H_2, D_2 and HD gases

Gas	T_0 (K)	$\tau_S p_0$ (s μPa)	$\tau_\Pi p_0$ (s μPa)	$\tau_q p_0$ (s μPa)
n-H_2	77.3	3.50	3360	4.70
	273	8.33	1380	12.2
	295.15	8.95	1150	13.1
	873.15	18.6	2510	27.6
	1073.15	21.0	3000	31.2
p-H_2	77.3	3.50	2710	4.64
	90.2	3.97	2030	5.16
	170	6.10	1010	6.93
	293	8.82	894	12.9
n-D_2	77.3	4.82	632	6.25
	273.15	11.8	1090	18.8
	295.15	12.6	990	20.0
	773.15	24.2	2640	35.0
	1073.15	30.4	3560	43.2
o-D_2	77.3	4.82	673	6.09
	90.2	5.50	562	6.81
	293	12.3	1040	18.1
HD	77.3	4.21	28.3	5.32
	293	10.8	91.9	15.5

the relaxation times as shown in (5.68). A point to be emphasized here is that, as the direct experiments to measure the bulk viscosity are usually difficult, the method for the evaluation of the bulk viscosity utilized here through analyzing the dispersion relation on the basis of the ET14 theory is quite useful. The values of ν thus evaluated are summarized in Table 7.1. See also the recent studies of the bulk viscosity [18–22].

7.4 Conclusion

Some remarks are made as follows:

1. There is a phenomenological theory of the dispersion relation for sound, the basic equations of which are composed of the relaxation equations for some nonequilibrium parameters and the Euler (or NSF) equations for the conservation laws [23, 24]. One crucial point is that the theory is based on the local equilibrium assumption. In this respect, this may be regarded as a theory in the framework of thermodynamics of irreversible processes [23]. Because of this, in the present chapter, we have compared the ET theory only with the NSF theory as a

representative one. In Chap. 11, the relationship between the simplified ET theory with six fields and the theory with one relaxation equation was studied in detail.

2. It is shown that the ET14 theory is applicable in a unified way to ultrasonic waves in rarefied polyatomic gases in a wide temperature range and also in a wide frequency range where the rotational and/or vibrational modes in a molecule play a role. It is expected that the theoretical predictions presented in this chapter will be useful in future when the ultrasonic waves with much higher frequencies are available in experiments.
3. In order to study the effect of the large value of the relaxation time τ_Π on various nonequilibrium phenomena such as shock wave phenomena, it seems to be appropriate to adopt a simpler model than the one adopted here. The theory with only 6 independent field variables (ρ, v_i, T, Π), which will be explained in Chap. 11, play an important role in such studies. See also Chap. 13. The results of this chapter are obtained in the papers [25] and [26].

References

1. I. Müller, T. Ruggeri, *Rational Extended Thermodynamics*, 4th edn. (Springer, New York, 1998)
2. T.G. Winter, G.L. Hill, High-temperature ultrasonic measurements of rotational relaxation in hydrogen, deuterium, nitrogen, and oxygen. J. Acoust. Soc. Am. **42**, 848 (1967)
3. E.J. Rhodes Jr., The velocity of sound in hydrogen when rotational degrees of freedom fail to be excited. Phys. Rev. **70**(11), 932 (1946)
4. C. Sluijter, H. Knaap, J. Beenakker, Determination of rotational relaxation times of hydrogen isotopes by sound absorption measurements at low temperatures. I. Physica **30**, 745 (1964)
5. E.S. Stewart, J.L. Stewart, Rotational dispersion in the velocity, attenuation, and reflection of ultrasonic waves in hydrogen and deuterium. J. Acoust. Soc. Am. **24**, 194 (1952)
6. E.M. Lifshitz, L.P. Pitaevskii, *Statistical Physics, 3rd Edition Part 1* (Pergamon Press, Oxford, 1980)
7. E.M. Lifshitz, L.P. Pitaevskii, *Statistical Physics, Part 2* (Pergamon Press, Oxford, 1980)
8. A.A. Radzig, B.M. Smirnov, *Reference Data on Atoms, Molecules, and Ions* (Springer, Berlin/Heidelberg/New York/Tokyo, 1985)
9. C.G. Sluijter, R.M. Jonkman, Sound absorption measurements in para hydrogen at 170 °K. Physica **30**, 1670 (1964)
10. H.J.M. Hanley, R.D. McCarty, H. Interman, The viscosity and thermal conductivity of dilute gaseous hydrogen from 15 to 5000 K. J. Res. Nat. Bur. Stand. Sect. A **74**, 331 (1970)
11. M.J. Assael, S. Mixafendi, W.A. Wakeham, The viscosity of normal deuterium in the limit of zero density. J. Phys. Chem. Ref. Data **16**, 189 (1987)
12. S.C. Saxena, W.K. Saxena, Thermal conductivity data for hydrogen and deuterium in the range 100–1100 degrees C. J. Phys. A **3**, 309 (1970)
13. W.P. Mason (ed.), *Physical Acoustics, Principles and Methods*, vol. II-Part A (Academic, New York/London, 1965)
14. W.G. Vincenti, C.H. Kruger Jr., *Introduction to Physical Gas Dynamics* (Wiley, New York/London/Sydney, 1965)
15. S. Chapman, T.G. Cowling, *The Mathematical Theory of Non-uniform Gases* (Cambridge University Press, Cambridge, 1991)
16. B.C. Eu, Y.G. Ohr, Generalized hydrodynamics, bulk viscosity, and sound wave absorption and dispersion in dilute rigid molecular gases. Phys. Fluids **13**(3), 744 (2001)

17. T. Arima, S. Taniguchi, T. Ruggeri, M. Sugiyama, Extended thermodynamics of real gases with dynamic pressure: an extension of Meixner's theory. Phys. Lett. A **376**, 2799 (2012)
18. G. Emanuel, Bulk viscosity of a dilute polyatomic gas. Phys. Fluids A **2**(12), 2252 (1990)
19. W.E. Meador, G.A. Miner, L.W. Townsend, Bulk viscosity as a relaxation parameter: fact or fiction? Phys. Fluids A **8**(1), 258 (1996)
20. G. Emanuel, Bulk viscosity as a relaxation parameter: fact or fiction? [Phys. Fluids 8, 258 (1996)]. Phys. Fluids A **8**(7), 1984 (1996)
21. G. Emanuel, Bulk viscosity in the Navier-Stokes equations. Int. J. Eng. Sci. **36**, 1313 (1998)
22. R.E. Graves, B.M. Argrow, Bulk viscosity: past to present. J. Thermophys. Heat Transf. **13**(3), 337 (1999)
23. S.R. de Groot, P. Mazur, *Non-equilibrium Thermodynamics* (North-Holland, Amsterdam, 1963)
24. H.J. Bauer, Phenomenological theory of the relaxation phenomena in gases, in *Physical Acoustics II Part A*, ed. by W.P. Mason (Academic, New York/London, 1965), pp. 47–131
25. T. Arima, S. Taniguchi, T. Ruggeri, M. Sugiyama, Dispersion relation for sound in rarefied polyatomic gases based on extended thermodynamics. Continuum Mech. Thermodyn. **25**, 727 (2013)
26. T. Arima, S. Taniguchi, T. Ruggeri, M. Sugiyama, A study of linear waves based on extended thermodynamics for rarefied polyatomic gases. Acta Appl. Math. **132**, 15 (2014)

Chapter 8
Shock Wave in a Polyatomic Gas

Abstract In this chapter we study the shock wave structure in a rarefied polyatomic gas by using the ET14 theory. We show how the ET14 theory can overcome the difficulties encountered in the previous approaches: Bethe-Teller approach and Gilbarg-Paolucci approach.

Firstly, the predictions derived from the ET14 theory are shown and compared with the results from the NSF theory. Secondly, the Bethe-Teller theory is reexamined in the light of the ET14 theory. Lastly, comparison between the theoretical predictions derived from the ET14 theory and the experimental data is made, where we show a very good agreement. We are able to explain in a unified manner the three different shock wave profiles Types A, B and C for increasing Mach number.

8.1 Introduction

In Sect. 1.6.4.2, characteristic features of the shock wave structure in a rarefied polyatomic gas have been briefly explained. In order to understand such structure, two different approaches by Bethe-Teller and by Gilbarg-Paolucci were proposed many years ago and are still in activity.

(I) Bethe-Teller approach [1]: In their celebrated theory, at the very beginning, the internal degrees of freedom of a molecule are assumed to be classified into two parts; the one part is composed of the "active" degrees of freedom that relax instantaneously, and the other part is composed of the "inert" degrees of freedom that relax slowly with a finite relaxation time. Except for a hydrogen gas, the translational and rotational modes are regarded as the "active" degrees of freedom but the vibrational modes are considered as "inert" degrees of freedom. In order to analyze the thin layer Δ shown in Fig. 1.1, the system of the Euler equations is adopted. The Rankine-Hugoniot relations [see (3.34)] of the system for the jumps of the physical quantities at Δ are derived under the following assumption: The internal energy due to the "inert" degrees of freedom is unchanged in the thin layer Δ

T. Ruggeri, M. Sugiyama, *Rational Extended Thermodynamics beyond the Monatomic Gas*, DOI 10.1007/978-3-319-13341-6_8

because only "active" degrees of freedom are able to adjust to such an instantaneous change. Therefore, the thin layer in this theory is just the jump discontinuity with no thickness. While, in order to analyze the relaxation process in the thick layer Ψ in Fig. 1.1, a variant of the Euler system with an additional linear relaxation equation for the internal vibrational modes is adopted.

This approach can describe the shock wave structure of Type C. The jumps of the physical quantities at the thin layer Δ can be calculated without using any adjustable parameters. The agreement between the theoretical predictions and the existing experimental data seems to be well.

It should be emphasized, however, that the basis of the Bethe-Teller theory is not clear enough. The assumption of the classification of the internal degrees of freedom should be regarded as a rough approximation even though it seems to be plausible intuitively. As the classification is not so much clear-cut in reality, it may introduce some arbitrariness into the theory. And furthermore two different systems of equations are adopted in their theory for analyzing the thin and thick layers separately. The compatibility between the two systems is, however, unclear from both mathematical and physical points of view. It is highly preferable, of course, to have one unified system of equations from which all Types A, B and C can be derived in a fully consistent way.

(II) Gilbarg-Paolucci approach [2]: This is basing on the system of the Navier-Stokes Fourier (NSF) equations. They studied, as a typical example, the shock wave structure in a rarefied carbon dioxide (CO_2) gas by adopting a very large value of the bulk viscosity. Although they could predict a thick shock wave structure, the shock profiles are always symmetric (Type A). No asymmetric shock wave structure (Type B) nor thin layer (Type C) could be explained by this theory.

One crucial point to be noted is that, because the NSF theory is constructed with the assumption of the local equilibrium (see Sect. 1.1.3), the theory is, in general, unsatisfactory for analyzing highly nonequilibrium phenomena such as shock wave phenomena.

Fortunately we have alternative theories as discussed below:

For rarefied monatomic gases, there already exist theories which can describe the phenomena out of local equilibrium, that is, the kinetic theory with the use of the Boltzmann equation (the Chapman-Enskog method [3] and the moment method [4]), and the theories of extended thermodynamics (ET) [5] and of molecular extended thermodynamics with the closure by the maximum entropy principle [6, 7]. These theories can indeed describe the structure of strong shock waves in a rarefied monatomic gas [5, 8]. Numerical techniques for solving the Boltzmann equation, such as the Direct Simulation Monte Carlo (DSMC) method [9], have also been developed, and their usefulness has been confirmed through the comparison between their predictions and the experimental data.

For rarefied polyatomic gases, the kinetic theory (the Chapman-Enskog method [3] and the moment method [10–13]) has been developed. Numerical methods for solving the Boltzmann equation have also been developed [9]. However, as the appropriate modeling of the collision term in the Boltzmann equation between two polyatomic molecules is very complicated, some simplifications are usually introduced into the modeling. It is therefore not self-evident that the numerical results thus obtained is compatible with the second law of thermodynamics (the entropy principle).

However, we know well that there is still another theory, that is, the ET14 theory of polyatomic gases explained in Chap. 5. It is remarkable that the ET theory is totally free from the difficulties mentioned just above [21].

8.2 Basic Equations

In this section, we summarize the basic equations for the present analysis.

8.2.1 Equations of State, Internal Energy, and Sound Velocity

We study a shock wave in a rarefied polyatomic gas with thermal and caloric equations of state given by (5.63). The functional form of the specific internal energy ε is determined through the specific heat c_v:

$$\varepsilon(T) = \frac{k_B}{m}\int_{T_R}^{T} \hat{c}_v(\xi)\, d\xi, \tag{8.1}$$

where $\hat{c}_v \equiv (m/k_B)c_v$ is the dimensionless specific heat and T_R is an inessential reference temperature. The velocity of sound is expressed as (7.9).

8.2.2 Balance Equations

As we analyze one-dimensional (plane) shock waves propagating along the x-axis, the vectorial and tensorial quantities are expressed in the form (7.2). Then, from (5.59) and (5.65), the system of equations in the present problem is given by

$$
\begin{aligned}
&\frac{\partial \rho}{\partial t} + \frac{\partial}{\partial x}(\rho v) = 0, \\
&\frac{\partial \rho v}{\partial t} + \frac{\partial}{\partial x}(p + \Pi - \sigma + \rho v^2) = 0, \\
&\frac{\partial}{\partial t}(2\rho\varepsilon + \rho v^2) + \\
&\quad + \frac{\partial}{\partial x}\left\{2\rho\varepsilon v + 2(p + \Pi - \sigma)v + \rho v^3 + 2q\right\} = 0, \\
&\frac{\partial}{\partial t}\left\{3(p + \Pi) + \rho v^2\right\} + \\
&\quad + \frac{\partial}{\partial x}\left\{(5p + 5\Pi - 2\sigma)v + \rho v^3 + \frac{5}{1 + \hat{c}_v}q\right\} = -\frac{3\Pi}{\tau_\Pi}, \\
&\frac{\partial}{\partial t}(p + \Pi - \sigma + \rho v^2) + \\
&\quad + \frac{\partial}{\partial x}\left\{3(p + \Pi - \sigma)v + \rho v^3 + \frac{3}{1 + \hat{c}_v}q\right\} = \frac{\sigma}{\tau_S} - \frac{\Pi}{\tau_\Pi}, \\
&\frac{\partial}{\partial t}\left\{2\rho\varepsilon v + 2(p + \Pi - \sigma)v + \rho v^3 + 2q\right\} + \\
&\quad + \frac{\partial}{\partial x}\left\{2\rho\varepsilon v^2 + 5(p + \Pi - \sigma)v^2 + \rho v^4 + \right. \\
&\qquad \left. + 2\left(\varepsilon + \frac{k_B}{m}T\right)p + 2\left(\varepsilon + 2\frac{k_B}{m}T\right)(\Pi - \sigma) + \frac{10 + 4\hat{c}_v}{1 + \hat{c}_v}qv\right\} \\
&\qquad\quad = -2\left\{\frac{q}{\tau_q} + \left(\frac{\Pi}{\tau_\Pi} - \frac{\sigma}{\tau_S}\right)v\right\}.
\end{aligned}
\tag{8.2}
$$

From (2.5), characteristic velocities λ of the hyperbolic system (8.2) evaluated in an equilibrium state are given by

$$
\begin{aligned}
\frac{\lambda}{c} = 0, \quad 0, \quad &\pm\sqrt{\frac{\hat{c}_v\left(7 + 4\hat{c}_v - \sqrt{37 + 32\hat{c}_v + 4\hat{c}_v^2}\right)}{2(1 + \hat{c}_v)^2}}, \\
&\pm\sqrt{\frac{\hat{c}_v\left(7 + 4\hat{c}_v + \sqrt{37 + 32\hat{c}_v + 4\hat{c}_v^2}\right)}{2(1 + \hat{c}_v)^2}}.
\end{aligned}
\tag{8.3}
$$

As is explained in Sect. 3.3, these velocities play an essential role in the study of shock wave propagation. The ET theory gives a differential system of hyperbolic type and, as a consequence, it predicts the shock wave structure with discontinuous part when the Mach number becomes large. According to the Theorem 3.1 (Sub-shock formation) of Boillat and Ruggeri in Sect. 3.4, a sub-shock emerges when the shock velocity s exceeds the maximum characteristic velocity $\lambda_{\max}$ of the hyperbolic system.

8.3 Setting of the Problem

In this section, conditions that we adopt for the present analysis are summarized. The parameters are fixed and the numerical method for the computation is explained.

As the differential system is Galilean invariant we can consider, without loss of generality, that the shock wave is stationary using the coordinate system moving with the shock wave, that is, the co-moving coordinate system. Both *the unperturbed state* (the state at $x = -\infty$ before and far from a shock wave) and *the perturbed state* (the state at $x = \infty$ after and far from a shock wave) are assumed to be in thermal equilibrium.

8.3.1 Dimensionless Form of the Field Equations

For convenience we introduce the following dimensionless quantities:

$$
\begin{aligned}
&\hat{\rho} \equiv \frac{\rho}{\rho_0}, \quad \hat{v} \equiv \frac{v}{c_0}, \quad \hat{T} \equiv \frac{T}{T_0},\\
&\hat{\sigma} \equiv \frac{\sigma}{\rho_0 \frac{k_B}{m} T_0}, \quad \hat{\Pi} \equiv \frac{\Pi}{\rho_0 \frac{k_B}{m} T_0}, \quad \hat{q} \equiv \frac{q}{\rho_0 \frac{k_B}{m} T_0 c_0},\\
&\hat{x} \equiv \frac{x}{\tau_\Pi(\rho_0, T_0) c_0}, \quad \hat{t} \equiv \frac{t}{\tau_\Pi(\rho_0, T_0)},\\
&\hat{\tau}_\Pi \equiv \frac{\tau_\Pi(\rho, T)}{\tau_\Pi(\rho_0, T_0)}, \quad \hat{\tau}_S \equiv \frac{\tau_S(\rho, T)}{\tau_\Pi(\rho_0, T_0)}, \quad \hat{\tau}_q \equiv \frac{\tau_q(\rho, T)}{\tau_\Pi(\rho_0, T_0)},
\end{aligned}
\tag{8.4}
$$

where the quantities with subscript 0 represent the quantities in the unperturbed state. The balance equations (8.2) are now rewritten in terms of the dimensionless quantities as follows:

$$
\begin{aligned}
&\frac{\mathrm{d}}{\mathrm{d}\hat{x}}(\hat{\rho}\hat{v}) = 0,\\
&\frac{\mathrm{d}}{\mathrm{d}\hat{x}}\left\{\frac{1}{\gamma_0}\left(\hat{\rho}\hat{T} + \hat{\Pi} - \hat{\sigma}\right) + \hat{\rho}\hat{v}^2\right\} = 0,\\
&\frac{\mathrm{d}}{\mathrm{d}\hat{x}}\left\{\frac{2}{\gamma_0}\left(\hat{\rho}\hat{v}\frac{1}{T_0}\int_{T_R}^{T}\hat{c}_v(\xi)\,d\xi + (\hat{\rho}\hat{T} + \hat{\Pi} - \hat{\sigma})\hat{v} + \hat{q}\right) + \hat{\rho}\hat{v}^3\right\} = 0,\\
&\frac{\mathrm{d}}{\mathrm{d}\hat{x}}\left\{\frac{1}{\gamma_0}\left((5\hat{\rho}\hat{T} + 5\hat{\Pi} - 2\hat{\sigma})\hat{v} + \frac{5}{\hat{c}_v(T)+1}\hat{q}\right) + \hat{\rho}\hat{v}^3\right\} = -\frac{3}{\gamma_0}\frac{\hat{\Pi}}{\hat{\tau}_\Pi},\\
&\frac{\mathrm{d}}{\mathrm{d}\hat{x}}\left\{\frac{3}{\gamma_0}\left((\hat{\rho}\hat{T} + \hat{\Pi} - \hat{\sigma})\hat{v} + \frac{1}{\hat{c}_v(T)+1}\hat{q}\right) + \hat{\rho}\hat{v}^3\right\} = \frac{1}{\gamma_0}\left(\frac{\hat{\sigma}}{\hat{\tau}_S} - \frac{\hat{\Pi}}{\hat{\tau}_\Pi}\right),\\
&\frac{\mathrm{d}}{\mathrm{d}\hat{x}}\left\{\frac{1}{\gamma_0}\left\{\left(\frac{1}{\gamma_0}\left(\hat{\rho}\hat{T} + \hat{\Pi} - \sigma\right) + \hat{\rho}\hat{v}^2\right)\frac{2}{T_0}\int_{T_R}^{T}\hat{c}_v(\xi)\,d\xi +\right.\right.\\
&\quad\left.\left. + \frac{2}{\gamma_0}\hat{T}\left(\hat{\rho}\hat{T} + 2\hat{\Pi} - 2\hat{\sigma}\right) + 5\left(\hat{\rho}\hat{T} + \hat{\Pi} - \hat{\sigma}\right)\hat{v}^2 + \frac{4\hat{c}_v(T)+10}{\hat{c}_v(T)+1}\hat{q}\hat{v}\right\} + \hat{\rho}\hat{v}^4\right\}\\
&\quad = -\frac{2}{\gamma_0}\left\{\frac{\hat{q}}{\hat{\tau}_q} + \left(\frac{\hat{\Pi}}{\hat{\tau}_\Pi} - \frac{\hat{\sigma}}{\hat{\tau}_S}\right)\hat{v}\right\},
\end{aligned}
\tag{8.5}
$$

where $\gamma_0 \equiv \gamma(T_0)$. As the conservation laws $(8.5)_{1-3}$ can be easily integrated, we can simplify the balance equations as follows:

$$
\begin{aligned}
&\hat{\rho} = \frac{M_0}{\hat{v}},\\
&\frac{1}{\gamma_0}\left(\frac{M_0\hat{T}}{\hat{v}} + \hat{\Pi} - \hat{\sigma}\right) + M_0\hat{v} = \frac{1}{\gamma_0} + M_0^2,\\
&\frac{2}{\gamma_0}\left(\frac{M_0}{T_0}\int_{T_0}^{T}\hat{c}_v(\xi)\,d\xi + M_0\hat{T} + \hat{\Pi}\hat{v} - \hat{\sigma}\hat{v} + \hat{q}\right) + M_0\hat{v}^2 = \frac{2}{\gamma_0}M_0 + M_0^3,\\
&\frac{\mathrm{d}}{\mathrm{d}\hat{x}}\left\{\frac{1}{\gamma_0}\left(5M_0\hat{T} + 5\hat{\Pi}\hat{v} - 2\hat{\sigma}\hat{v} + \frac{5}{\hat{c}_v(T)+1}\hat{q}\right) + M_0\hat{v}^2\right\} = -\frac{3}{\gamma_0}\frac{\hat{\Pi}}{\hat{\tau}_\Pi},\\
&\frac{\mathrm{d}}{\mathrm{d}\hat{x}}\left\{\frac{3}{\gamma_0}\left(\left(M_0\hat{T} + \hat{\Pi}\hat{v} - \hat{\sigma}\hat{v}\right) + \frac{1}{\hat{c}_v(T)+1}\hat{q}\right) + M_0\hat{v}^2\right\} = \frac{1}{\gamma_0}\left(\frac{\hat{\sigma}}{\hat{\tau}_S} - \frac{\hat{\Pi}}{\hat{\tau}_\Pi}\right),
\end{aligned}
$$

$$\frac{\mathrm{d}}{\mathrm{d}\hat{x}}\left\{\frac{1}{\gamma_0}\left\{\left(\frac{1}{\gamma_0}+M_0^2\right)\frac{2}{T_0}\int_{T_R}^{T}\hat{c}_v(\xi)\,d\xi+\frac{2}{\gamma_0}\hat{T}\left(\frac{M_0\hat{T}}{\hat{v}}+2\hat{\Pi}-2\hat{\sigma}\right)+\right.\right.$$
$$\left.\left.+5\left(M_0\hat{T}+\hat{\Pi}\hat{v}-\hat{\sigma}\hat{v}\right)\hat{v}+\frac{4\hat{c}_v(T)+10}{\hat{c}_v(T)+1}\hat{q}\hat{v}\right\}+M_0\hat{v}^3\right\} \tag{8.6}$$
$$=-\frac{2}{\gamma_0}\left\{\frac{\hat{q}}{\hat{\tau}_q}+\left(\frac{\hat{\Pi}}{\hat{\tau}_\Pi}-\frac{\hat{\sigma}}{\hat{\tau}_S}\right)\hat{v}\right\},$$

where M_0 represents the Mach number in the unperturbed state expressed by

$$M_0 \equiv \frac{v_0}{c_0}. \tag{8.7}$$

8.3.2 Boundary Conditions: Rankine-Hugoniot Conditions for the System of the Euler Equations

The boundary conditions for the basic system of equations expressed above are determined as follows: Inserting $\hat{\Pi}=0$, $\hat{\sigma}=0$ and $\hat{q}=0$ into $(8.6)_{1-3}$, we obtain the expressions for the quantities in the perturbed state:

$$\begin{aligned}
&\hat{\rho}_1=\frac{M_0}{\hat{v}_1},\\
&\frac{1}{\gamma_0}\left(\frac{M_0\hat{T}_1}{\hat{v}_1}-1\right)+M_0\left(\hat{v}_1-M_0\right)=0,\\
&\frac{2}{\gamma_0}M_0\left(\hat{T}_1+\frac{1}{T_0}\int_{T_0}^{T_1}\hat{c}_v(\xi)\,d\xi-1\right)+M_0\left(\hat{v}_1^2-M_0^2\right)=0,
\end{aligned} \tag{8.8}$$

where the quantities with subscript 1 are those in the perturbed state. These relations express the Rankine-Hugoniot (RH) conditions for the system of the Euler equations.

8.3.3 Parameters

In order to compare the theoretical predictions with experimental data, we will focus our study on the experimental data for the shock wave structure in a rarefied CO_2 gas at $T = 295$ K and $p = 69$ mmHg in the unperturbed state [15].

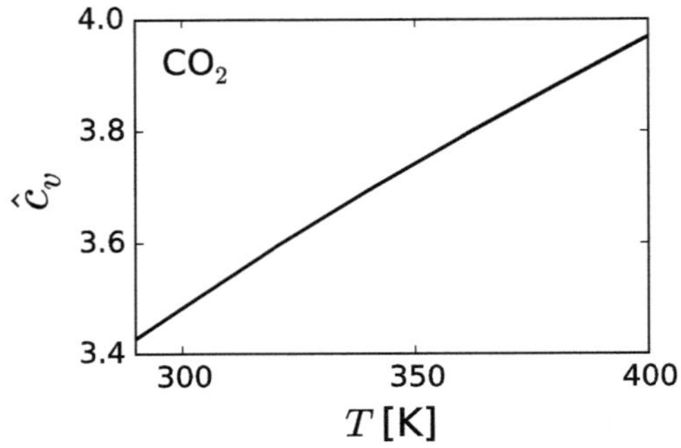

Fig. 8.1 Dependence of the dimensionless specific heat of a rarefied CO_2 gas on the temperature [16]

Table 8.1 Experimental values of the dimensionless specific heat $\hat{c}_v$ [16], sound velocity c_0 [16], heat conductivity κ [17] and shear viscosity μ [17] of a rarefied CO_2 gas at $T = 295\,\mathrm{K}$ and $p = 69\,\mathrm{mmHg}$

	$\hat{c}_v$	c_0 (m/s)	κ (W/(m K))	μ (Pa s)
CO_2	3.45	269	1.68×10^{-2}	1.5×10^{-5}

We have determined the dependence of the specific heat (8.1) on the temperature, which is shown in Fig. 8.1, by inserting the data on the temperature dependence of the sound velocity [16] into (7.9). The values of the dimensionless specific heat $\hat{c}_v$, sound velocity c_0, heat conductivity κ [17] and shear viscosity μ [17] in the unperturbed state are summarized in Table 8.1. We note that the rotational modes are completely excited and the vibrational modes are partially excited at this temperature. From (8.3), the maximum characteristic velocity at $T = 295\,\mathrm{K}$ is estimated as $\lambda_{max}/c \approx 1.74$. Therefore we recognize that, from the Theorem 3.1 (Sub-shock formation), the shock wave structure predicted by the present analysis is continuous up to $M_0 \approx 1.74$.

For a rarefied CO_2 gas, the temperature dependence of the phenomenological coefficients was already estimated by both the kinetic theoretical considerations and the experimental data [2] as follows:

$$\mu \propto T^n, \nu \propto T^n, \kappa \propto T^n \hat{c}_v(T), \tag{8.9}$$

where the exponent n was estimated as $n = 0.935$. We adopt the same temperature dependence in the present analysis. We have confirmed that the shock wave structure studied below depends weakly on the value of the exponent n.

Table 8.2 Relaxation times for a CO_2 gas at $T = 295$ K and $p = 69$ mmHg. The relaxation times τ_S and τ_q are obtained from the experimental data shown in Table 8.1. Only the relaxation time for the dynamic pressure τ_Π is remained as a fitting parameter

	$\tau_S(\rho_0, T_0)$ (s)	$\tau_q(\rho_0, T_0)$ (s)	$\tau_\Pi(\rho_0, T_0)$ (s)
CO_2	1.6×10^{-9}	2.2×10^{-9}	2.2×10^{-5}

By substituting (8.9) into the relations (5.68), we have the following dependence of the relaxation times on the mass density and the temperature:

$$\begin{aligned} \hat{\tau}_\Pi &= \frac{1}{\hat{\rho}\hat{T}^{1-n}} \frac{5-3\gamma_0}{5-3\gamma(T)}, \\ \hat{\tau}_S &= \frac{\tau_S(\rho_0, T_0)}{\tau_\Pi(\rho_0, T_0)} \frac{1}{\hat{\rho}\hat{T}^{1-n}}, \\ \hat{\tau}_q &= \frac{\tau_q(\rho_0, T_0)}{\tau_\Pi(\rho_0, T_0)} \frac{1}{\hat{\rho}\hat{T}^{1-n}} \frac{\gamma_0}{\gamma(T)}. \end{aligned} \tag{8.10}$$

Inserting the values of the phenomenological coefficients in Table 8.1 into the relations (5.68), we obtain the values of the relaxation times τ_S and τ_q for the shear stress and the heat flux in the unperturbed state as shown in Table 8.2. The remaining undetermined parameter is only the relaxation time τ_Π for the dynamic pressure, which is proportional to the bulk viscosity ν. Because of the lack of knowledge of reliable data on ν due to the difficulty in its experimental measurements, as was already done in Chap. 7, we will use τ_Π in the unperturbed state as a fitting parameter. As will be explained below, the value of τ_Π in the unperturbed state is determined by the comparison of the theoretical prediction with the experimental data for $M_0 = 1.47$. See also Figs. 8.5 and 8.6 below. It is noticeable that the value of τ_Π is larger, with different order of magnitude, than the other two relaxation times.

8.3.4 Numerical Methods

We solve numerically the system of balance equations (8.6) under the boundary conditions (8.8) by adopting the methods proposed by Weiss [5, 18].

We introduce the $N + 1$ grid points such that the range $[-\hat{L}/2, \hat{L}/2]$ in the $\hat{x}$-axis is discretized with constant intervals $\Delta\hat{x} = \hat{L}/N$ as follows:

$$\hat{x}^i = -\frac{\hat{L}}{2} + \frac{\hat{L}}{N} i \quad \text{for } i = 0, 1, \cdots, N, \tag{8.11}$$

where superscript i represents the number of the grid point.

Because the mass density $\hat{\rho}$ is already expressed by other variables in $(8.6)_1$, we need to solve the system $(8.6)_{2-6}$ for $\mathbf{u} = (\hat{v}, \hat{T}, \hat{\Pi}, \hat{\sigma}, \hat{q})$. The boundary conditions (8.8) give

$$\mathbf{u}^0 = \mathbf{u}_0, \qquad \mathbf{u}^N = \mathbf{u}_1, \tag{8.12}$$

where $\mathbf{u}^i$ represents $\mathbf{u}|_{x=x^i}$, $\mathbf{u}_0 = (M_0, 1, 0, 0, 0)$ and $\mathbf{u}_1 = (\hat{v}_1, \hat{T}_1, 0, 0, 0)$. For the conservation laws $(8.6)_{2,3}$ expressed as $\mathbf{F}(\mathbf{u}) = \mathbf{F}(\mathbf{u}_0)$ with $\mathbf{F}$ being the general flux, we have

$$\mathbf{F}(\mathbf{u}^i) = \mathbf{F}(\mathbf{u}_0) \quad \text{for } i = 1, 2, \cdots, N-1. \tag{8.13}$$

Replacing the differentiation in the balance equations $(8.6)_{4-6}$, which we express as $d\mathbf{F}(\mathbf{u})/d\hat{x} = \mathbf{P}(\mathbf{u})$ briefly with $\mathbf{P}$ being the general production, by the central difference, we get

$$\frac{\mathbf{F}(\mathbf{u}^{i+1}) - \mathbf{F}(\mathbf{u}^{i-1})}{2\Delta\hat{x}} = \mathbf{P}(\mathbf{u}^i) \quad \text{for } i = 1, 2, \cdots, N-1. \tag{8.14}$$

The nonlinear algebraic equations (8.13) and (8.14) with the condition (8.12) may be solved with the help of numerical solvers equipped with softwares for numerical computations. In the present analysis, we have constructed numerical codes by adopting the numerical solver implemented in the Mathematica based on the Newton's method. The computation starts from an appropriate initial guess, e.g.,

$$\mathbf{u}^i = \begin{cases} \mathbf{u}_0 & \text{for } i = 0, 1, \cdots, \frac{N}{2}, \\ \mathbf{u}_1 & \text{for } i = \frac{N}{2}+1, \frac{N}{2}+2, \cdots, N, \end{cases} \tag{8.15}$$

and the iterative calculations are repeated until the numerical solution converges to the one that satisfies the system (8.6) and the boundary conditions (8.8) within the appropriate accuracy we have set; eight digits of the precision in the present analysis. We have chosen $\Delta\hat{x}$ small enough and have confirmed that the dependence of the profiles in Figs. 8.2, 8.3, 8.4, 8.5, and 8.6 on $\Delta\hat{x}$ is negligibly small.

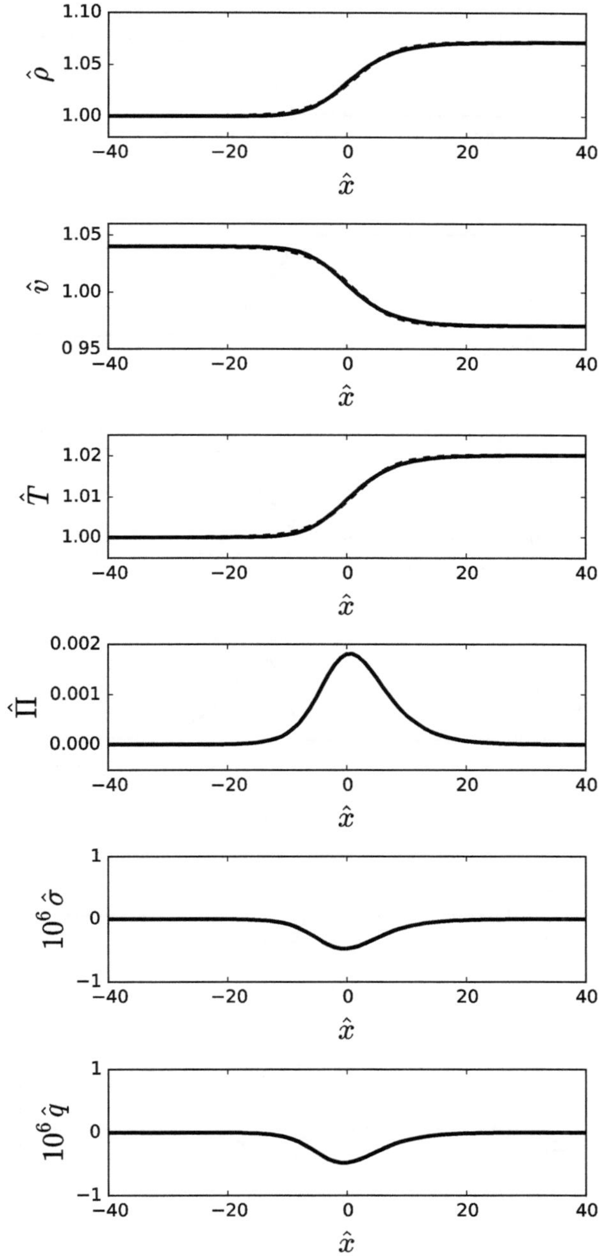

Fig. 8.2 Type A: Profiles of the dimensionless mass density, velocity, temperature, dynamic pressure, shear stress and heat flux predicted by the ET theory (*solid curves*). Profiles of the dimensionless mass density, the dimensionless velocity and the temperature predicted by the NSF theory (*dashed curves*) are also shown. $M_0 = 1.04$. The conditions for the numerical calculations are $\hat{L} = 100$ and $N = 100$

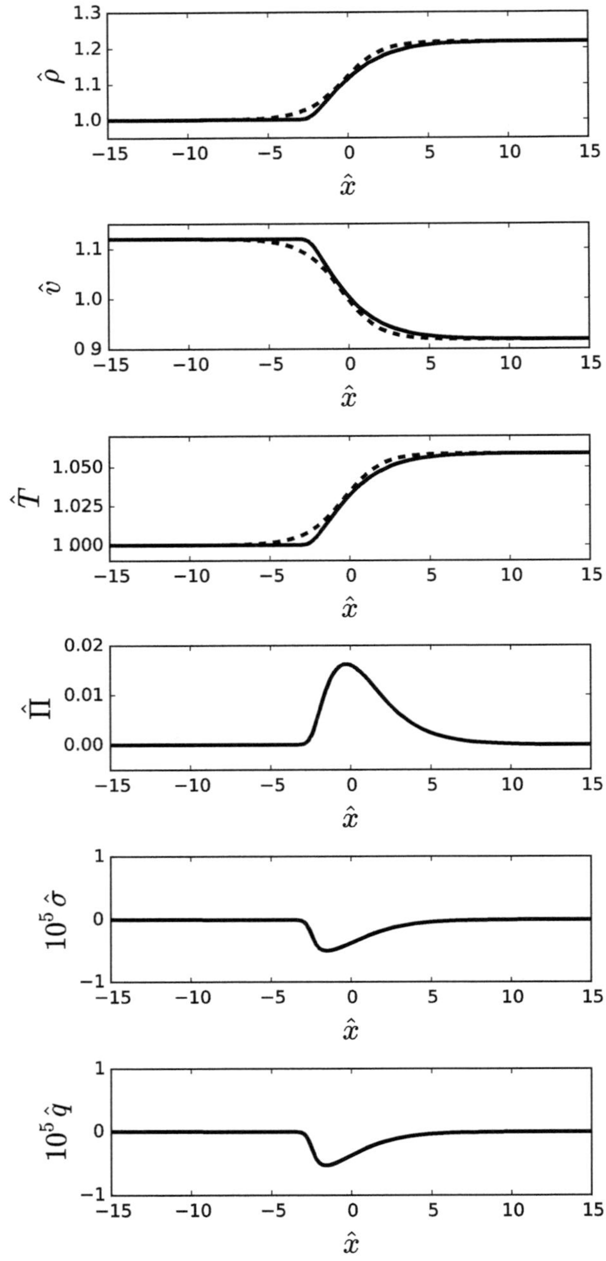

Fig. 8.3 Type B: Shock wave structure predicted by the ET theory (*solid curves*) and by the NSF theory (*dashed curves*). $M_0 = 1.12$. The numerical conditions are $\hat{L} = 50$, $N = 5000$ for the ET theory and $N = 100$ for the NSF theory

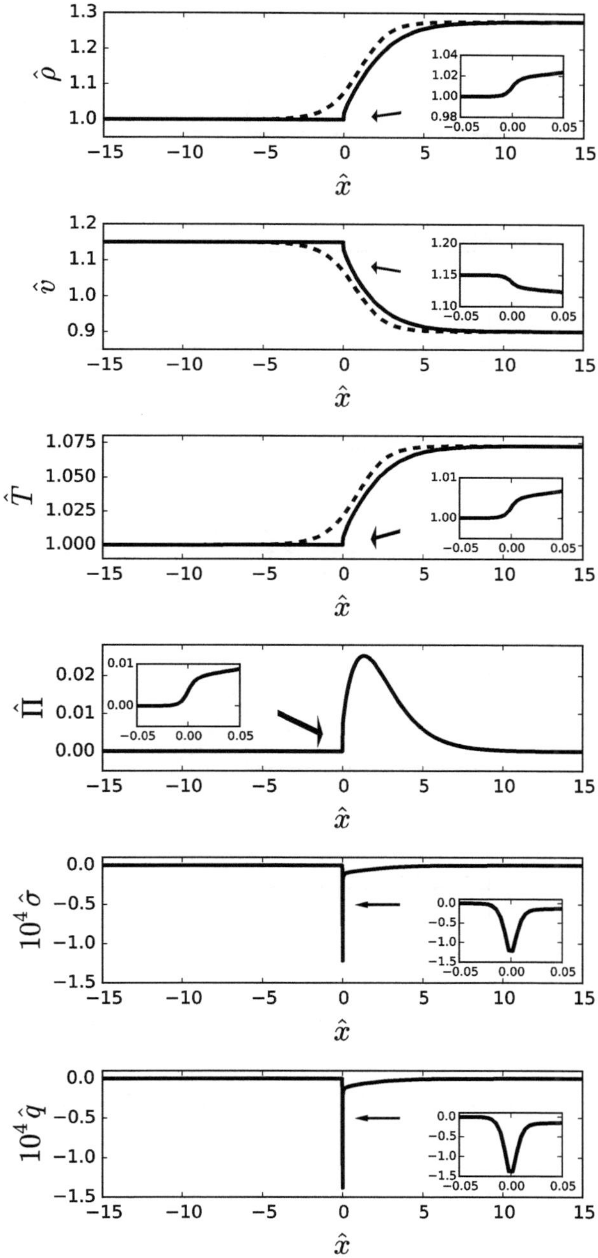

Fig. 8.4 Type C: Shock wave structure predicted by the ET theory (*solid curves*) and by the NSF theory (*dashed curves*). $M_0 = 1.15$. The numerical conditions are $\hat{L} = 40$, $N = 10{,}000$ for the ET theory and $N = 100$ for the NSF theory

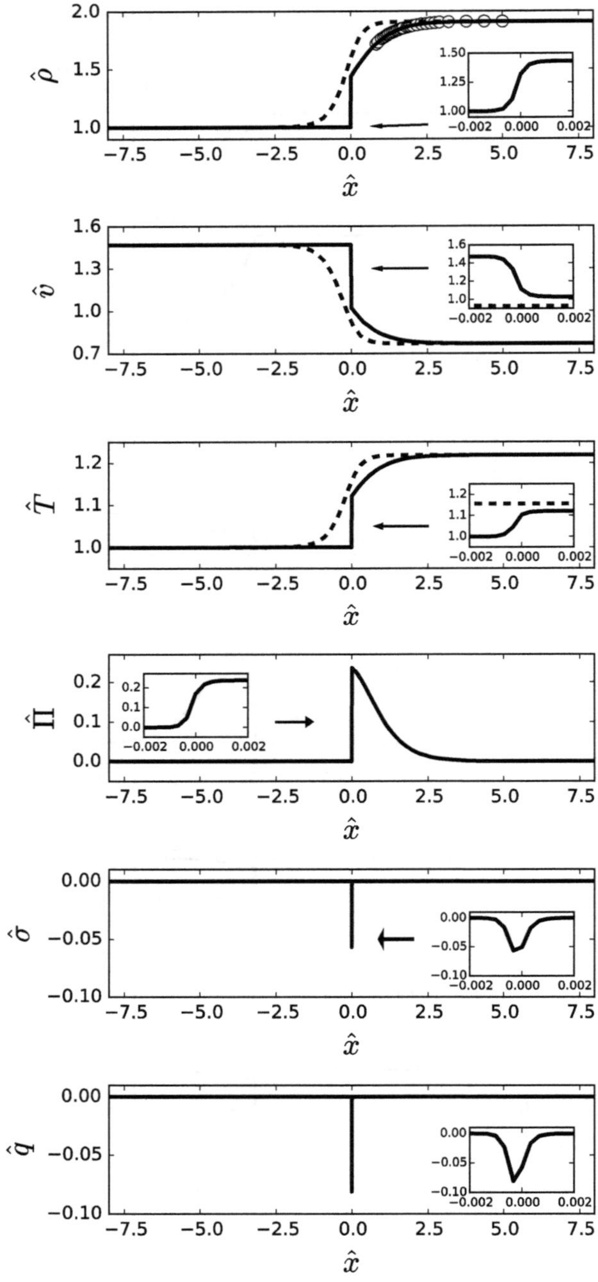

Fig. 8.5 Type C: Shock wave structure predicted by the ET theory (*solid curves*) and by the NSF theory (*dashed curves*). The experimental data [15] in the thick layer are also shown by *circles*. $M_0 = 1.47$. The numerical conditions are $\hat{L} = 20$, $N = 60{,}000$ for the ET theory and $N = 100$ for the NSF theory

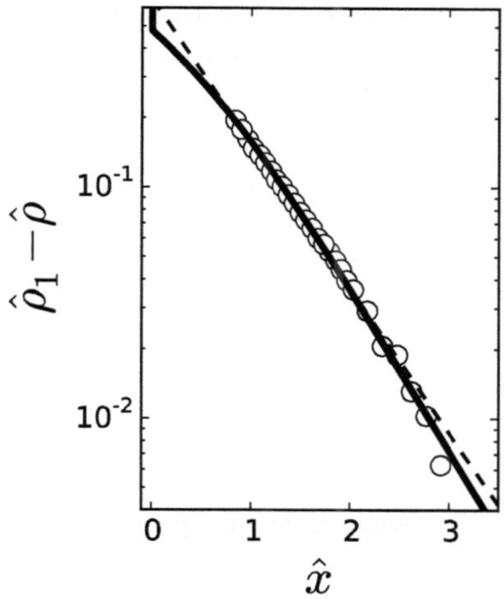

Fig. 8.6 Profile of the mass density difference $\hat{\rho}_1 - \hat{\rho}$ predicted by the ET theory (*solid curve*) and the experimental data (*circles*) [15]. The *dotted line* shows the exponential decay

8.4 Navier-Stokes Fourier Theory

As both results obtained by the ET14 theory and by the NSF theory are compared with each other in the next section, we here summarize also the NSF system of equations. It is obtained as the first approximation of the ET14 system (5.59) by using the Maxwellian iteration (see Sect. 5.3.6):

$$
\begin{aligned}
&\frac{\partial \rho}{\partial t} + \frac{\partial}{\partial x}(\rho v) = 0, \\
&\frac{\partial \rho v}{\partial t} + \frac{\partial}{\partial x}\left(p + \Pi - \sigma + \rho v^2\right) = 0, \\
&\frac{\partial}{\partial t}\left(2\rho\varepsilon + \rho v^2\right) + \frac{\partial}{\partial x}\left\{2\rho\varepsilon v + 2(p + \Pi - \sigma)v + \rho v^3 + 2q\right\} = 0, \\
&\Pi = -\left(\frac{2}{3} - \frac{1}{\hat{c}_v}\right) p\tau_\Pi \frac{\partial v}{\partial x}, \\
&\sigma = \frac{4}{3} p\tau_S \frac{\partial v}{\partial x}, \\
&q = -\left(1 + \hat{c}_v\right) \frac{k_B}{m} p\tau_q \frac{\partial T}{\partial x}.
\end{aligned}
\tag{8.16}
$$

From Eq. $(8.16)_{4-6}$, we obtain the relationship between relaxation times and the phenomenological coefficients (5.68). Note that the system (8.16) is of parabolic type although the original ET system (5.59) is hyperbolic.

The dimensionless form of the conservation laws are the same as $(8.6)_{1-3}$. While the dimensionless constitutive relations are expressed by

$$\begin{aligned} \hat{\Pi} &= -\left(\frac{2}{3} - \frac{1}{\hat{c}_v(T)}\right)\frac{M_0\hat{T}}{\hat{v}}\hat{\tau}_\Pi\frac{\mathrm{d}\hat{v}}{\mathrm{d}\hat{x}}, \\ \hat{\sigma} &= \frac{4}{3}\frac{M_0\hat{T}}{\hat{v}}\hat{\tau}_S\frac{\mathrm{d}\hat{v}}{\mathrm{d}\hat{x}}, \\ \hat{q} &= -\left(1+\hat{c}_v(T)\right)\frac{1}{\gamma_0}\frac{M_0\hat{T}}{\hat{v}}\hat{\tau}_q\frac{\mathrm{d}\hat{T}}{\mathrm{d}\hat{x}}. \end{aligned} \tag{8.17}$$

8.5 Shock Wave Structure

In this section, we show and emphasize that all of the three types of the shock wave structure, Types A, B, and C, can be described naturally within the ET14 theory.

8.5.1 *Type A: Nearly Symmetric Shock Wave Structure*

The nearly symmetric shock wave structure appears in a small Mach number region just above unity. The typical example of the shock wave structure of Type A is obtained at $M_0 = 1.04$ as shown in Fig. 8.2. We have depicted the profiles of all independent variables; the mass density, the velocity, the temperature, the dynamic pressure, the shear stress and the heat flux. We can confirm that the shock wave structure is indeed nearly the same as the one predicted by the NSF theory.

We notice that the thickness of a shock wave is very large even at several centimeters order because of the large characteristic length estimated as $\tau_\Pi(\rho_0, T_0)c_0 = 0.60\,\mathrm{cm}$. And the dimensionless dynamic pressure is also several orders larger than the dimensionless shear stress and heat flux. These features are, of course, due to the fact that the relaxation time for the dynamic pressure τ_Π, which is proportional to the bulk viscosity ν, is much larger than the other two relaxation times τ_S and τ_q that are, respectively, proportional to the shear viscosity μ and the heat conductivity κ.

Because of the large thickness and the small Mach number, i.e., small gradients of physical quantities, a shock wave is not so much far from local equilibrium. Therefore the predictions from the ET14 theory and the predictions from the NSF theory are similar to each other. It is this Type A that Gilbarg and Paolucci [2] studied.

8.5.2 Type B: Asymmetric Shock Wave Structure

When the Mach number increases further, the gradient of the physical quantities in the shock wave structure near the unperturbed state becomes much steeper than the gradient near the perturbed state. The shock wave structure now becomes evidently asymmetric. The NSF theory cannot describe such asymmetric profiles. Typical shock wave structure of Type B is shown in Fig. 8.3 where $M_0 = 1.12$.

From Fig. 8.3, we can see that the dimensionless shear stress and the heat flux are still several orders smaller than the dimensionless dynamic pressure in the whole range of the shock wave structure. Therefore we may conclude that the dynamic pressure plays much more important role in the global structure of a shock wave of Types A and B than the shear stress and the heat flux.

8.5.3 Type C: Shock Wave Structure Composed of Thin and Thick Layers

When the Mach number increases further more, the shock wave structure changes from a single-layer asymmetric structure (Type B) to a structure composed of thick and thin layers (Type C). Typical examples of Type C are shown in Fig. 8.4 with $M_0 = 1.15$ and Fig. 8.5 with $M_0 = 1.47$. It is this Type C that Bethe and Teller mainly studied. The NSF theory again cannot describe such shock wave structures with two layers.

We notice from Figs. 8.4 and 8.5 clearly that the thickness of the thin layer is finite although it is still much smaller than that of the thick layer, the thickness of which is at several centimeters order. Therefore we can analyze the detailed structure in the thin layer, which is impossible to be addressed by the Bethe-Teller theory. For example, as shown in Figs. 8.4 and 8.5, we understand the detailed profiles of the dissipative quantities in the thin layer.

We see that the shear stress and the heat flux are negligibly small everywhere except for the thin-layer region. On the other hand, the dynamic pressure is large in both the thick and thin layers. Therefore we may say that, in the thin layer with finite thickness, all dissipative quantities together play a crucial role, while, in the thick layer, only the dynamic pressure seems to be essential.

Within the present theory, as is pointed out above, the continuous shock wave structure is obtained until $M_0 \approx 1.74$. If we want to study the shock wave structure at larger Mach numbers than 1.74, we need the ET theory with more independent variables, which will be discussed in Chap. 10.

8.5.4 Critical Mach Numbers for the Transitions Between the Types A-B and B-C

We have estimated numerically the critical Mach numbers for the transition of the type of the shock wave structure: The asymmetric character of Type B becomes evident when the Mach number is around 1.08. And, for the transition between Type B and Type C, we have the critical Mach number $M_0 \approx 1.14$. Note that these values of the Mach number are merely rough indications because the boundary between two different types cannot be clearly defined.

In the Bethe-Teller theory, from the stability analysis of the discontinuous part of the shock wave structure [nowadays known as the Lax condition (see Sect. 3.3.2.1)], the critical Mach number between Type B and Type C was estimated as $M_0 \approx 1.04$. This value does not agree with the value mentioned above, but is not so far from it.

8.5.5 Reexamination of the Bethe-Teller Theory

Let us summarize the features of the Bethe-Teller theory in the light of the ET14 theory.

(A) As explained above, the Bethe-Teller theory describes the shock wave structure of Type C by adopting the two systems of equations under the assumption that the internal degrees of freedom of a molecule can be divided into two parts, that is, "inert" part and "active" part. One system is applied to analyze a thin layer and the other system to a thick layer. The compatibility of the two system of equations is, however, not self-evident. In the ET14 theory, on the other hand, a single system of equations can describe all Types A, B and C without any ambiguity. There is no compatibility problem.

(B) The thin layer is a jump discontinuity with zero thickness. While, in the ET14 theory, the thin layer has a structure with finite thickness.

(C) The thick layer is described essentially by a relaxation equation with a finite relaxation time. If necessary, the theory may be generalized so as to have several relaxation equations with different relaxation times. Usually the relaxation equation is assumed to be linear. The ET theory includes, in a natural way, the relaxation mechanism of the internal degrees of freedom. In this respect, see also Chap. 11.

(D) The critical Mach number between Type B and Type C can be estimated by the stability analysis. Its predicted values by the Bethe-Teller theory and by the ET theory are not far from each other.

(E) There is a qualitative difference between the Bethe-Teller theory and the ET theory in the temperature profile. The temperature just after the discontinuous jump derived from the Bethe-Teller theory may have the possibility to be larger than the temperature in the perturbed state (i.e., so-called temperature overshoot), while the temperature profile derived from the ET theory is always smaller than the perturbed temperature.

Experiments to observe the temperature profile, however, seem to be extremely difficult because a shock wave has a very steep and rapid change in space and time. There is another difficulty from a theoretical point of view. We should be careful about the definition of the temperature in nonequilibrium. We will discuss this point later in Sect. 13.5 and Chap. 15.

8.6 Comparison with Experimental Data

The experiments of the shock wave structure in a CO_2 gas at the room temperature and the atmospheric pressure indicate that the shock profiles with no thin layer are obtained at least in the range $1 < M_0 < 1.04$ [19]. The present result is consistent with this.

The experimental results at $M_0 = 1.134$ and $M_0 = 1.16$ are available [20, 21]. We have confirmed that our theoretical predictions shown in Fig. 8.4 are qualitatively the same as the shock profiles obtained by the experiments. Quantitative comparison is, however, impossible because only the interferograms are shown in the papers.

The experimental data on the mass density profile [15] and the theoretical mass density profile derived from the ET14 theory at $M_0 = 1.47$ are shown in Fig. 8.5. Note that only the experimental data in the thick layer are reported in the paper, in which the authors said that the accurate measurement in the region near the thin layer was impossible because the change of physical quantities is so steep. In order to study in more detail, Fig. 8.5 is shown in a different way: the single logarithmic plot of the profile of the mass density difference $\hat{\rho}_1 - \hat{\rho}$ as shown in Fig. 8.6. We can see that the agreement between the theoretical prediction and the experimental data is excellent. It is also remarkable that the ET theory seems to explain the deviation of the experimental data $\hat{\rho}_1 - \hat{\rho}$ from the the dotted line in Fig. 8.6, i.e., from the purely exponential decay.

Unfortunately, only the experimental data of the mass density profile at $M_0 = 1.47$ are available at present. More detailed experimental studies of the shock wave structure are highly expected.

8.7 Conclusion

We have shown that the ET14 theory can describe the shock wave structure in a polyatomic gas excellently. We have found the fact that the dynamic pressure Π is essentially important in the shock wave structure but the shear stress and the heat flux are not so important everywhere except for the inside of a thin layer. Therefore it is natural to expect that, by neglecting all dissipative fluxes but the dynamic pressure, we can study the shock wave structure properly. In fact, in Chap. 13, we will show that the ET theory with 6 independent fields (ET6) can describe the shock wave structure of Types A to C reasonably well.

References

1. H.A. Bethe, E. Teller, Deviations from thermal equilibrium in shock waves, Reprinted by Engineering Research Institute, University of Michigan, Ann Arbor
2. D. Gilbarg, D. Paolucci, The structure of shock waves in the continuum theory of fluids. J. Ration. Mech. Anal. **2**, 617 (1953)
3. S. Chapman, T.G. Cowling, *The Mathematical Theory of Non-uniform Gases* (Cambridge University Press, Cambridge, 1991)
4. H. Grad, On the kinetic theory of rarefied gases. Commun. Pure Appl. Math. **2**(4), 331 (1949)
5. I. Müller, T. Ruggeri, *Rational Extended Thermodynamics*, 2nd edn. (Springer, New York, 1998)
6. I. Müller, T. Ruggeri, *Extended Thermodynamics*. Springer Tracts in Natural Philosophy, vol. 37, I edn. (Springer, New York, 1993)
7. W. Dreyer, Maximization of the entropy in non-equilibrium. J. Phys. A Math. Gen. **20**, 6505 (1987)
8. H. Struchtrup, *Macroscopic Transport Equations for Rarefied Gas Flow, Approximation Methods in Kinetic Theory* (Springer, Berlin/Heidelberg, 2005)
9. G. Bird, *Molecular Gas Dynamics and the Direct Simulation of Gas Flows* (Clarendon Press, Oxford, 1994)
10. V.M. Zhdanov, The kinetic theory of a polyatomic gas. Soviet Phys. JETP **26**, 1187 (1968)
11. F.J. MacCormack, Kinetic equations for polyatomic gases: the 17-moment approximation. Phys. Fluids **11**, 2533 (1968)
12. F.J. MacCormack, Kinetic moment equations for a gas of polyatomic molecules with many internal degrees of freedom. Phys. Fluids **13**, 1446 (1970)
13. F. Mallinger, Generalization of the Grad theory to polyatomic gases. INRIA Research Report (1998), p. 3581
14. S. Taniguchi, T. Arima, T. Ruggeri, M. Sugiyama, Thermodynamic theory of the shock wave structure in a rarefied polyatomic gas: beyond the Bethe-Teller theory. Phys. Rev. E **89**, 013025 (2014)
15. N.H. Johannesen, H.K. Zienkiewicz, P.A. Blythe, J.H. Gerrard, Experimental and theoretical analysis of vibrational relaxation regions in carbon dioxide. J. Fluid Mech. **13**, 213 (1962)
16. Japan Society of Mechanical Engineers, *JSME Data Book, Thermophysical Properties of Fluids* (Japan Society of Mechanical Engineers, Tokyo, 1983)
17. W.M. Haynes, D.R. Lide (eds.), *CRC Handbook of Chemistry and Physics*, 91st edn. (CRC Press, Boca Raton, 2010)
18. W. Weiss, Continuous shock structure in extended thermodynamics. Phys. Rev. E **52**, R5760 (1995)
19. W.C. Griffith, A. Kenny, On fully-dispersed shock waves in carbon dioxide. J. Fluid Mech. **3**, 286 (1957)
20. W.C. Griffith, W. Bleakney, Shock waves in gases. Am. J. Phys. **22**, 597 (1954)
21. W. Griffith, D. Brickl, V. Blackman, Structure of shock waves in polyatomic gases. Phys. Rev. **102**, 1209 (1956)

Chapter 9
Light Scattering, Heat Conduction, and Fluctuation

Abstract In this chapter, we discuss briefly some other interesting applications of the ET14 theory: light scattering, stationary heat conduction, and fluctuating hydrodynamics. However, since these studies have just begun recently and is still quite primitive, this chapter should be read as an outline of these promising research fields.

9.1 Light Scattering

Experiments of light scattering in a gas afford us with precise information about irreversible processes in a gas even out of local equilibrium. These are a good test for checking the validity of a nonequilibrium thermodynamic theory.

In this section, light scattering in a rarefied polyatomic gas is studied by the ET14 theory [1]. The results obtained are compared with the results derived from the Navier-Stokes Fourier theory and also with some experimental data.

9.1.1 Introduction

Light scattering occurs due to the fluctuations in the mass density ρ through the dielectric constant $\epsilon(\rho)$. The intensity of scattered light is directly related to the dynamic structure factor [2]:

$$S(\boldsymbol{q},\omega) = \frac{1}{\pi}\left(\frac{\partial \epsilon}{\partial \rho}\right)^2 \Re\langle \delta\rho^*(\boldsymbol{q},0)\delta\hat{\rho}(\boldsymbol{q},s)\rangle_{s=i\omega}, \tag{9.1}$$

where $\boldsymbol{q}$ is the scattering vector, magnitude of which is $|\boldsymbol{q}| = (4\pi/\Lambda)\sin(\theta/2)$ with Λ and θ being the wavelength of the incident light and the scattering angle, ω is the shift in angular frequency, and $\langle \delta\rho^*(\boldsymbol{q},0)\delta\hat{\rho}(\boldsymbol{q},s)\rangle$ is the Laplace transform of the autocorrelation of the density fluctuations $\langle \delta\rho^*(\boldsymbol{q},0)\delta\rho(\boldsymbol{q},t)\rangle$ where $\langle\ \rangle$ denotes the thermal average and $\delta\rho(\boldsymbol{q},t)$ is the Fourier transform of the mass density fluctuation $\delta\rho(\boldsymbol{x},t)$.

T. Ruggeri, M. Sugiyama, *Rational Extended Thermodynamics beyond the Monatomic Gas*, DOI 10.1007/978-3-319-13341-6_9

Usually the dynamic structure factor has been studied theoretically by means of two approaches: the hydrodynamic approach [3–5] based on the Navier-Stokes-Fourier (NSF) theory, and the kinetic theory of gases based on a special model [6–12].

The analyses of light scattering in monatomic gases on the basis of the RET theory were also made in detail [2, 13]. The RET theory is shown to be a better theory to describe light scattering than the NSF theory. The analysis of the light scattering in polyatomic gases by using the RET theory is, however, not well developed until now [1].

9.1.2 Basic Equations

9.1.2.1 ET14 Theory

The basic equations for the present analysis are given by the linearized system of equation (7.1) with the thermal and caloric equations of state given by (5.63). This linear system can be decomposed into two uncoupled systems with independent variables:

system-L $\rho,\ \psi = \frac{\partial v_i}{\partial x_i},\ T,\ \sigma = \frac{\partial^2 \sigma_{\langle ij \rangle}}{\partial x_i \partial x_j},\ \Pi,\ Q = \frac{\partial q_i}{\partial x_i},$

system-T rot $\boldsymbol{v},\ \epsilon_{ijk} \frac{\partial^2 \sigma_{\langle jn \rangle}}{\partial x_k \partial x_n},$ rot $\boldsymbol{q},$

where ϵ_{ijk} is the Levi-Civita symbol. Here and hereafter, we denote the fluctuations $\delta\rho$ and δT simply as ρ and T.

For the present purpose, we adopt the system-L because only this system affects the density fluctuations. The system-L is expressed as follows:

$$
\begin{aligned}
&\frac{\partial \rho}{\partial t} + \rho_0 \psi = 0, \\
&\frac{\partial \psi}{\partial t} + \frac{p_0}{\rho_0^2} \Delta\rho + \frac{k_B}{m} \Delta T - \frac{1}{\rho_0}\sigma + \frac{1}{\rho_0}\Delta\Pi = 0, \\
&\frac{\partial T}{\partial t} + \frac{T_0}{\hat{c}_v}\psi + \frac{T_0}{p_0 \hat{c}_v} Q = 0, \\
&\frac{\partial \Pi}{\partial t} + \left(\frac{2}{3} - \frac{1}{\hat{c}_v}\right) p_0 \psi + \frac{2\hat{c}_v - 3}{3\hat{c}_v(1+\hat{c}_v)} Q = -\frac{1}{\tau_\Pi}\Pi, \\
&\frac{\partial \sigma}{\partial t} - \frac{4}{3} p_0 \Delta\psi - \frac{4}{3(1+\hat{c}_v)} \Delta Q = -\frac{1}{\tau_S}\sigma, \\
&\frac{\partial Q}{\partial t} + (1+\hat{c}_v)\frac{k_B}{m} p_0 \Delta T - \frac{p_0}{\rho_0}\sigma + \frac{p_0}{\rho_0}\Delta\Pi = -\frac{1}{\tau_q} q_i.
\end{aligned}
\tag{9.2}
$$

The Fourier (space) and Laplace (time) transformed basic equations provide an algebraic system:

$$\begin{pmatrix} s & \rho_0 & 0 & 0 & 0 & 0 \\ -q^2\dfrac{p_0}{\rho_0^2} & s & -q^2\frac{k_B}{m} & -\dfrac{q^2}{\rho_0} & -\dfrac{1}{\rho_0} & 0 \\ 0 & \dfrac{T_0}{\hat{c}_v} & s & 0 & 0 & \dfrac{T_0}{p_0\hat{c}_v} \\ 0 & \left(\dfrac{2}{3}-\dfrac{1}{\hat{c}_v}\right)p_0 & 0 & s+\dfrac{1}{\tau_\Pi} & 0 & \dfrac{2\hat{c}_v-3}{3\hat{c}_v(1+\hat{c}_v)} \\ 0 & \dfrac{4}{3}q^2 p_0 & 0 & 0 & s+\dfrac{1}{\tau_S} & \dfrac{4q^2}{3(1+\hat{c}_v)} \\ 0 & 0 & -q^2(1+\hat{c}_v)\dfrac{k_B}{m}p_0 & -q^2\dfrac{p_0}{\rho_0} & -\dfrac{p_0}{\rho_0} & s+\dfrac{1}{\tau_q} \end{pmatrix}$$

$$\begin{pmatrix} \hat{\rho}(\boldsymbol{q},s) \\ \hat{\psi}(\boldsymbol{q},s) \\ \hat{T}(\boldsymbol{q},s) \\ \hat{\sigma}(\boldsymbol{q},s) \\ \hat{\Pi}(\boldsymbol{q},s) \\ \hat{Q}(\boldsymbol{q},s) \end{pmatrix} = \begin{pmatrix} \rho(\boldsymbol{q},0) \\ \psi(\boldsymbol{q},0) \\ T(\boldsymbol{q},0) \\ \sigma(\boldsymbol{q},0) \\ \Pi(\boldsymbol{q},0) \\ Q(\boldsymbol{q},0) \end{pmatrix}, \tag{9.3}$$

where $\hat{X}(\boldsymbol{q},s)$ denotes the Fourier and Laplace transform of a generic quantity $X(\boldsymbol{x},t)$, and $X(\boldsymbol{q},0)$ denotes the initial value of spatial Fourier transform of $X(\boldsymbol{x},t)$.

Denoting the coefficient matrix of (9.3) as $\boldsymbol{A}$, we obtain the density as

$$\begin{aligned} \hat{\rho}(\boldsymbol{q},s) =& (\boldsymbol{A}^{-1})_{11}\rho(\boldsymbol{q},0) + (\boldsymbol{A}^{-1})_{12}\psi(\boldsymbol{q},0) + (\boldsymbol{A}^{-1})_{13}T(\boldsymbol{q},0) \\ &+ (\boldsymbol{A}^{-1})_{14}\Pi(\boldsymbol{q},0) + (\boldsymbol{A}^{-1})_{15}\sigma(\boldsymbol{q},0) + (\boldsymbol{A}^{-1})_{16}Q(\boldsymbol{q},0). \end{aligned} \tag{9.4}$$

When we consider the adiabatic system with volume V_G which contains the scattering volume V, we obtain

$$\langle \rho^*(\boldsymbol{q},0)\hat{\rho}(\boldsymbol{q},s)\rangle = (\boldsymbol{A}^{-1})_{11}\rho^2 k_B T\kappa_T V Q_0, \tag{9.5}$$

where κ_T is the isothermal compressibility and the function Q_0 is defined as follows:

$$Q_0 = 1 - \frac{1}{VV_G}\left(\int_V e^{-iq\cdot x}\mathrm{d}x\right)^2. \tag{9.6}$$

It is useful to introduce the following dimensionless quantities:

$$x = \frac{\omega}{v_0 q}, \quad y = \frac{p_0}{\mu q v_0}, \tag{9.7}$$

where $v_0 = \sqrt{2k_B T_0/m}$. Usually the case of large y is referred to as the hydrodynamic region and the case of small y as the kinetic region. By using x and y, the relative intensity of $S(\boldsymbol{q},\omega)$ can be expressed as a function of x, y, $\hat{c}_v$, τ_{qs} and τ_{ps}, where $\tau_{qs} = \tau_q/\tau_S$ and $\tau_{ps} = \tau_\Pi/\tau_S$. In fact, the essential part of $S(\boldsymbol{q},\omega)$ is obtained as follows:

$$\frac{S(\boldsymbol{q},\omega)}{\rho^2 k_B T \kappa_T V Q_0 \frac{1}{\pi}\left(\frac{\partial\epsilon}{\partial\rho}\right)^2} = \Re\left(\frac{N}{D}\right),$$

where

$$\begin{aligned}
N(x,y) =& -12ix^5\hat{c}_v(\hat{c}_v+1)\tau_{\Pi S}\tau_{qS} - 12x^4 y\hat{c}_v(\hat{c}_v+1)(\tau_{\Pi S}\tau_{qS}+\tau_{\Pi S}+\tau_{qS})\\
&+x^3\left\{18i\hat{c}_v(\hat{c}_v+2)\tau_{\Pi S}\tau_{qS} + 12iy^2\hat{c}_v(\hat{c}_v+1)(\tau_{\Pi S}+\tau_{qS}+1)\right\}\\
&+x^2\left\{12y^3\hat{c}_v(\hat{c}_v+1)\right.\\
&\left.+y\left[2(\hat{c}_v+2)\tau_{qS}(5\hat{c}_v\tau_{\Pi S}+7\hat{c}_v+3)+12\hat{c}_v(\hat{c}_v+1)\tau_{\Pi S}\right]\right\}\\
&+x\left\{-2iy^2(\hat{c}_v+1)\left[3(\hat{c}_v+2)\tau_{qS}+2\hat{c}_v\tau_{\Pi S}+4\hat{c}_v+3\right]\right.\\
&\left.+3i(1-2\hat{c}_v)\hat{c}_v\tau_{\Pi S}\tau_{qS}\right\}\\
&-6y^3(\hat{c}_v+1) - y\hat{c}_v\tau_{qS}\left[(2\hat{c}_v-3)\tau_{\Pi S}+4\hat{c}_v\right],\\
D(x,y) =& 12x^6\hat{c}_v(\hat{c}_v+1)\tau_{\Pi S}\tau_{qS} - 12ix^5 y\hat{c}_v(\hat{c}_v+1)(\tau_{\Pi S}\tau_{qS}+\tau_{\Pi S}+\tau_{qS})\\
&+x^4\left\{-6\hat{c}_v(4\hat{c}_v+7)\tau_{\Pi S}\tau_{qS} - 12y^2\hat{c}_v(\hat{c}_v+1)(\tau_{\Pi S}+\tau_{qS}+1)\right\}\\
&+x^3\left\{12y^3 i\hat{c}_v(\hat{c}_v+1)\right.\\
&\left.+2iy\left[\tau_{qS}(\hat{c}_v(8\hat{c}_v+13)\tau_{\Pi S}+10\hat{c}_v(\hat{c}_v+2)+6)+9\hat{c}_v(\hat{c}_v+1)\tau_{\Pi S}\right]\right\}\\
&+x^2\left\{2y^2(\hat{c}_v+1)\left[6(\hat{c}_v+1)\tau_{qS}+5\hat{c}_v\tau_{\Pi S}+7\hat{c}_v+3\right]\right.\\
&\left.+9\hat{c}_v(\hat{c}_v+1)\tau_{\Pi S}\tau_{qS}\right\}\\
&+x\left\{-6iy^3(\hat{c}_v+1)^2 - iy(\hat{c}_v+1)\tau_{qS}(5\hat{c}_v\tau_{\Pi S}+7\hat{c}_v+3)\right\}\\
&-3y^2(\hat{c}_v+1)^2\tau_{qS}.
\end{aligned}$$

9.1.2.2 Navier-Stokes Fourier Theory

In the case of the NSF theory, the Fourier-Laplace transformed field equations are as follows:

$$\begin{pmatrix} s & \rho_0 & 0 \\ -\dfrac{k_B}{m}\dfrac{q^2 T_0}{\rho_0} & s + \dfrac{1}{\rho_0}\left(\dfrac{4\mu}{3} + \nu\right) q^2 & -\dfrac{k_B}{m} q^2 \\ 0 & \dfrac{T_0}{\hat{c}_v} & s + \dfrac{m}{k_B}\dfrac{\kappa q^2}{\hat{c}_v \rho_0} \end{pmatrix} \begin{pmatrix} \hat{\rho}(\boldsymbol{q}, s) \\ \hat{\psi}(\boldsymbol{q}, s) \\ \hat{T}(\boldsymbol{q}, s) \end{pmatrix} = \begin{pmatrix} \rho(\boldsymbol{q}, 0) \\ \psi(\boldsymbol{q}, 0) \\ T(\boldsymbol{q}, 0) \end{pmatrix}. \tag{9.8}$$

The dynamic structure factor can be derived from this equation in a similar way.

9.1.3 *Comparison with Experimental Data for CO_2*

For polyatomic gases (both rarefied and dense), there are many experiments of Rayleigh-Brillouin Scattering basing on the traditional (called spontaneous) [14–21] and new (called coherent) scattering techniques [22–24].

Here we compare the theoretical prediction of ET14 with the laser scattering experiment of Greytak and Benedek [14, 15] in a carbon dioxide gas. The experiments were achieved under an approximately fixed pressure $p_0(= \frac{k_B}{m}\rho_0 T_0) = 1$ [atm], and the parameter y is set as $y = 11.1$ (hydrodynamic region) and $y = 1.01$ (kinetic region) by changing the detecting angle. The wavelength of the laser is $\Lambda = 6328$ [Å]. As thermodynamic parameters, we use $\mu = 1.46 \times 10^{-5}$ [$\mathrm{kg\,m^{-1}\,s^{-1}}$] and $\kappa = 1.31$ [$\mathrm{W\,K^{-1}\,m^{-1}}$] which are adopted in [23] with Eucken relation, therefore $\tau_{qS} = 1.36$.

In the present comparison, the instrumental function $f_{inst}(\omega)$ is approximated by the Lorentzian with a full-width at half-maximum of about 28 MHz for the case of $y = 11.1$ and 210 MHz for the case of $y = 1.01$, and is convolved with the theoretical predictions:

$$S_{convolved}(\omega) = \int_{-\infty}^{\infty} S(q, \omega') f_{inst}(\omega' - \omega) \mathrm{d}\omega. \tag{9.9}$$

As indicated in [15, 16, 23, 24], since the frequencies in light scattering experiments are much larger than the characteristic frequency of the vibrational modes, the modes seem to freeze. Therefore, in contrast with the study of shock waves in Chap. 8, the effect of the relaxation time of the dynamic pressure (or the bulk viscosity) is not considerably larger than that of the shear stress and the heat flux. This consideration indicates that we need to adopt the specific heat without the vibrational modes, i.e., $\hat{c}_v = 2.5$. Here we adopt these assumptions as a working hypothesis, although further study of these is evidently necessary.

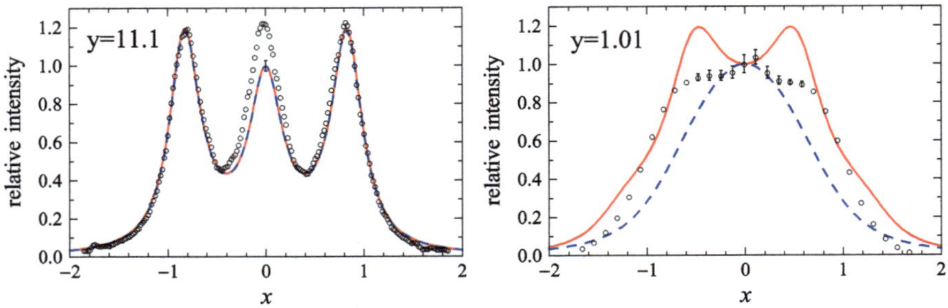

Fig. 9.1 Comparison of the theoretical prediction of relative intensity of the dynamic structure factor derived from ET (*solid line*) and NSF (*dashed line*) with the experimental data (*circle*) [14, 15]. *Left* and *right figures* show the cases with $y = 11.1$ and 1.01, respectively

In Fig. 9.1, we show the comparison of $S(\boldsymbol{q}, \omega)$ derived from ET14 and NSF with experimental data. The theoretical predictions are regularized to be 1 at $x = 0$.

In the hydrodynamic region, the experiment was made at scattering angle $\theta = 10.6$ [°] (therefore $y = 11.1$), $T = 297.9$ [K] and $p_0 = 770$ [mmHg]. In this region, the predictions by ET14 and NSF give almost no difference from each other and show good agreement with experimental data. The ratio of the relaxation times is estimated as $\tau_{\Pi S} = 0.8$ in such a way that the difference between the experiment and theories becomes minimum. Then the relaxation time of the dynamic pressure is 1.9×10^{-10} [s] and the bulk viscosity in this case is 5.2×10^{-6} [kg m^{-1} s^{-1}].

For the kinetic region, the experiment was made at scattering angle $\theta = 169.4$ [°] (therefore $y = 1.01$), $T = 298.1$ [K] and $p_0 = 750$ [mmHg]. The relaxation time of the dynamic pressure is same as the one estimated in hydrodynamic region. From Fig. 9.1, also for the kinetic region, the ET14 theory shows better agreement with experimental data than NSF except for small x. To study more details of the kinetic region, as is expected, the theory with more moments [25] (see Chap. 10) will be useful.

9.2 Heat Conduction

In this section, the effect of the dynamic pressure on stationary heat conduction in a rarefied polyatomic gas at rest confined in a bounded domain in planar or radial (cylindrical and spherical) geometry is studied by the ET14 theory [26]. The effect is observable only in a polyatomic gas because the dynamic pressure vanishes identically in a monatomic gas and is intrinsically related to the internal degrees of freedom of a polyatomic molecule. The effect in the case of a para-hydrogen gas (p-H_2) is explained briefly as a typical example. As discussed in Sect. 4.2.1, we know that, for monatomic gases, the ET and NSF theories carry mutually different results in such a heat transfer problem [27–32].

The present analysis is a typical example in the problem of a bounded domain discussed in Sect. 4.2.2.

9.2.1 *Basis of the Present Analysis*

The basic system of field equations is given by (5.59) and (5.65) with the thermal and caloric equations of state (5.63). We study the one-dimensional heat conduction problem in the planar, radially symmetric cylindrical and spherical cases. All the quantities will be described in the *physical components* [33].

9.2.1.1 Basic System of Equations

For convenience, we use the following dimensionless quantities:

$$\begin{aligned}
&\hat{x} = \frac{x}{L}, \quad \hat{c}_v' = T_0 \frac{\mathrm{d}\hat{c}_v}{\mathrm{d}T}, \quad \hat{T} = \frac{T}{T_0}, \quad \hat{p} = \frac{p}{p_0}, \\
&\hat{\sigma}_{\langle ij \rangle} = \frac{\sigma_{\langle ij \rangle}}{p_0}, \quad \hat{\Pi} = \frac{\Pi}{p_0}, \quad \hat{q} = \frac{q_1}{p_0\sqrt{k_B T_0/m}}, \\
&\hat{\tau}_S = \frac{\tau_S \sqrt{k_B T_0/m}}{L}, \hat{\tau}_\Pi = \frac{\tau_\Pi \sqrt{k_B T_0/m}}{L}, \hat{\tau}_q = \frac{\tau_q \sqrt{k_B T_0/m}}{L},
\end{aligned} \tag{9.10}$$

where x represents the position along the axis normal to the plates in the planar case or the radius in the radial geometry, L is the distance between the two boundaries, and the index 0 denotes the value at a reference state.

The basic system of equations can be rewritten as follows, in which the mass conservation law is identically satisfied:

$$\begin{aligned}
&\frac{\mathrm{d}}{\mathrm{d}\hat{x}}(\hat{p} + \hat{\Pi} - \hat{\sigma}_{\langle 11 \rangle}) - \frac{j(j+1)}{2\hat{x}}\hat{\sigma}_{\langle 11 \rangle} - \frac{j(j-2)}{\hat{x}}\hat{\sigma}_{\langle 22 \rangle} = 0, \\
&\frac{\mathrm{d}\hat{q}}{\mathrm{d}\hat{x}} + \frac{j}{\hat{x}}\hat{q} = 0, \\
&\frac{2}{1+\hat{c}_v}\frac{\mathrm{d}\hat{q}}{\mathrm{d}\hat{x}} - \frac{4}{3}\frac{\hat{c}_v'}{(1+\hat{c}_v)^2}\hat{q}\frac{\mathrm{d}\hat{T}}{\mathrm{d}\hat{x}} = \frac{\hat{\sigma}_{\langle 11 \rangle}}{\hat{\tau}_S}, \\
&\frac{2(1-\delta_{j0})}{1+\hat{c}_v}\frac{\hat{q}}{\hat{x}} + \frac{2}{3}\frac{\hat{c}_v'}{(1+\hat{c}_v)^2}\hat{q}\frac{\mathrm{d}\hat{T}}{\mathrm{d}\hat{x}} = \frac{\hat{\sigma}_{\langle 22 \rangle}}{\hat{\tau}_S}, \\
&\frac{5}{3}\frac{\hat{c}_v'}{(1+\hat{c}_v)^2}\hat{q}\frac{\mathrm{d}\hat{T}}{\mathrm{d}\hat{x}} = \frac{\hat{\Pi}}{\hat{\tau}_\Pi}, \\
&\left\{(1+\hat{c}_v)\hat{p} + (2+\hat{c}_v)(\hat{\Pi} - \hat{\sigma}_{\langle 11 \rangle})\right\}\frac{\mathrm{d}\hat{T}}{\mathrm{d}\hat{x}} - \hat{T}\frac{\mathrm{d}\hat{p}}{\mathrm{d}\hat{x}} = -\frac{\hat{q}}{\hat{\tau}_q},
\end{aligned} \tag{9.11}$$

where the index j is 0, 1 and 2, respectively, for the planar case, the cylindrical case and the spherical case, and δ denotes the Kronecker symbol.

9.2.1.2 Reduced Basic System of Equations

From the basic system (9.11), several relations are obtained:

$$\begin{aligned}\hat{q} &= \frac{Q_0}{\hat{x}^j},\\ \hat{\sigma}_{\langle 11\rangle} &= -\frac{2j}{1+\hat{c}_v}\frac{Q_0\hat{\tau}_S}{\hat{x}^{j+1}} - \frac{4}{5}\frac{\hat{\tau}_S}{\hat{\tau}_\Pi}\hat{\Pi},\\ \hat{\sigma}_{\langle 22\rangle} &= \frac{j+\delta_{j1}}{1+\hat{c}_v}\frac{Q_0\hat{\tau}_S}{\hat{x}^{j+1}} + \frac{2}{5}\frac{\hat{\tau}_S}{\hat{\tau}_\Pi}\hat{\Pi},\end{aligned} \tag{9.12}$$

where Q_0 is a constant determined by the boundary condition. We note that, in the planar and spherical cases, $\sigma_{\langle 22\rangle} = \sigma_{\langle 33\rangle}$.

Hereafter, in order to set apart the effect of the dynamic pressure, we neglect the T dependence of the relaxation times assuming that these have constant values determined by the temperature range under consideration. Essential features in the present analysis are not changed by this assumption, which was also adopted in Sect. 4.2.1. Then, the following relation is obtained form $(9.11)_1$ and (9.12):

$$\hat{p} + \left(1 + \frac{4}{5}\frac{\hat{\tau}_S}{\hat{\tau}_\Pi}\right)\hat{\Pi} = P_0, \tag{9.13}$$

where P_0 is a constant determined by the boundary condition. The quantities $\hat{\Pi}$, $\hat{\sigma}_{\langle 11\rangle}$, $\hat{\sigma}_{\langle 22\rangle}$ and $\hat{q}$ are determined by the solution of the differential equations for $\hat{T}$ and $\hat{p}$:

$$\begin{aligned}(A-\hat{p})\frac{\mathrm{d}\hat{T}}{\mathrm{d}\hat{x}} - \hat{T}\frac{\mathrm{d}\hat{p}}{\mathrm{d}\hat{x}} &= -\frac{Q_0}{\hat{\tau}_q\hat{x}^j},\\ \left(1 + \frac{4}{5}\frac{\hat{\tau}_S}{\hat{\tau}_\Pi}\right)\hat{c}_v'\frac{\mathrm{d}\hat{T}}{\mathrm{d}\hat{x}} &= B\left(P_0 - \hat{p}\right),\end{aligned} \tag{9.14}$$

where

$$\begin{aligned}A &= (2+\hat{c}_v)\left(P_0 + \frac{2j}{1+\hat{c}_v}\frac{Q_0\hat{\tau}_S}{\hat{x}^{j+1}}\right),\\ B &= \frac{3}{5}(1+\hat{c}_v)^2\frac{\hat{x}^j}{Q_0\hat{\tau}_\Pi}.\end{aligned} \tag{9.15}$$

9.2.1.3 Navier-Stokes Fourier Theory

For reference, we write down the NSF basic equation:

$$(1+\hat{c}_v)\hat{p}\frac{\mathrm{d}\hat{T}}{\mathrm{d}\hat{x}} = -\frac{\hat{q}}{\hat{\tau}_q} \tag{9.16}$$

with $\hat{p} = P_0$ and $\hat{q} = Q_0/\hat{x}^j$. The viscous stress $\sigma_{<ij>}$ and the dynamic pressure Π vanish identically.

9.2.2 *Boundary Conditions*

We assume, for simplicity, a permeable plate at $\hat{x} = 0$ and an impermeable one at $\hat{x} = 1$. At the permeable plate, we can impose both the density ρ_0 and the temperature T_0. While, at the impermeable plate, a fixed heat flux Q_0 is applied.

Although these boundary data are sufficient to determine completely the solution of the NSF system, we need one more condition for the ET system of equations. This is the so-called "non-controllable boundary data problem" and several approaches have been introduced to overcome it. See the discussions in Sect. 4.2.2.

In the present study we have determined it as follows: We tried to assign different values of $\sigma_{<11>}$ at one plate. We found that the corresponding solutions of $\sigma_{<11>}$ exhibit, in general, thin boundary layers near the plate. We then adopt, as an appropriate boundary value, the one for which no boundary layer is observed. A similar phenomenon was also observed in a different context [34]. We remark that different boundary values for $\sigma_{<11>}$ create minimally different solutions for T and Π, but create significant differences for q and for p.

9.2.3 *Effect of the Dynamic Pressure*

From $(9.11)_5$, we notice that

$$\hat{c}_v' = 0 \qquad \longrightarrow \qquad \hat{\Pi} = 0, \tag{9.17}$$

therefore the polyatomic effect ($\hat{\Pi} \neq 0$) is observable only in the case with $\hat{c}_v' \neq 0$, which is the case that we can not see in rarefied monatomic gases. From (9.13), the pressure $\hat{p}$ is no more constant but depends on $\hat{x}$.

Let us divide this case into the following two cases:

(a) Planar case: It is evident that, even in the planar case, the ET theory predicts the existence of the polyatomic effect.
(b) Cylindrical and spherical cases: From the numerical analysis [26], we see that the radial geometry enhances the polyatomic effect significantly.

Two remarks are made below:

Remark 9.1 Two quantities $\hat{c}_v'$ and $\hat{\Pi}$ have qualitatively different characters from each other: $\hat{c}_v'$ is, by definition, an equilibrium quantity, while $\hat{\Pi}$ is a nonequilibrium one. However, both quantities are intrinsically related to the internal degrees of freedom of a polyatomic molecule. In stationary heat conduction in a rarefied polyatomic gas, we have found that both quantities play an important role simultaneously.

Remark 9.2 In the NSF theory (9.16), the dynamic pressure Π vanishes identically even in the case with $\hat{c}_v' \neq 0$. From an extended thermodynamics point of view, this fact reveals an inconsistency in the NSF theory of a rarefied polyatomic gas.

Before closing this subsection, we discuss the case with $\hat{c}_v' = 0$ for the sake of completeness. When the specific heat $\hat{c}_v$ is constant, we have

$$\Pi = 0, \quad \hat{p} = P_0 \tag{9.18}$$

in all geometries. There is no polyatomic effect. Then, $(9.14)_2$ is identically satisfied, and the temperature profile is governed by

$$\frac{\mathrm{d}\hat{T}}{\mathrm{d}\hat{x}} = -\frac{Q_0}{(A - P_0)\hat{\tau}_q \hat{x}^j}. \tag{9.19}$$

We discuss this case further as follows:

(i) $\hat{c}_v = 3/2$: This case corresponds to a monatomic gas or a polyatomic gas in a low temperature range where no internal degrees of freedom of a molecule are excited. Equation (9.19) is exactly the same as that studied in Sect. 4.2.1, where it was shown that the temperature profiles in the planar case predicted by the ET theory and the NSF theory coincide with each other, and that unphysical singularities of the temperature on the axis of the cylinder and at the center of the sphere predicted by the NSF theory can be removed by the ET theory.
(ii) $\hat{c}_v = \textit{constant} > 3/2$: This case corresponds to a polyatomic gas with excited internal degrees of freedom. However, except for the value of $\hat{c}_v$, the analytical expression of the temperature profile is the same as that of the case (i) above.

9.2.4 An Example: Polyatomic Effect in a Para-Hydrogen Gas

By taking the importance of the dynamic pressure $\hat{\Pi}$ and the relationship (9.17) into consideration, the numerical study of the polyatomic effect in a para-hydrogen (p-H_2) gas was made as a typical example. See its temperature dependence of the specific heat in Fig. 7.2, where $\hat{c}_v' \neq 0$ except for one special point.

It is shown that, both in the planar case and in the radial case, the polyatomic effect due to the presence of the dynamic pressure Π appears. The presence of Π affects both the heat flux and the mass density, which may be detected by the experimental techniques described in [35]. The radial geometry enhances significantly the difference between the ET and NSF predictions although the difference of the temperature profile is still small. For details of the numerical results, see [26].

9.3 Fluctuating Hydrodynamics of ET14

9.3.1 Introduction

Landau and Lifshitz developed the theory of fluctuating hydrodynamics for viscous, heat-conducting fluids with constitutive equations of Navier-Stokes and Fourier type [36–38] basing on thermodynamics of irreversible processes (TIP). They introduced additional stochastic flux terms (generalized random forces) into the constitutive equations of the viscous stress and the heat flux by applying the fluctuation-dissipation theorem [39–41]. See also review articles on fluctuating hydrodynamics [42–44].

In recent years, the Landau-Lifshitz (LL) theory has been applied to, in particular, nano-technology [45, 46] and molecular biology [47, 48]. Numerical analyses of the fluctuations by using the theory have been made extensively [49–55]. The fluctuating-hydrodynamic approach can also contribute to the study of fluctuations in nonequilibrium states [44, 56, 57]. However, as TIP rests essentially on the local equilibrium assumption, it is highly probable that TIP may no longer be valid for highly nonequilibrium cases such as the cases where nanoflows are involved, or the cases where rarefied gases play a role.

The purpose of the present section is to summarize briefly the theory of fluctuating hydrodynamics based on ET through the study of the ET14 theory as a representative case. See also [58, 59] for fluctuating hydrodynamics based on ET with 13-field theory of rarefied monatomic gases.

9.3.2 Theory of Fluctuating Hydrodynamics Based on ET

The basic equations in the present study are the linearized equations of ET14 for a rarefied polyatomic gas (7.1) but in a polytropic case with $\hat{c}_v = D/2$.

Let us now try to introduce the random forces into ET. Following the general theory [60], we can have the expressions for the productions P_{jj}, $P_{\langle ij\rangle}$ and Q_i in terms of Π, $\sigma_{\langle ij\rangle}$, q_i and the Gaussian white random forces $\mathfrak{r}$, $\mathfrak{r}_{\langle ij\rangle}$, $\mathfrak{s}_i$ (see (5.38)):

$$P_{jj} = -\frac{3}{\tau_\Pi}\Pi + \mathfrak{r}, \tag{9.20}$$

$$P_{\langle ij\rangle} = \frac{1}{\tau_S}\sigma_{\langle ij\rangle} + \mathfrak{r}_{\langle ij\rangle},$$

$$Q_i = -\frac{2}{\tau_q}q_i + \mathfrak{s}_i.$$

The means of the random forces $\mathfrak{r}$, $\mathfrak{r}_{\langle ij\rangle}$ and $\mathfrak{s}_i$ vanish. And their correlations are given by

$$\langle \mathfrak{r}(\boldsymbol{x},t)\mathfrak{r}(\boldsymbol{x}',t')\rangle = k_B\frac{12\frac{D-3}{D}\frac{k_B}{m}\rho_0T_0^2}{\tau_\Pi}\delta(\boldsymbol{x}-\boldsymbol{x}')\delta(t-t'),$$

$$\langle \mathfrak{r}_{\langle ij\rangle}(\boldsymbol{x},t)\mathfrak{r}_{\langle mn\rangle}(\boldsymbol{x}',t')\rangle = k_B\frac{2\frac{k_B}{m}\rho_0T_0^2}{\tau_S}$$

$$\times(\delta_{im}\delta_{jn} + \delta_{in}\delta_{jm} - \frac{2}{3}\delta_{ij}\delta_{mn})\delta(\boldsymbol{x}-\boldsymbol{x}')\delta(t-t'), \tag{9.21}$$

$$\langle \mathfrak{s}_i(\boldsymbol{x},t)\mathfrak{s}_j(\boldsymbol{x}',t')\rangle = k_B\frac{4(D+2)\left(\frac{k_B}{m}\right)^2\rho_0T_0^3}{\tau_q}\delta_{ij}\delta(\boldsymbol{x}-\boldsymbol{x}')\delta(t-t'),$$

$$\langle \mathfrak{r}(\boldsymbol{x},t)\mathfrak{r}_{\langle ij\rangle}(\boldsymbol{x}',t')\rangle = \langle \mathfrak{r}(\boldsymbol{x},t)\mathfrak{s}_m(\boldsymbol{x}',t')\rangle = \langle \mathfrak{r}_{\langle ij\rangle}(\boldsymbol{x},t)\mathfrak{s}_m(\boldsymbol{x}',t')\rangle = 0,$$

where brackets $\langle\ \ \rangle$ in the left-hand side stand for the statistical average at the reference equilibrium state.

Field equations (7.1) with (9.20) and (9.21) constitute *the basic system of equations for fluctuating hydrodynamics based on ET14*. The relaxation times can be evaluated by experiments or kinetic-theoretical analyses.

9.3.3 Two Subsystems of the Stochastic Field Equations

The system of equations obtained above may be decomposed into two uncoupled subsystems, that is, the subsystem composed of longitudinal modes (System-L) and the subsystem of transverse modes (System-T). (See also Sect. 9.1.2.1.)

9.3.3.1 System-L

The relevant quantities of the system are given by

$$\begin{aligned}&\rho,\ T,\ \psi\left(\equiv\frac{\partial v_i}{\partial x_i}\right),\quad \Pi,\ \tau\left(\equiv\frac{\partial^2\sigma_{\langle ij\rangle}}{\partial x_i\partial x_j}\right),\ \varphi\left(\equiv\frac{\partial q_i}{\partial x_i}\right),\\ &\mathfrak{u}\left(\equiv\frac{1}{3}\mathfrak{r}\right),\quad \mathfrak{v}\left(\equiv-\frac{\partial^2\mathfrak{r}_{\langle ij\rangle}}{\partial x_i\partial x_j}\right),\quad \text{and}\ \mathfrak{w}\left(\equiv\frac{1}{2}\frac{\partial\mathfrak{s}_i}{\partial x_i}\right).\end{aligned} \tag{9.22}$$

The spatial Fourier transform of the system is the system of the rate-type differential equations in the space of the wave number $\boldsymbol{k}$ and time t ($\boldsymbol{k}t$-representation) as follows:

$$\begin{aligned}&\frac{\partial\rho(\boldsymbol{k},t)}{\partial t}+\rho_0\psi(\boldsymbol{k},t)=0,\\
&\frac{\partial\psi(\boldsymbol{k},t)}{\partial t}-\frac{\frac{k_B}{m}T_0}{\rho_0}k^2\rho(\boldsymbol{k},t)-\frac{k_B}{m}k^2T(\boldsymbol{k},t)-\frac{1}{\rho_0}k^2\Pi(\boldsymbol{k},t)-\frac{1}{\rho_0}\tau(\boldsymbol{k},t)=0,\\
&\frac{\partial T(\boldsymbol{k},t)}{\partial t}+\frac{2}{D}T_0\psi(\boldsymbol{k},t)+\frac{m}{k_B}\frac{2}{D\rho_0}\varphi(\boldsymbol{k},t)=0,\\
&\frac{\partial\Pi(\boldsymbol{k},t)}{\partial t}+\frac{4(D-3)}{3D(D+2)}\varphi(\boldsymbol{k},t)+\frac{2(D-3)}{3D}\frac{k_B}{m}\rho_0T_0\psi(\boldsymbol{k},t)=-\frac{1}{\tau_\Pi}\Pi(\boldsymbol{k},t)+\mathfrak{u}(\boldsymbol{k},t),\\
&\frac{\partial\tau(\boldsymbol{k},t)}{\partial t}+\frac{8}{3(D+2)}k^2\varphi(\boldsymbol{k},t)+\frac{4}{3}\frac{k_B}{m}\rho_0T_0k^2\psi(\boldsymbol{k},t)=-\frac{1}{\tau_S}\tau(\boldsymbol{k},t)+\mathfrak{v}(\boldsymbol{k},t),\\
&\frac{\partial\varphi(\boldsymbol{k},t)}{\partial t}-\frac{k_B}{m}T_0k^2\Pi(\boldsymbol{k},t)-\frac{k_B}{m}T_0\tau(\boldsymbol{k},t)-\frac{D+2}{2}\left(\frac{k_B}{m}\right)^2\rho_0T_0k^2T(\boldsymbol{k},t)\\
&\qquad=-\frac{1}{\tau_q}\varphi(\boldsymbol{k},t)+\mathfrak{w}(\boldsymbol{k},t),\end{aligned} \tag{9.23}$$

where $\rho(\boldsymbol{k},t)$ is the spatial Fourier transform of $\rho(\boldsymbol{x},t)$.

From Eq. (9.21), the quantities $\mathfrak{v}(\boldsymbol{k},t)$ and $\mathfrak{w}(\boldsymbol{k},t)$ are the Gaussian white random forces with null means and correlations:

$$\begin{aligned}&\langle\mathfrak{u}(\boldsymbol{k},t)\mathfrak{u}(\boldsymbol{k}',t')\rangle=k_B\frac{\frac{D-3}{D}\frac{k_B}{m}\rho_0T_0^2}{6\pi^3\tau_\Pi}\delta(\boldsymbol{k}+\boldsymbol{k}')\delta(t-t'),\\
&\langle\mathfrak{v}(\boldsymbol{k},t)\mathfrak{v}(\boldsymbol{k}',t')\rangle=k_B\frac{\frac{k_B}{m}\rho_0T_0^2}{3\pi^3\tau_S}k^4\delta(\boldsymbol{k}+\boldsymbol{k}')\delta(t-t'),\\
&\langle\mathfrak{w}(\boldsymbol{k},t)\mathfrak{w}(\boldsymbol{k}',t')\rangle=k_B\frac{(D+2)\left(\frac{k_B}{m}\right)^2\rho_0T_0^3}{8\pi^3\tau_q}k^2\delta(\boldsymbol{k}+\boldsymbol{k}')\delta(t-t'),\\
&\langle\mathfrak{u}(\boldsymbol{k},t)\mathfrak{v}(\boldsymbol{k}',t')\rangle=\langle\mathfrak{u}(\boldsymbol{k},t)\mathfrak{w}(\boldsymbol{k}',t')\rangle=\langle\mathfrak{v}(\boldsymbol{k},t)\mathfrak{w}(\boldsymbol{k}',t')\rangle=0.\end{aligned} \tag{9.24}$$

9.3.3.2 System-T

The relevant quantities of the system are given by

$$\omega_i\,(\equiv (\mathrm{rot}\boldsymbol{v})_i)\,,\quad \sigma_i\left(\equiv \epsilon_{ijk}\frac{\partial^2\sigma_{\langle kn\rangle}}{\partial x_j\partial x_n}\right),\quad \pi_i(\equiv (\mathrm{rot}\boldsymbol{q})_i),$$
$$\mathfrak{x}_i\left(\equiv -\epsilon_{ijk}\frac{\partial^2\mathfrak{r}_{\langle kn\rangle}}{\partial x_j\partial x_n}\right),\quad \text{and}\quad \mathfrak{y}_i\left(\equiv \frac{1}{2}(\mathrm{rot}\mathfrak{s})_i\right). \tag{9.25}$$

The field equations in the $\boldsymbol{k}t$-representation are as follows:

$$\frac{\partial\omega_i(\boldsymbol{k},t)}{\partial t} - \frac{1}{\rho_0}\sigma_i(\boldsymbol{k},t) = 0,$$
$$\frac{\partial\sigma_i(\boldsymbol{k},t)}{\partial t} + \frac{2}{D+2}k^2\pi_i(\boldsymbol{k},t) + \frac{k_B}{m}\rho_0T_0k^2\omega_i(\boldsymbol{k},t) = -\frac{1}{\tau_S}\sigma_i(\boldsymbol{k},t) + \mathfrak{x}_i(\boldsymbol{k},t),$$
$$\frac{\partial\pi_i(\boldsymbol{k},t)}{\partial t} - \frac{k_B}{m}T_0\sigma_i(\boldsymbol{k},t) = -\frac{1}{\tau_q}\pi_i(\boldsymbol{k},t) + \mathfrak{y}_i(\boldsymbol{k},t). \tag{9.26}$$

Note that, for given $\mathfrak{x}_i$ and $\mathfrak{y}_i$, the equations for the set of variables $(\omega_i, \sigma_i, \pi_i)$ with the same suffix i can be solved separately from those with the different suffix $j(\neq i)$. In view of Eq. (9.21), $\mathfrak{x}_i$ and $\mathfrak{y}_i$ are the Gaussian white random forces with null means and correlations:

$$\langle\mathfrak{x}_i(\boldsymbol{k},t)\mathfrak{x}_m(\boldsymbol{k}',t')\rangle = k_B\frac{\frac{k_B}{m}\rho_0T_0^2}{4\pi^3\tau_S}k^4\left(\delta_{im} - \frac{k_ik_m}{k^2}\right)$$
$$\times\delta(\boldsymbol{k}+\boldsymbol{k}')\delta(t-t'),$$
$$\langle\mathfrak{y}_i(\boldsymbol{k},t)\mathfrak{y}_m(\boldsymbol{k}',t')\rangle = k_B\frac{(D+2)\left(\frac{k_B}{m}\right)^2\rho_0T_0^3}{8\pi^3\tau_q}k^2\left(\delta_{im} - \frac{k_ik_m}{k^2}\right)$$
$$\times\delta(\boldsymbol{k}+\boldsymbol{k}')\delta(t-t'),$$
$$\langle\mathfrak{x}_i(\boldsymbol{k},t)\mathfrak{y}_m(\boldsymbol{k}',t')\rangle = 0. \tag{9.27}$$

9.3.4 Relationship to the Landau-Lifshitz Theory

In what follows, we adopt the coarse-graining approximation where the fast modes are eliminated [58], and show explicitly the coarse-grained solutions for the System-L and System-T. We will see that these solutions are just the ones in the LL theory.

9.3.4.1 System-L

We have the following relation up to the leading term with respect to τ_Π, τ_S and τ_q

$$\begin{bmatrix} \Pi(\boldsymbol{k},t) \\ \tau(\boldsymbol{k},t) \\ \varphi(\boldsymbol{k},t) \end{bmatrix} = \begin{bmatrix} -\frac{2(D-3)}{3D}\frac{k_B}{m}\rho_0 T_0 \tau_\Pi \psi(\boldsymbol{k},t) + \mathfrak{f}(\boldsymbol{k},t) \\ -\frac{4}{3}\frac{k_B}{m}\rho_0 T_0 \tau_S k^2 \psi(\boldsymbol{k},t) + \mathfrak{g}(\boldsymbol{k},t) \\ \frac{D+2}{2}\left(\frac{k_B}{m}\right)^2 \rho_0 T_0 \tau_q k^2 T(\boldsymbol{k},t) + \mathfrak{h}(\boldsymbol{k},t) \end{bmatrix}. \tag{9.28}$$

The Gaussian white random forces $\mathfrak{g}$ and $\mathfrak{h}$ have null means and correlations:

$$\begin{aligned} \langle \mathfrak{f}(\boldsymbol{k},t)\mathfrak{f}(\boldsymbol{k}',t')\rangle &= \frac{1}{6\pi^3}\frac{D-3}{D}k_B\frac{k_B}{m}\rho_0 T_0^2 \tau_\Pi \delta(\boldsymbol{k}+\boldsymbol{k}')\delta(t-t'), \\ \langle \mathfrak{g}(\boldsymbol{k},t)\mathfrak{g}(\boldsymbol{k}',t')\rangle &= \frac{1}{3\pi^3}k_B\frac{k_B}{m}\rho_0 T_0^2 \tau_S k^4 \delta(\boldsymbol{k}+\boldsymbol{k}')\delta(t-t'), \\ \langle \mathfrak{h}(\boldsymbol{k},t)\mathfrak{h}(\boldsymbol{k}',t')\rangle &= \frac{D+2}{8\pi^3}k_B\left(\frac{k_B}{m}\right)^2 \rho_0 T_0^3 \tau_q k^2 \delta(\boldsymbol{k}+\boldsymbol{k}')\delta(t-t'), \\ \langle \mathfrak{f}(\boldsymbol{k},t)\mathfrak{g}(\boldsymbol{k}',t')\rangle &= \langle \mathfrak{f}(\boldsymbol{k},t)\mathfrak{h}(\boldsymbol{k}',t')\rangle = \langle \mathfrak{g}(\boldsymbol{k},t)\mathfrak{h}(\boldsymbol{k}',t')\rangle = 0. \end{aligned} \tag{9.29}$$

9.3.4.2 System-T

We obtain the following relations in a similar way as above

$$\begin{bmatrix} \sigma_i(\boldsymbol{k},t) \\ \pi_i(\boldsymbol{k},t) \end{bmatrix} = \begin{bmatrix} -\frac{k_B}{m}\rho_0 T_0 \tau_S k^2 \omega_i(\boldsymbol{k},t) + \mathfrak{k}_i(\boldsymbol{k},t) \\ \mathfrak{l}_i(\boldsymbol{k},t) \end{bmatrix}. \tag{9.30}$$

Note that there is no deterministic part in $\pi_i(\boldsymbol{k},t)$, therefore, only the random force plays a role. The correlations between the zero-mean Gaussian white random forces are given by

$$\begin{aligned} \langle \mathfrak{k}_i(\boldsymbol{k},t)\mathfrak{k}_m(\boldsymbol{k}',t')\rangle &= \frac{1}{4\pi^3}k_B\frac{k_B}{m}\rho_0 T_0^2 \tau_S k^4 \left(\delta_{im} - \frac{k_i k_m}{k^2}\right) \\ &\quad \times \delta(\boldsymbol{k}+\boldsymbol{k}')\delta(t-t'), \\ \langle \mathfrak{l}_i(\boldsymbol{k},t)\mathfrak{l}_m(\boldsymbol{k}',t')\rangle &= \frac{D+2}{8\pi^3}k_B\left(\frac{k_B}{m}\right)^2 \rho_0 T_0^3 \tau_q k^2 \left(\delta_{im} - \frac{k_i k_m}{k^2}\right) \\ &\quad \times \delta(\boldsymbol{k}+\boldsymbol{k}')\delta(t-t'), \\ \langle \mathfrak{k}_i(\boldsymbol{k},t)\mathfrak{l}_m(\boldsymbol{k}',t')\rangle &= 0. \end{aligned} \tag{9.31}$$

The Relationship Between the Present Theory and the LL Theory We can now confirm that the expressions in (9.28), (9.29), (9.30) and (9.31) are exactly the same as those derived from the LL theory where the viscosities ν, μ and the heat conductivity κ are identified by the relations (5.76). Thus we have proved that the LL theory can be derived from the ET14 theory by using the coarse-graining approximation, and that the LL theory is included in the present theory as a limiting case.

The ET14 theory and the LL theory belong to the two different levels of description of fluctuating hydrodynamics. As we analyzed above, the rapidly changing deterministic modes (fast modes) in ET have been consistently re-normalized into the random forces in the LL theory. Therefore, from a physical point of view, the delta functions appeared in the correlations have their own validity range depending on the spatio-temporal resolution of their description level.

9.3.5 Conclusion

In the present section, we have summarized the theory of fluctuating hydrodynamics based on ET. And we have made clear the link between the two levels of description of fluctuating hydrodynamics, that is, the ET14 theory and the LL theory.

Generally speaking, there are many such levels. As explained in Sect. 2.4, Boillat and Ruggeri found the hierarchy structure of ET and the important concept called principal subsystem of field equations. Each subsystem gives us one level of description with different resolution from each other. And, in a similar way as above, we can develop the corresponding fluctuating hydrodynamics basing on a given subsystem. Detailed discussions are omitted here for simplicity.

References

1. T. Arima, S. Taniguchi, M. Sugiyama, Light scattering in rarefied polyatomic gases based on extended thermodynamics, in *Proceedings of the 34th Symposium on Ultrasonic Electronics* (2013), pp. 15–16
2. I. Müller, T. Ruggeri, *Rational Extended Thermodynamics*, 2nd edn. (Springer, New York, 1998)
3. L.P. Kadanoff, P.C. Martin, Hydrodynamic equations and correlation functions. Ann. Phys. **24**, 419 (1963)
4. R.D. Mountain, Spectral distribution of scattered light in a simple fluid. Rev. Mod. Phys. **38**, 205 (1966)
5. R.D. Mountain, Density fluctuations in fluids having an internal degree of freedom. J. Res. Natl. Bur. Stand. **72**, 95 (1968)
6. S. Yip, M. Nelkin, Application of a kinetic model to time-dependent density correlations in fluids. Phys. Rev. **135**, 1241 (1964)
7. A. Sugawara, S. Yip, Kinetic model analysis of light scattering by molecular gases. Phys. Fluids **10**, 1911 (1967)
8. R.C. Desai, L.R. Kapral, Translational hydrodynamics and light scattering from molecular fluids. Phys. Rev. A **6**, 2377 (1972)

9. C.D. Boley, R.C. Desai, G. Tenti, Kinetic models and brillouin scattering in a molecular gas. Can. J. Phys. **50**, 2158 (1972)
10. G. Tenti, C.D. Boley, R.C. Desai, On the Kinetic model description of Rayleigh–Brillouin scattering from molecular gases. Can. J. Phys. **52**, 285 (1974)
11. W. Marques Jr., Light scattering and sound propagation in polyatomic gases with classical degrees of freedom. Contin. Mech. Thermodyn. **16**, 517 (2004)
12. A.S. Fernandes, W. Marques Jr., Kinetic model analysis of time-dependent problems in polyatomic gases. Phys. A: Stat. Mech. Appl. **373**, 97 (2007)
13. W. Weiss, I. Müller, Light scattering and extended thermodynamics. Contin. Mech. Thermodyn. **7**, 123 (1995)
14. T.J. Greytak, G.B. Benedek, Spectrum of light scattered from thermal fluctuations in gases. Phys. Rev. Lett. **17**, 179 (1966)
15. T.J. Greytak, Ph.D. thesis, Massachusetts Institute of Technology, 1967
16. Q.H. Lao, P.E. Schoen, B. Chu, Rayleigh–Brillouin scattering of gases with internal relaxation. J. Chem. Phys. **64**, 3547 (1976)
17. E.H. Hara, A.D. May, H.P.F. Knaap, Rayleigh–Brillouin scattering in compressed H_2, D_2, and HD. Can. J. Phys. **49**, 420 (1971)
18. E.H. Hara, A.D. May, Rayleigh–Brillouin scattering in compressed H_2, D_2, and HD II. Can. J. Phys. **50**, 1227 (1972)
19. M. Hubert, A.D. May, The Rayleigh–Brillouin spectrum of normal and parahydrogen: a test of model solutions of the Wang–Chang Uhlenbeck equation. Can. J. Phys. **53**, 343 (1975)
20. Z. Gu, M.O. Vieitez, E.J. van Duijn, W. Ubachs, A Rayleigh-Brillouin scattering spectrometer for ultraviolet wavelengths. Rev. Sci. Instrum. **83**, 053112 (2012)
21. Z. Gu, W. Ubachs, W. van de Water, Rayleigh–Brillouin scattering of carbon dioxide. Opt. Lett. **39**, 3301 (2014)
22. X. Pan, M.N. Shneider, R.B. Miles, Coherent Rayleigh-Brillouin scattering in molecular gases. Phys. Rev. A **69**, 033814 (2004)
23. X. Pan, M.N. Shneider, R.B. Miles, Power spectrum of coherent Rayleigh-Brillouin scattering in carbon dioxide. Phys. Rev. A **71**, 045801 (2005)
24. M.O. Vieitez, E.J. van Duijn, W. Ubachs, B. Witschas, A. Meijer, A.S. de Wijn, N.J. Dam, W. van de Water, Coherent and spontaneous Rayleigh-Brillouin scattering in atomic and molecular gases and gas mixtures. Phys. Rev. A **82**, 043836 (2010)
25. T. Arima, A. Mentrelli, T. Ruggeri, Molecular extended thermodynamics of rarefied polyatomic gases and wave velocities for increasing number of moments. Ann. Phys. **345**, 111 (2014)
26. T. Arima, E. Barbera, F. Brini, M. Sugiyama, The role of the dynamic pressure in stationary heat conduction of a rarefied polyatomic gas. Phys. Lett. A **378**, 2695 (2014)
27. I. Müller, T. Ruggeri, Stationary heat conduction in radially, symmetric situations—an application of extended thermodynamics. J. Non-Newtonian Fluid Mech. **119**, 139 (2004)
28. E. Barbera, F. Brini, On stationary heat conduction in 3D symmetric domains: an application of extended thermodynamics. Acta Mech. **215**, 241 (2010)
29. E. Barbera, F. Brini, G. Valenti, Some non-linear effects of stationary heat conduction in 3D domains through extended thermodynamics. Eur. Phys. Lett. **98**, 54004 (2012)
30. E. Barbera, F. Brini, Heat transfer in gas mixtures: advantages of an extended thermodynamics approach. Phys. Lett. A **375**(4), 827 (2011)
31. E. Barbera, F. Brini, Heat transfer in a binary gas mixture between two parallel plates: an application of linear extended thermodynamics. Acta Mech. **220**, 87 (2011)
32. E. Barbera, F. Brini, Heat transfer in multi-component gas mixtures described by extended thermodynamics. Meccanica **47**(3), 655 (2012)
33. C. Truesdell, The physical components of vektors and tensors. Z. Angew. Math. Mech. **33**, 345 (1953)
34. E. Barbera, I. Müller, Heat conduction in a non-inertial frame, in *Rational Continua, Classical and New*, ed. by P. Podio-Guidugli, M. Brocato (Springer, Milano, 2002), pp. 1–10

35. W.P. Teagan, G.S. Springer, Heat-transfer and density-distribution measurements between parallel plates in the transition regime. Phys. Fluids **11**, 497 (1968)
36. L.D. Landau, E.M. Lifshitz, Hydrodynamic fluctuations. Soviet Phys. JETP **5**, 512 (1957)
37. L.D. Landau, E.M. Lifshitz, *Fluid Mechanics* (Pergamon, London, 1958)
38. E.M. Lifshitz, L.P. Pitaevskii, *Statistical Physics, Part 2* (Pergamon, Oxford, 1980)
39. H.B. Callen, T.A. Welton, Irreversibility and generalized noise. Phys. Rev. **83**, 34 (1951)
40. M.S. Green, Brownian motion in a gas of noninteracting molecules. J. Chem. Phys. **19**, 1036 (1951)
41. R. Kubo, The fluctuation-dissipation theorem and Brownian motion, in *1965 Tokyo Summer Lectures in Theoretical Physics*, ed. by R. Kubo (Part I. Many-Body Theory) 1966 Shokabo (Tokyo) and Benjamin (New York), pp. 1–16
42. R.F. Fox, Gaussian stochastic processes in physics. Phys. Rep. **48**, 179 (1978)
43. M. Kac, J. Logan, *Fluctuation Phenomena, Studies in Statistical Mechanics*, ed. by E.W. Montroll, J.L. Lebowitz, Chap. 1, vol. VII (North-Holland, Amsterdam, 1979)
44. J.M. Ortiz de Zarate, J.V. Sengers, *Hydrodynamic Fluctuations* (Elsevier, Amsterdam, 2006)
45. G. Karniadakis, A. Beskok, N. Aluru, *Microflows and Nanoflows: Fundamentals and Simulation* (Springer, New York, 2005)
46. W. Kang, U. Landman, Universality crossover of the pinch-off shape profiles of collapsing liquid nanobridges in vacuum and gaseous environments. Phys. Rev. Lett. **98**, 064504 (2007)
47. R.D. Astumian, P. Hänggi, Brownian motors. Phys. Today **55**, 33 (2002)
48. G. Oster, Darwin's motors. Nature (London) **417**, 25 (2002)
49. A.L. Garcia, M. Malek-Mansour, G.C. Lie, E. Clementi, Numerical integration of the fluctuating hydrodynamic equations. J. Stat. Phys. **47**, 209 (1987)
50. A.L. Garcia, C. Penland, Fluctuating hydrodynamics and principal oscillation pattern analysis. J. Stat. Phys. **64**, 1121 (1991)
51. N. Sharma, N.A. Patankar, Direct numerical simulation of the Brownian motion of particles by using fluctuating hydrodynamic equations. J. Comput. Phys. **201**, 466 (2004)
52. G. De Fabritiis, R. Delgado-Buscalioni, P.V. Coveney, Multiscale modeling of liquids with molecular specificity. Phys. Rev. Lett. **97**, 134501 (2006)
53. G. De Fabritiis, M. Serrano, R. Delgado-Buscalioni, P.V. Coveney, Fluctuating hydrodynamic modeling of fluids at the nanoscale. Phys. Rev. E **75**, 026307 (2007)
54. J.B. Bell, A.L. Garcia, S.A. Williams, Numerical methods for the stochastic Landau-Lifshitz Navier-Stokes equations. Phys. Rev. E **76**, 016708 (2007)
55. J.B. Bell, A.L. Garcia, S.A. Williams, Computational fluctuating fluid dynamics. ESAIM: M2AN **44**, 1085 (2010)
56. R. Schmitz, Fluctuations in nonequilibrium fluids. Phys. Rep. **171**, 1 (1988)
57. P. Español, Stochastic differential equations for non-linear hydrodynamics. Physica A **248**, 77 (1998)
58. A. Ikoma, T. Arima, S. Taniguchi, N. Zhao, M. Sugiyama, Fluctuating hydrodynamics for a rarefied gas based on extended thermodynamics. Phys. Lett. A **375**, 2601 (2011)
59. T. Arima, A. Ikoma, S. Taniguchi, M. Sugiyama, N. Zhao, Fluctuating hydrodynamics based on extended thermodynamics. Note Mat. **32**, 227 (2012)
60. E.M. Lifshitz, L.P. Pitaevskii, *Statistical Physics, Part 1*, 3rd edn. (Pergamon, Oxford, 1980)

Part V
Maximum Entropy Principle and Nesting Theories of Many Moments

Chapter 10
Molecular ET Theory of Rarefied Polyatomic Gas

Abstract Molecular extended thermodynamics of rarefied polyatomic gases is characterized by the system composed of two hierarchies of balance equations for the moments of a distribution function. The internal degrees of freedom of a molecule are properly taken into account in the distribution function. By the reasoning of physical relevance, the truncation orders of the two hierarchies are proved to be not independent of each other. And the two closure procedures based on the maximum entropy principle (MEP) and on the entropy principle are also proved to be equivalent to each other.

Characteristic velocities of a hyperbolic system of the balance equations for a polyatomic gas are compared to those obtained for a monatomic gas. The lower bound estimate for the maximum equilibrium characteristic velocity established for a monatomic gas is proved to be valid also for a rarefied polyatomic gas, that is, the estimate is independent of the degrees of freedom of a molecule. As a consequence, also for polyatomic gases, when the number of moments increases the maximum characteristic velocity becomes unbounded.

10.1 Introduction

We discussed, in Sect. 1.5.5, the role of the MEP and, in Sect. 4.3, we gave a survey of molecular ET. We proved the equivalence of the closures via MEP and via the entropy principle in the case of monatomic gases. In Chap. 6, we described the closure via MEP of the 14-field theory of the new ET valid for rarefied polyatomic gases following the results given in [1]. We proved that the binary hierarchy of moment equations obtained with the distribution function is consistent with the binary hierarchy presented in Chap. 5. In particular, it was shown that the *momentum-like* hierarchy is related to the usual moments of the distribution function, and the *energy-like* hierarchy is related to the moments of an additional continuous variable representing the internal energy of a molecule.

The purpose of the present chapter is to give a contribution to the development and understanding of the new ET theory of rarefied polyatomic gases for any number of moments. These results are based on the papers of Arima et al. [2, 3].

T. Ruggeri, M. Sugiyama, *Rational Extended Thermodynamics beyond the Monatomic Gas*, DOI 10.1007/978-3-319-13341-6_10

In particular, by the physical arguments (namely, Galilean invariance and the requirement that the characteristic velocities depend on the degrees of freedom of a molecule), the relation between the orders of truncation of the *momentum-like* and *energy-like* hierarchies are investigated, leading to the conclusion that once the order of truncation of one hierarchy is chosen, the order of truncation of the other one is automatically prescribed.

The closure of the system is achieved by means of MEP, and it is proved that, also in the present general case, this MEP approach is equivalent to the approach with the requirement that the truncated system satisfies the entropy principle with the convex entropy density. In this way, it is shown that the system is symmetric when written in terms of the *main field* components.

And the characteristic velocities in an equilibrium state are analyzed. These velocities play an important role in processes such as the propagation of acceleration waves [4, 5] (see Sect. 3.2), the determination of the phase velocity of linear waves in the high-frequency limit [6, 7] (see Sect. 3.1), and the *subshock* formation [8] (see Theorem 3.1). With regard to this, it will be discussed how the characteristic velocities of the system depend on the internal degrees of freedom and on the orders of the truncation of hierarchies. In particular, the two limit cases of monatomic gases and of a gas with infinite internal degrees of freedom are investigated.

Finally, using convexity arguments and sub-characteristic conditions for principal subsystems (see Theorem 2.3), we prove that the lower bound estimate for the maximum characteristic velocity established for monatomic gases by Boillat and Ruggeri [9] [see (4.35)] still holds and is independent of the degrees of freedom of a molecule. Therefore, also for polyatomic gases, the maximum characteristic velocity tends to be unbounded when the orders of the hierarchies tend to infinity!

10.2 MEP Closure for Rarefied Polyatomic Gases with Many Moments

The discussion is focused on gases characterized by the thermal and caloric equations of state (1.28), that is, rarefied polytropic and polyatomic gases.

We consider now the same binary hierarchy of the 14-moment theory but for a generic number of moments truncated for the F-series at the index of truncation N and for the G-series at the index M:

$$\begin{aligned}
&\partial_t F + \partial_i F_i = 0, \\
&\partial_t F_{k_1} + \partial_i F_{ik_1} = 0, \\
&\partial_t F_{k_1k_2} + \partial_i F_{ik_1k_2} = P_{k_1k_2}, \qquad && \partial_t G_{kk} + \partial_i G_{ikk} = 0, \\
&\vdots && \partial_t G_{kkj_1} + \partial_i G_{kkij_1} = Q_{kkj_1},
\end{aligned} \tag{10.1}$$

$$\vdots \qquad\qquad\qquad \vdots$$

$$\partial_t F_{k_1 k_2 \dots k_N} + \partial_i F_{ik_1 k_2 \dots k_N} = P_{k_1 k_2 \dots k_N} \quad \vdots$$

$$\partial_t G_{kkj_1 j_2 \dots j_M} + \partial_i G_{kkij_1 j_2 \dots j_M} = Q_{kkj_1 j_2 \dots j_M}.$$

Definition 10.1 ((N, M)-**System**) The above system can be rewritten in a simple form by using the multi-index notations defined below:

$$\begin{aligned} &\partial_t F_A + \partial_i F_{iA} = P_A, \\ &(0 \leqslant A \leqslant N) \qquad \partial_t G_{llA'} + \partial_i G_{lliA'} = Q_{llA'}, \\ &\qquad\qquad\qquad \left(0 \leqslant A' \leqslant M\right) \end{aligned} \tag{10.2}$$

which we call (N, M)-system.

The moments of the F-series are the usual ones:

$$\begin{aligned} F_A &= \int_{R^3} \int_0^\infty mfc_A I^\alpha \, dI d\mathbf{c}, \qquad F_{iA} = \int_{R^3} \int_0^\infty mfc_i c_A I^\alpha \, dI d\mathbf{c}, \\ P_A &= \int_{R^3} \int_0^\infty mQ(f) c_A I^\alpha \, dI d\mathbf{c}, \end{aligned} \tag{10.3}$$

while the moments of the G-series are expressed with the additional variable I as

$$\begin{aligned} G_{llA'} &= \int_{R^3} \int_0^\infty mf \left(c^2 + \frac{2I}{m}\right) c_{A'} I^\alpha \, dI d\mathbf{c}, \\ G_{lliA'} &= \int_{R^3} \int_0^\infty mf \left(c^2 + \frac{2I}{m}\right) c_i c_{A'} I^\alpha \, dI d\mathbf{c}, \\ Q_{llA'} &= \int_{R^3} \int_0^\infty mQ(f) \left(c^2 + \frac{2I}{m}\right) c_{A'} I^\alpha \, dI d\mathbf{c}. \end{aligned} \tag{10.4}$$

The following multi-index notations are introduced for the sake of compactness:

$$F_A = \begin{cases} F & \text{for } A = 0 \\ F_{i_1 \cdots i_A} & \text{for } 1 \leqslant A \leqslant N \end{cases}, \qquad F_{iA} = \begin{cases} F_i & \text{for } A = 0 \\ F_{i i_1 \cdots i_A} & \text{for } 1 \leqslant A \leqslant N \end{cases},$$

$$P_A = \begin{cases} 0 & \text{for } A = 0 \\ 0 & \text{for } A = 1 \\ P_{i_1 \cdots i_A} & \text{for } 2 \leqslant A \leqslant N \quad (\text{ with } P_{ll} = 0) \end{cases},$$

$$G_{llA'} = \begin{cases} G_{ll} & \text{for } A' = 0 \\ G_{lli_1\cdots i_{A'}} & \text{for } 1 \leqslant A' \leqslant M \end{cases}, \qquad G_{lliA'} = \begin{cases} G_{lli} & \text{for } A' = 0 \\ G_{llii_1\cdots i_{A'}} & \text{for } 1 \leqslant A' \leqslant M \end{cases},$$

$$Q_{llA'} = \begin{cases} 0 & \text{for } A' = 0 \\ Q_{lli_1\cdots i_{A'}} & \text{for } 1 \leqslant A' \leqslant M \end{cases},$$

and

$$c_A = \begin{cases} 1 & \text{for } A = 0 \\ c_{i_1} \cdots c_{i_A} & \text{for } 1 \leqslant A \leqslant N \end{cases},$$

where the indices i and $i_1 \leqslant i_2 \leqslant \cdots \leqslant i_A$ assume the values $1, 2, 3$. The truncation order N of the F-hierarchy (*momentum-like* hierarchy) and the order M of the G-hierarchy (*energy-like* hierarchy) are a priori independent of each other. It is worth noting that the first and second equations of the F-hierarchy represent the conservation laws of mass and momentum, respectively ($P \equiv 0, P_i \equiv 0$), while the first equation of the G-hierarchy represents the conservation law of energy ($Q_{ll} \equiv 0$), and that, in each of the two hierarchies, the flux in one equation appears as the density in the following equation—a feature in common with the single hierarchy of monatomic gases.

The Euler 5-moment system (4.3) is a particular case of (10.2) with $N = 1$, $M = 0$. And the 14-moment system (5.2) is another particular case of (10.2) with $N = 2$, $M = 1$.

Instead of the (N, M)-system that includes all tensors of F_A and $G_{llA'}$, it is possible to construct the system that includes all the tensors except that the last ones, i.e., N-order and/or M-order tensorial equations, are replaced by the part of them as independent variables. Then similarly to the Definition 4.1, we have the following:

Definition 10.2 A system (10.1) is called (N^-, M)-system if, for the last balance equation in the F-hierarchy, we consider only the trace with respect to the two indexes, $F_{k_1k_2\ldots k_{N-2}ll}$, instead of the full N-order tensor $F_{k_1k_2\ldots k_{N-2}k_{N-1}k_N}$. Similar definition is valid for the G-hierarchy and we have an (N, M^-)-system. If, instead of 2 indexes, we have the contraction with respect to the 2 couples of 2 indexes, we add another $-$: (N^{--}, M) etc.

For example, in the classical kinetic approach for rarefied polyatomic gases, the 17-moment theory is proposed [10, 11]. This theory corresponds to the ET theory with the densities; $F, F_i, F_{ij}, G_{ll}, F_{lli}, G_{lli}$. In this case, in the last tensor of F-hierarchy, only the trace is adopted as an independent field, and therefore all components of the G-hierarchy have the corresponding components in the F-hierarchy. As this 17-moment system adopts the trace with respect to the indexes i_{N-1} and i_N of the N-order tensorial equation with $N = 3$, the 17-moment system is denoted as $(3^-, 1)$-system. Similarly, the monatomic 13-moment system is denoted as (3^-)-system.

Remark Definition 10.2 does not cover all possible cases of physically plausible systems. For example, the system that adopts the traceless part of the highest order tensor of F- and/or G-hierarchies is also possible. Another possibility is that the system adopts the trace part not only of the highest-order tensor of F- and/or G-hierarchies but also of the second-highest-order tensor. However, in the present study, we skip such systems and pay attention solely to the (N, M)-system with or without $-$, because, in our opinion, these systems are most interesting from a physical point of view. From (4.23) and (4.24), the number of moments for an (N, M)-system is given by

$$n_{(N,M)} = \frac{1}{6}(N+1)(N+2)(N+3) + \frac{1}{6}(M+1)(M+2)(M+3), \tag{10.5}$$

and, for an (N^-, M)-system,

$$n_{(N^-,M)} = \frac{1}{6}N(N^2+6N-1) + \frac{1}{6}(M+1)(M+2)(M+3). \tag{10.6}$$

10.2.1 Galilean Invariance

An important issue to be addressed concerns the relation between the orders N and M of the two hierarchies (10.2). In the treatment outlined above, the orders N and M have been considered to be independent of each other. However, from the physical point of view, one should clarify whether some restrictions on the truncation procedure of the two hierarchies exist or not.

In the spirit of ET, the application of the universal principles—in the present case, the Galilean invariance—suggests that in order to have a physically acceptable model, the orders of truncation N and M cannot be chosen independently. Indeed, we can prove the following theorem that gives the first restriction on the relation between the two indexes N and M:

Theorem 10.1 *In order to be Galilean invariant for the* (N, M)*-system* (10.2)*, it must be* $M \leqslant N-1$.

Proof Recalling the definition of the peculiar velocity $C_i = c_i - v_i$, and defining the velocity-independent *internal* moments (so-called Galilean tensors) $\hat{F}_A$ and $\hat{G}_{llA'}$ as follows:

$$\hat{F}_A = \int_{R^3}\int_0^\infty mf\, C_A I^\alpha \, dI d\mathbf{c},$$

$$\hat{G}_{llA'} = \int_{R^3}\int_0^\infty mf\left(C^2 + \frac{2I}{m}\right) C_{A'} I^\alpha \, dI d\mathbf{c},$$

we can write

$$F_A = \sum_{k=0}^{A} \binom{A}{k} \hat{F}_{(i_1 \cdots i_k} v_{i_{k+1}} \cdots v_{i_A)}, \\ = X_{AB} \hat{F}_B, \tag{10.7}$$

where X_{AB} is a component of the matrix (2.51), and

$$G_{llA'} = \sum_{k=0}^{A'} \binom{A'}{k} \left(\hat{G}_{ll(i_1 \cdots i_k} v_{i_{k+1}} \cdots v_{i_{A'})} + 2 v_l \hat{F}_{l(i_1 \cdots i_k} v_{i_{k+1}} \cdots v_{i_{A'})} \right. \\ \left. + v^2 \hat{F}_{(i_1 \cdots i_k} v_{i_{k+1}} \cdots v_{i_{A'})} \right), \\ = X_{A'B'} \left(\hat{G}_{llB'} + 2 v_l \hat{F}_{lB'} + v^2 \hat{F}_{B'} \right). \tag{10.8}$$

It is noticeable that $G_{llA'}$ depends not only on the Galilean tensors of the G-hierarchy, but also on those of the F-hierarchy. Similarly, the fluxes F_{iA}, $G_{lliA'}$ and the productions P_A, $Q_{llA'}$ are expressed, respectively, by these internal quantities $\hat{F}_{iA}$, $\hat{G}_{lliA'}$ and $\hat{P}_A$, $\hat{Q}_{llA'}$, which are evaluated by imposing $v_{i_k} = 0$ for all k.

The structure of the terms (10.7) and (10.8) guarantees that the results proved in [12] can be extended to the present case and, when $M > N - 1$, unknown moments of the F-hierarchy appear in a higher order moment of the G-hierarchy. For example, in the case $M = N$, $G_{lli_1 \cdots i_N}$, which is the highest-order moment of the G-hierarchy, includes $\hat{F}_{li_1 \cdots i_N}$, and the corresponding moment $F_{li_1 \cdots i_N}$ is not included in the F-hierarchy. Therefore we conclude that the system cannot be Galilean invariant whenever $M > N - 1$. The 14 moment theory, namely $(2, 1)$-system and the Euler $(1, 0)$-system are of course Galilean invariant.

It should be noted that the requirement of the Galilean invariance for (N, M)-system; $M \leqslant N - 1$ is satisfied also for (N^-, M), (N, M^-) and (N^-, M^-) except for $(N^-, N-1)$-system.

10.2.2 Closure of the (N, M)-System via the Maximum Entropy Principle

To close the (N, M)-system (10.2) with $M \leqslant N - 1$, the maximum entropy principle is applied: The actual distribution function $f_{(N,M)}$ is the one that maximizes the entropy h^0 defined by

$$h^0 = -k_B \int_{R^3} \int_0^{\infty} f \log f \, I^{\alpha} \, dI d\mathbf{c}, \tag{10.9}$$

under the constraints that the moments F_A and $G_{llA'}$ are given by $(10.3)_1$ and $(10.4)_1$.

The variational problem from which the distribution function $f_{(N,M)}$ is obtained is connected to the functional:

$$\mathscr{L}_{(N,M)}(f) =$$
$$-k_B \int_{R^3}\int_0^\infty f \log f\, I^\alpha\, dI d\mathbf{c} + \sum_{A=0}^{N} \lambda_A \left(F_A - \int_{R^3}\int_0^\infty mfc_A I^\alpha\, dI d\mathbf{c} \right) +$$
$$+ \sum_{A'=0}^{M} \mu_{A'} \left(G_{llA'} - \int_{R^3}\int_0^\infty mf\left(c^2 + \frac{2I}{m}\right) c_{A'} I^\alpha dI d\mathbf{c} \right),$$

where λ_A and $\mu_{A'}$ are the Lagrange multipliers. Then the distribution function is obtained as follows [2, 3]:

Theorem 10.2 *The distribution function $f_{(N,M)}$ that maximizes the functional $\mathscr{L}_{(N,M)}$ is (we omit from now the symbol of summation in A and in A′):*

$$f_{(N,M)} = \exp\left(-1 - \frac{m}{k_B}\chi_{(N,M)}\right), \qquad \chi_{(N,M)} = \lambda_A c_A + \left(c^2 + \frac{2I}{m}\right)\mu_{A'} c_{A'}. \tag{10.10}$$

By inserting (10.10) into $(10.3)_1$ and $(10.4)_1$, the Lagrange multipliers λ_A and $\mu_{A'}$ are evaluated in terms of the densities F_A and $G_{llA'}$. Finally, plugging (10.10) into the last flux and production terms, the system can be closed. We will give more details in the following sections.

10.2.3 Closure of the (N, M)-System via the Entropy Principle

As discussed above, an alternative approach to achieve the closure of the system makes use of the entropy principle. In this case, it is required that all the solutions of (10.2) satisfy the entropy principle (2.8). The condition (2.12) can now be written as

$$\partial_t h^0 + \partial_i h^i - u'_A\left(\partial_t F_A + \partial_i F_{iA} - P_A\right) - v'_{A'}\left(\partial_t G_{llA'} + \partial_i G_{lliA'} - Q_{llA'}\right) = \Sigma \geqslant 0,$$

where u'_A and $v'_{A'}$ are the main field components. Treating h^0, h^i and Σ as constitutive functions of F_A and $G_{llA'}$, we obtain [see (2.13)]

$$dh^0 = u'_A dF_A + v'_{A'} dG_{A'}, \quad dh_i = u'_A dF_{iA} + v'_{A'} dG_{iA'}, \quad \Sigma = u'_A P_A + v'_{A'} Q_{llA'} \geqslant 0.$$

Recalling (10.3) and (10.4), we can write

$$dh^0 = \int_{R^3}\int_0^\infty m\chi_{(N,M)} I^\alpha \, df \, dI d\mathbf{c}, \qquad dh^i = \int_{R^3}\int_0^\infty mc_i \, \chi_{(N,M)} I^\alpha \, df \, dI d\mathbf{c},$$
$$\Sigma = \int_{R^3}\int_0^\infty m\chi_{(N,M)} \, Q(f) \, I^\alpha \, dI d\mathbf{c}, \qquad \chi_{(N,M)} = u'_A c_A + \left(c^2 + \frac{2I}{m}\right) v'_{A'} c_{A'}. \tag{10.11}$$

On the other hand, recalling that the entropy density h^0 is expressed as (10.9) by using the distribution function, and similarly the entropy flux and production can be expressed as

$$h^i = -k_B \int_{R^3}\int_0^\infty c_i f \log f \, I^\alpha \, dI d\mathbf{c}, \qquad \Sigma = -k_B \int_{R^3}\int_0^\infty Q(f) \log f \, I^\alpha \, dI d\mathbf{c}, \tag{10.12}$$

we can write the quantities dh^0 and dh^i as

$$\begin{aligned} dh^0 &= -k_B \int_{R^3}\int_0^\infty (\log f + 1) \, I^\alpha \, df \, dI d\mathbf{c}, \\ dh_i &= -k_B \int_{R^3}\int_0^\infty c_i(\log f + 1) \, I^\alpha \, df \, dI d\mathbf{c}. \end{aligned} \tag{10.13}$$

By comparing (10.11)–(10.13) it is easy to have:

Theorem 10.3 *The distribution function $f_{(N,M)}$ that satisfies the entropy principle is given by*

$$f_{(N,M)} = \exp\left(-1 - \frac{m}{k_B}\chi_{(N,M)}\right). \tag{10.14}$$

As a consequence of Theorems 10.2 and 10.3, it is concluded that:

Statement 2 *The MEP and the entropy principle are equivalent with respect to the closure also for rarefied polyatomic gases. In addition, the Lagrange multipliers of MEP coincide with the main field of the entropy principle: $u'_A \equiv \lambda_A$ and $v'_{A'} \equiv \mu_{A'}$.*

For the case of a generic entropy functional $h^0 = h^0(f)$ valid for any gas including degenerate gas, the equivalence of the entropy principle and MEP was proved in [3] in a similar way in the case of monatomic gases (see Sect. 4.3).

10.2.4 Closure and Symmetric Hyperbolic Form

As discussed in Sect. 4.3 in the case of rarefied monatomic gases, the system (10.2) can be written in a symmetric hyperbolic form in terms of the main field $(u'_A, v'_{A'})$. In fact:

Theorem 10.4 *The system* (10.2) *may be written as follows:*

$$\begin{pmatrix} J^0_{AB} & J^1_{AB'} \\ J^1_{A'B} & J^2_{A'B'} \end{pmatrix} \partial_t \begin{pmatrix} u'_B \\ v'_{B'} \end{pmatrix} + \begin{pmatrix} J^0_{iAB} & J^1_{iAB'} \\ J^1_{iA'B} & J^2_{iA'B'} \end{pmatrix} \partial_i \begin{pmatrix} u'_B \\ v'_{B'} \end{pmatrix} = \begin{pmatrix} P_A \\ Q_{llA'} \end{pmatrix}. \tag{10.15}$$

The coefficient matrix of time-derivatives of the fields is negative-definite, and all other matrices are symmetric. Therefore the system is symmetric hyperbolic.

Proof From (10.14) dropping the subscript (N, M) for simplicity, we have

$$\partial_t f = -\frac{m}{k_B} f\, \partial_t \chi = -\frac{m}{k_B} f \left(c_A \partial_t u'_A + \left(c^2 + \frac{2I}{m}\right) c_{A'} \partial_t v'_{A'} \right),$$

and then

$$\begin{aligned} \partial_t F_A &= J^0_{AB} \partial_t u'_B + J^1_{AB'} \partial_t v'_{B'}, \qquad & \partial_t G_{llA'} &= J^1_{BA'} \partial_t u'_B + J^2_{A'B'} \partial_t v'_{B'}, \\ \partial_i F_{iA} &= J^0_{iAB} \partial_i u'_B + J^1_{iAB'} \partial_i v'_{B'}, \qquad & \partial_i G_{lliA'} &= J^1_{iBA'} \partial_i u'_B + J^2_{iA'B'} \partial_i v'_{B'}, \end{aligned} \tag{10.16}$$

where

$$\begin{aligned} J^0_{AB} &= -\frac{m^2}{k_B} \int_{R^3} \int_0^\infty f c_A c_B I^\alpha \, dI d\mathbf{c}, \\ J^0_{iAB} &= -\frac{m^2}{k_B} \int_{R^3} \int_0^\infty f c_i c_A c_B I^\alpha \, dI d\mathbf{c}, \\ J^1_{AB'} &= -\frac{m^2}{k_B} \int_{R^3} \int_0^\infty f c_A c_{B'} \left(c^2 + \frac{2I}{m}\right) I^\alpha \, dI d\mathbf{c}, \\ J^1_{iAB'} &= -\frac{m^2}{k_B} \int_{R^3} \int_0^\infty f c_i c_A c_{B'} \left(c^2 + \frac{2I}{m}\right) I^\alpha \, dI d\mathbf{c}, \\ J^2_{A'B'} &= -\frac{m^2}{k_B} \int_{R^3} \int_0^\infty f c_{A'} c_{B'} \left(c^2 + \frac{2I}{m}\right)^2 I^\alpha \, dI d\mathbf{c}, \\ J^2_{iA'B'} &= -\frac{m^2}{k_B} \int_{R^3} \int_0^\infty f c_i c_{A'} c_{B'} \left(c^2 + \frac{2I}{m}\right)^2 I^\alpha \, dI d\mathbf{c}. \end{aligned} \tag{10.17}$$

Inserting (10.16) into (10.2), we obtain (10.15), and this completes the proof.

10.3 Closure in the Neighborhood of a Local Equilibrium State and Principal Subsystems

As seen in the monatomic gas case (Sect. 4.5), we have also in this case the problem of the convergence of the moment integrals. Therefore the distribution function (10.10) obtained as a solution of the variational problem is expanded in the neighborhood of a local equilibrium state:

$$f \approx f_E \left[1 - \frac{m}{k_B} \left(\tilde{u}'_A c_A + \left(c^2 + \frac{2I}{m} \right) \tilde{v}'_{A'} c_{A'} \right) \right], \quad \tilde{u}'_A = u'_A - u'^E_A, \; \tilde{v}'_{A'} = v'_{A'} - v'^E_{A'}, \tag{10.18}$$

where u'^E_A and $v'^E_{A'}$ are the main field components evaluated in the local equilibrium state. The equilibrium distribution function f_E, obtained by Pavić et al. [1] (see also [13]), is given by (6.9).

All the Lagrange multipliers (main field) in equilibrium vanish except those corresponding to the hydrodynamic variables ρ, v_i, ε. By inserting (10.18) into $(10.3)_1$ and $(10.4)_1$, a linear algebraic system that permits to evaluate the main field $\left(u'_A, v'_{A'}\right)$ in terms of the densities F_A and $G_{llA'}$ is obtained:

$$\begin{pmatrix} J^{0|E}_{AB} & J^{1|E}_{AB'} \\ J^{1|E}_{A'B} & J^{2|E}_{A'B'} \end{pmatrix} \begin{pmatrix} \tilde{u}'_B \\ \tilde{v}'_{B'} \end{pmatrix} = \begin{pmatrix} F_A - F^E_A \\ G_{llA'} - G^E_{llA'} \end{pmatrix}, \tag{10.19}$$

where the superscript "E"denotes the quantities evaluated in an equilibrium state by using the local equilibrium distribution function f_E given by (6.9). Plugging (6.9) into (10.17), we obtain the following useful relations that express the quantities $J^{0|E}_{AB}$, $J^{1|E}_{AB'}$ and $J^{2|E}_{A'B'}$ by the corresponding quantities $J^{\mathscr{M}}_{AB}$ of the monatomic gas case defined in (4.41) and the parameter α (the degrees of freedom of a molecule):

$$\begin{aligned} J^{0|E}_{AB} &= J^{\mathscr{M}}_{AB}, \qquad J^{1|E}_{AB'} = J^{\mathscr{M}}_{iiAB'} + 2c_s^2(1+\alpha)J^{\mathscr{M}}_{AB'}, \\ J^{2|E}_{A'B'} &= J^{\mathscr{M}}_{iijjA'B'} + 4c_s^2(1+\alpha)\left(J^{\mathscr{M}}_{iiA'B'} + c_s^2(2+\alpha)J^{\mathscr{M}}_{A'B'}\right), \end{aligned} \tag{10.20}$$

where

$$c_s = \sqrt{\frac{k_B}{m}T}.$$

Once $\left(\tilde{u}'_B, \tilde{v}'_{B'}\right)$ is calculated as a solution of (10.19) in terms of the densities F_A and $G_{llA'}$, the main field $\left(u'_A, v'_{A'}\right)$ is obtained from $(10.18)_2$ by considering the fact that, from the general Theorem 2.4, all the components of main field in equilibrium vanish except for the first five components corresponding to the Lagrange multipliers of the conservation laws of mass, momentum and energy (u', u'_i, v'_{ll}). Then inserting the solution of (10.19) $\left(\tilde{u}'_B, \tilde{v}'_{B'}\right)$ into $(10.18)_1$, we obtain all the fluxes $(10.3)_{2,3}$ and the productions $(10.4)_{2,3}$ in terms of the densities. Thus the closure is completed.

10.3.1 14-Moment System and Its Principal Subsystems

In the case of the 14-moment system ($N = 2$, $M = 1$), the results coincide with the ones described in Chap. 6 and the closed system in physical variables is given in (1.29).

According to the theory of principal subsystems by Boillat and Ruggeri [14] presented in Sect. 2.4, the following systems can be obtained as *principal subsystems* of the 14-moment system:

The 11-moment system ($N = 2$, $M = 0$): Setting $q_i = 0$ and neglecting the equation of q_j in (1.29), the following principal subsystem is obtained

$$\begin{aligned}
&\partial_t \rho + \partial_i (\rho v_i) = 0, \\
&\partial_t \left\{\rho v_j\right\} + \partial_i \left\{\rho v_i v_j + (p+\Pi)\delta_{ij} - \sigma_{\langle ij\rangle}\right\} = 0, \\
&\partial_t \left\{\rho v^2 + 3(p+\Pi)\right\} + \partial_i \left\{\rho v^2 v_i + 5 v_i (p+\Pi) - 2 v_l \sigma_{\langle li\rangle}\right\} = -\frac{3\Pi}{\tau_\Pi}, \\
&\partial_t \left\{\rho v_{\langle j} v_{k\rangle} - \sigma_{\langle jk\rangle}\right\} + \partial_i \left\{\rho v_{\langle j} v_{k\rangle} v_i + 2 v_{\langle j}\delta_{k\rangle i}(p+\Pi) - 2 v_{\langle j}\sigma_{k\rangle i} - v_i \sigma_{\langle jk\rangle}\right\} = \frac{\sigma_{\langle jk\rangle}}{\tau_S}, \\
&\partial_t \left\{\rho v^2 + 2\rho\varepsilon\right\} + \partial_i \left\{(\rho v^2 + 2\rho\varepsilon + 2p + 2\Pi) v_i - 2\sigma_{\langle li\rangle} v_l\right\} = 0.
\end{aligned} \tag{10.21}$$

This is the system in which F, F_i, F_{ij} and G_{ll} are retained as field variables.

The 6-moment system ($N = 2^-$, $M = 0$): Setting $\sigma_{\langle ij\rangle} = 0$ and neglecting the equation of $\sigma_{\langle ij\rangle}$ in (10.21), the following principal subsystem is obtained

$$\begin{aligned}
&\partial_t \rho + \partial_i (\rho v_i) = 0, \\
&\partial_t \left\{\rho v_j\right\} + \partial_i \left\{\rho v_i v_j + (p+\Pi)\delta_{ij}\right\} = 0, \\
&\partial_t \left\{\rho v^2 + 3(p+\Pi)\right\} + \partial_i \left\{\rho v^2 v_i + 5 v_i (p+\Pi)\right\} = -\frac{3\Pi}{\tau_\Pi}, \\
&\partial_t \left\{\rho v^2 + 2\rho\varepsilon\right\} + \partial_i \left\{(\rho v^2 + 2\rho\varepsilon + 2p + 2\Pi) v_i\right\} = 0.
\end{aligned} \tag{10.22}$$

This is the system in which F, F_i, F_{ll} and G_{ll} appear as field variables.

The 5-moment system ($N = 1$, $M = 0$): The 5-moment system (well-known as the Euler system), which retains F, F_i, and G_{ll} as field variables, is obtained as a principal subsystem of the 14-moment system by setting $\Pi = 0$ in the 6-moment principal subsystem. The system reads as follows:

$$\partial_t \rho + \partial_i (\rho v_i) = 0,$$
$$\partial_t (\rho v_j) + \partial_i (\rho v_i v_j + p\delta_{ij}) = 0,$$
$$\partial_t (\rho v^2 + 2\rho\varepsilon) + \partial_i ((\rho v^2 + 2\rho\varepsilon + 2p)v_i) = 0.$$

10.3.2 Closure for Higher-Order Systems

As examples of the higher order system than 14-moment system, the closed system with 17, 18 and 30 moments are shown. For simplicity, the one-dimensional variables are displayed with the following notation[1]:

$$\begin{aligned} F_{p,q} &= \int_{R^3}\int_0^\infty mf(c_1)^p(c^2)^q I^\alpha \, dI d\mathbf{c}, \\ G_{p',q'} &= \int_{R^3}\int_0^\infty mf\left(c^2 + \frac{2I}{m}\right)(c_1)^{p'}(c^2)^{q'} I^\alpha \, dI d\mathbf{c}, \end{aligned} \tag{10.23}$$

where the indexes p, q, p', and q' are the non-negative integers satisfying:

$$0 \leqslant p + 2q \leqslant N, \qquad 0 \leqslant p' + 2q' \leqslant M \qquad (M \leqslant N - 1).$$

In these systems, the first 14 moments ($F_{0,0}, F_{1,0}, F_{2,0}, F_{0,1}, G_{0,0}, G_{1,0}$) are common, and these moments have the following form:

$$F_{0,0} = \rho, \quad F_{1,0} = \rho v, \quad F_{2,0} = p + \Pi - \sigma + \rho v^2, \quad F_{0,1} = 3(p + \Pi) + \rho v^2,$$
$$G_{0,0} = 2\rho\varepsilon + \rho v^2, \quad G_{1,0} = 2q + v((D+2)p + 2\Pi - 2\sigma) + \rho v^3.$$

In the following, we show the explicit form of the remaining densities and fluxes. The nonequilibrium parts of the Galilean tensors of higher order are denoted as $\tilde{F}_{p,q} = \hat{F}_{p,q} - \hat{F}_{p,q}|_E$ and $\tilde{G}_{p',q'} = \hat{G}_{p',q'} - \hat{G}_{p',q'}|_E$.

[1]Upon inspection it can be seen that in the one-dimensional case, for any N and M, $\{F_{i_1 i_2 \ldots i_A},\ 0 \leqslant A \leqslant N\}$ is mapped into $\{F_{p,q},\ 0 \leqslant p + 2q \leqslant N\}$, and $\{G_{lli_1 i_2 \ldots i_{A'}},\ 0 \leqslant A' \leqslant M\}$ is mapped into $\{G_{p',q'},\ 0 \leqslant p' + 2q' \leqslant M\}$.

10.3.2.1 17-Moment System ($N = 3^-, M = 1$)

The independent variables are $F_{0,0}$, $F_{1,0}$, $F_{2,0}$, $F_{0,1}$, $F_{1,1}$, $G_{0,0}$, $G_{1,0}$ and the constitutive functions are $F_{3,0}, F_{2,1}, G_{2,0}$. The densities and fluxes are obtained as follows:

$$\begin{aligned}
F_{3,0} &= \frac{3}{5}\tilde{F}_{1,1} + 3v(p + \Pi - \sigma) + \rho v^3, \\
F_{1,1} &= \tilde{F}_{1,1} + v(5p + 5\Pi - 2\sigma) + \rho v^3, \\
F_{2,1} &= \frac{p}{\rho}(5p + 10\Pi - 7\sigma) + \frac{16}{5}v\tilde{F}_{1,1} + v^2(8p + 8\Pi - 5\sigma) + \rho v^4, \\
G_{2,0} &= \frac{p}{\rho}\{(D+2)p + (D+4)(\Pi - \sigma)\} + v\left(\frac{6}{5}\tilde{F}_{1,1} + 4q\right) \\
&\quad + v^2\{(D+5)p + 5\Pi - 5\sigma\} + \rho v^4.
\end{aligned}$$

10.3.2.2 18-Moment System ($N = 3^-, M = 2^-$)

The independent variables are $F_{0,0}$, $F_{1,0}$, $F_{2,0}$, $F_{0,1}$, $F_{1,1}$, $G_{0,0}$, $G_{1,0}$, $G_{0,1}$ and the constitutive functions are $F_{3,0}, F_{2,1}, G_{2,0}, G_{1,1}$. The densities and fluxes are obtained as follows:

$$\begin{aligned}
F_{3,0} &= \frac{3}{5}\tilde{F}_{1,1} + 3v(p + \Pi - \sigma) + \rho v^3, \\
F_{1,1} &= \tilde{F}_{1,1} + v(5p + 5\Pi - 2\sigma) + \rho v^3, \\
F_{2,1} &= \frac{10\tilde{G}_{0,1} + 3\frac{p}{\rho}\{5(D+7)p + 30\Pi - 7(D+7)\sigma\}}{3(D+7)} \\
&\quad + \frac{16}{5}v\tilde{F}_{1,1} + v^2(8p + 8\Pi - 5\sigma) + \rho v^4, \\
G_{2,0} &= \frac{1}{3}\tilde{G}_{0,1} + v\left(\frac{6}{5}\tilde{F}_{1,1} + 4q\right) \\
&\quad + \frac{p}{\rho}\{(D+2)p - (D+4)\sigma\} + v^2\{(D+5)p + 5\Pi - 5\sigma\} + \rho v^4, \\
G_{0,1} &= \tilde{G}_{0,1} + 3(D+2)\frac{p^2}{\rho} + 2v(\tilde{F}_{1,1} + 2q) + v^2\{(D+7)p + 7\Pi - 4\sigma\} + \rho v^4,
\end{aligned}$$

$$G_{1,1} = \frac{p}{\rho}\left\{(D+6)\tilde{F}_{1,1} + 10q\right\}$$
$$+ v\frac{5(D+11)\tilde{G}_{0,1} + 3\frac{p}{\rho}\left\{5(D+7)(D+4)p + 60\Pi - 2(D+7)(D+11)\sigma\right\}}{3(D+7)}$$
$$+ v^2\left(\frac{27}{5}\tilde{F}_{1,1} + 6q\right) + v^3\left\{(D+11)p + 11\Pi - 8\sigma\right\} + \rho v^5.$$

The 18 fields system from phenomenological point of view was studied in [15].

10.3.2.3 30-Moment System ($N = 3, M = 2$)

The independent variables are $F_{0,0}$, $F_{1,0}$, $F_{2,0}$, $F_{0,1}$, $F_{3,0}$, $F_{1,1}$, $G_{0,0}$, $G_{1,0}$, $G_{2,0}$, $G_{0,1}$ and the constitutive functions are $F_{4,0}$, $F_{2,1}$, $G_{3,0}$, $G_{1,1}$. The densities and fluxes are obtained as follows:

$$F_{3,0} = \tilde{F}_{3,0} + 3v(p + \Pi - \sigma) + \rho v^3,$$

$$F_{1,1} = \tilde{F}_{1,1} + v(5p + 5\Pi - 2\sigma) + \rho v^3,$$

$$F_{4,0} = \frac{6\left\{(D+7)\tilde{G}_{2,0} - \tilde{G}_{0,1}\right\} + 3\frac{p}{\rho}(D+4)\left\{(D+7)p + 6\Pi\right\}}{(D+4)(D+7)}$$
$$+ 4v\tilde{F}_{3,0} + 6v^2(p + \Pi - \sigma) + \rho v^4,$$

$$F_{2,1} = \frac{(D-3)\tilde{G}_{0,1} + 7(D+7)\tilde{G}_{2,0} + 5\frac{p}{\rho}(D+4)\left\{(D+7)p + 6\Pi\right\}}{(D+4)(D+7)}$$
$$+ 2v(\tilde{F}_{1,1} + \tilde{F}_{3,0}) + v^2(8p + 8\Pi - 5\sigma) + \rho v^4,$$

$$G_{2,0} = \tilde{G}_{2,0} + (D+2)\frac{p^2}{\rho} + 2v(\tilde{F}_{3,0} + 2q) + v^2\left\{(D+5)p + 5\Pi - 5\sigma\right\} + \rho v^4,$$

$$G_{0,1} = \tilde{G}_{0,1} + 3(D+2)\frac{p^2}{\rho} + 2v(\tilde{F}_{1,1} + 2q) + v^2\left\{(D+7)p + 7\Pi - 4\sigma\right\} + \rho v^4,$$

$$G_{3,0} = \frac{p}{\rho}\left\{(D+6)\tilde{F}_{3,0} + 6q\right\}$$
$$+ 3v\frac{(D+7)(D+8)\tilde{G}_{2,0} - 4\tilde{G}_{0,1} + \frac{p}{\rho}(D+4)\left\{(D+4)(D+7)p + 12\Pi\right\}}{(D+4)(D+7)}$$
$$+ v^2(7\tilde{F}_{3,0} + 6q) + v^3((D+9)p + 9\Pi - 9\sigma) + \rho v^5,$$

$$G_{1,1} = \frac{p}{\rho}\left\{(D+6)\tilde{F}_{1,1} + 10q\right\}$$
$$+ v\frac{(D+11)\left\{(D+2)\tilde{G}_{0,1} + 2(D+7)\tilde{G}_{2,0}\right\} + 5\frac{p}{\rho}(D+4)\left\{(D+4)(D+7)p + 12\Pi\right\}}{(D+4)(D+7)}$$
$$+ v^2(3\tilde{F}_{1,1} + 4\tilde{F}_{3,0} + 6q) + v^3((D+11)p + 11\Pi - 8\sigma) + \rho v^5.$$

10.4 Characteristic Velocities of the (N, M)-System

The set of the characteristic velocities $\lambda_{(N,M)}$ of the system (10.15) in the propagation direction with a unit vector $\mathbf{n} \equiv (n^i)$ is the set of the roots of the characteristic polynomial $T_{(N,M)}$:

$$T_{(N,M)} = \det\left[\begin{pmatrix} J^0_{iAB} & J^1_{iAB'} \\ J^1_{iA'B} & J^2_{iA'B'} \end{pmatrix} n^i - \lambda_{(N,M)} \begin{pmatrix} J^0_{AB} & J^1_{AB'} \\ J^1_{A'B} & J^2_{A'B'} \end{pmatrix}\right] = 0.$$

In particular, the wave velocities for disturbances propagating in an equilibrium state are the solutions of the characteristic polynomial $T^E_{(N,M)}$:

$$T^E_{(N,M)} = \det\left[\begin{pmatrix} J^{0|E}_{iAB} & J^{1|E}_{iAB'} \\ J^{1|E}_{iA'B} & J^{2|E}_{iA'B'} \end{pmatrix} n^i - \lambda^E_{(N,M)} \begin{pmatrix} J^{0|E}_{AB} & J^{1|E}_{AB'} \\ J^{1|E}_{A'B} & J^{2|E}_{A'B'} \end{pmatrix}\right] = 0.$$

Hereafter, the analysis is restricted to the one-dimensional case for the sake of simplicity. Under this assumption, in the same manner as (10.23), the relations (10.20) can be rewritten as

$$\begin{aligned} J^{0|E}_{p+r,q+s} &= -\frac{m^2}{k_B}\int_{R^3}\int_0^\infty f^E c_1^{p+r}(c^2)^{q+s} I^\alpha \, dI d\mathbf{c} \\ &= J^{\mathscr{M}}_{p+r,q+s}, \end{aligned} \tag{10.24}$$

$$\begin{aligned} J^{1|E}_{p'+r',q'+s'} &= -\frac{m^2}{k_B}\int_{R^3}\int_0^\infty f^E c_1^{p'+q'}(c^2)^{r'+s'}\left(c^2 + \frac{2I}{m}\right) I^\alpha \, dI d\mathbf{c} \\ &= J^{\mathscr{M}}_{p'+r',q'+s'+1} + 2c_s^2(1+\alpha) J^{\mathscr{M}}_{p'+r',q'+s'}, \end{aligned} \tag{10.25}$$

$$\begin{aligned} J^{2|E}_{p'+r',q'+s'} &= -\frac{m^2}{k_B}\int_{R^3}\int_0^\infty f^E c_1^{p'+r'}(c^2)^{q'+s'}\left(c^2 + \frac{2I}{m}\right)^2 I^\alpha \, dI d\mathbf{c} \\ &= J^{\mathscr{M}}_{p'+r',q'+s'+2} + 4c_s^2(1+\alpha)\left(J^{\mathscr{M}}_{p'+r',q'+s'+1} + c_s^2(2+\alpha) J^{\mathscr{M}}_{p'+r',q'+s'}\right), \end{aligned} \tag{10.26}$$

where

$$0 \leqslant r + 2s \leqslant N, \qquad 0 \leqslant r' + 2s' \leqslant M \qquad (M \leqslant N-1),$$

and

$$J^{\mathcal{M}}_{p,q} = -\frac{m^2}{k_B}\int_{R^3} f^{\mathcal{M}} c_1^p (c^2)^q \, d\mathbf{c}$$
$$= -\frac{m}{k_B}\rho\left(\frac{k_B T}{m}\right)^{\frac{p+2q}{2}} \frac{2^{\frac{p}{2}+q}}{p+1}\Gamma\left(\frac{p+3}{2}+q\right)\frac{1+(-1)^p}{\sqrt{\pi}}.$$

Introducing the notation

$$\tilde{J}^{a|E}_{p,q} = -\lambda J^{a|E}_{p,q} + J^{a|E}_{p+1,q} \qquad (a = 0, 1, 2),$$

we rewrite the characteristic polynomial $T^E_{(N,M)}$ in one-dimensional case as

$$T^E_{(N,M)} = \det\begin{pmatrix} \tilde{J}^{0|E}_{p+r,q+s} & \tilde{J}^{1|E}_{p'+r,q'+s} \\ \tilde{J}^{1|E}_{p+r',q+s'} & \tilde{J}^{2|E}_{p'+r',q'+s'} \end{pmatrix} =$$
$$= \det\begin{pmatrix} \tilde{J}^{\mathcal{M}}_{p+r,q+s} & \tilde{J}^{\mathcal{M}}_{p'+r,q'+s+1} \\ \tilde{J}^{\mathcal{M}}_{p+r',q+s'+1} & \tilde{J}^{\mathcal{M}}_{p'+r',q'+s'+2} + 4c_s^4(1+\alpha)\tilde{J}^{\mathcal{M}}_{p'+r',q'+s'} \end{pmatrix} = 0, \tag{10.27}$$

where we make use of the properties of the determinant and of the relations (10.24), (10.25) and (10.26).

10.4.1 Characteristic Velocities of the 14-, 11-, 6- and 5-Moment Systems

For simplicity, the set $\hat{\lambda}^E_{(N,M)}$ of the dimensionless characteristic velocities is introduced, starting from the set $\lambda^E_{(N,M)}$, as follows:

$$\hat{\lambda}^E_{(N,M)} = \left\{\frac{\lambda}{c_0}, \ \forall\lambda \in \lambda^E_{(N,M)}\right\},$$

where

$$c_0 = \sqrt{\frac{5}{3}\frac{k_B}{m}T_0}$$

is the sound velocity of a monatomic gas at the temperature T_0.

The explicit form of the dimensionless characteristic velocities in the case of the 14-, 11-, 6- and 5-moment systems are derived.

14-moment system ($N = 2$, $M = 1$): The characteristic polynomial (10.27) is

$$T^{E}_{(2,1)} = \det \mathscr{J}^{E} = 0, \qquad \mathscr{J}^{E} = \begin{pmatrix} \tilde{J}^{0|E}_{0,0} & \tilde{J}^{0|E}_{1,0} & \tilde{J}^{0|E}_{0,1} & \tilde{J}^{0|E}_{2,0} & \tilde{J}^{1|E}_{0,0} & \tilde{J}^{1|E}_{0,1} \\ \tilde{J}^{0|E}_{1,0} & \tilde{J}^{0|E}_{2,0} & \tilde{J}^{0|E}_{1,1} & \tilde{J}^{0|E}_{3,0} & \tilde{J}^{1|E}_{1,0} & \tilde{J}^{1|E}_{1,1} \\ \tilde{J}^{0|E}_{0,1} & \tilde{J}^{0|E}_{1,1} & \tilde{J}^{0|E}_{0,2} & \tilde{J}^{0|E}_{2,1} & \tilde{J}^{1|E}_{0,1} & \tilde{J}^{1|E}_{0,2} \\ \tilde{J}^{0|E}_{2,0} & \tilde{J}^{0|E}_{3,0} & \tilde{J}^{0|E}_{2,1} & \tilde{J}^{0|E}_{4,0} & \tilde{J}^{1|E}_{2,0} & \tilde{J}^{1|E}_{2,1} \\ \tilde{J}^{1|E}_{0,0} & \tilde{J}^{1|E}_{1,0} & \tilde{J}^{1|E}_{0,1} & \tilde{J}^{1|E}_{2,0} & \tilde{J}^{2|E}_{0,0} & \tilde{J}^{2|E}_{0,1} \\ \tilde{J}^{1|E}_{0,1} & \tilde{J}^{1|E}_{1,1} & \tilde{J}^{1|E}_{0,2} & \tilde{J}^{1|E}_{2,1} & \tilde{J}^{2|E}_{0,1} & \tilde{J}^{2|E}_{0,2} \end{pmatrix}, \tag{10.28}$$

and the equilibrium characteristic velocities in one-dimensional case are calculated as follows:

$$\hat{\lambda}^{E}_{(2,1)} = \left\{ 0 \text{ (multiplicity 2)}, \ \pm\sqrt{\frac{3}{5}}\sqrt{\frac{2D + 7 \pm \sqrt{D^2 + 16D + 37}}{D+2}} \right\}. \tag{10.29}$$

The 11-moment system ($N = 2$, $M = 0$): The characteristic polynomial of the 11-moment system is obtained as the determinant of the matrix $\mathscr{J}^{E}$ after removing the last row and the last column. The characteristic velocities are given by

$$\hat{\lambda}^{E}_{(2,0)} = \left\{ 0 \text{ (multiplicity 3)}, \ \pm\sqrt{\frac{9}{5}} \right\}. \tag{10.30}$$

The 6-moment system ($N = 2^-$, $M = 0$): In this case, the characteristic polynomial is obtained as the determinant of the matrix $\mathscr{J}^{E}$ after removing the third and last row and column. In this case the characteristic velocities are given by

$$\hat{\lambda}^{E}_{(2^-,0)} = \{ 0 \text{ (multiplicity 2)}, \ \pm 1 \}. \tag{10.31}$$

The 5-moment system ($N = 1$, $M = 0$): The characteristic polynomial of the 5-moment system (Euler system) is the determinant of the matrix $\mathscr{J}^{E}$ after removing the last three rows and columns. The characteristic velocities are given by

$$\hat{\lambda}^{E}_{(1,0)} = \left\{ 0, \ \pm\sqrt{\frac{3}{5}}\sqrt{\frac{D+2}{D}} \right\}. \tag{10.32}$$

10.4.2 Systems with D-Independent Characteristic Velocities

By analyzing the features of the matrix $\mathscr{J}^E$, by means of (10.20), it seems that, as expected, the equilibrium characteristic velocities $\lambda^E_{(N,M)}$ depend on the parameter α, and thus on the degrees of freedom of a molecule D. Nonetheless, as seen, for example, in (10.30) for the case of the 11-moment system and in (10.31) for the case of the 6-moment system, this is not always the case. Upon inspection of the features of the matrix $\mathscr{J}^E$, the following theorem holds:

Theorem 10.5 *If $M < N-1$, the equilibrium characteristic velocities of an (N,M)-system, $\lambda^E_{(N,M)}$, are independent of the degrees of freedom of a molecule D. More specifically, the equilibrium characteristic velocities coincide with those of an (N)-system, $\lambda^E_{(N)}$,* and *those of an (M)-system, $\lambda^E_{(M)}$ of monatomic gases (Fig. 10.1):*

$$\lambda^E_{(N,M)} = \lambda^E_{(N)} \cup \lambda^E_{(M)}.$$

In particular, the maximum equilibrium characteristic velocity of the (N,M)-system, $\lambda^{E,\max}_{(N,M)}$, is independent of the order M, and coincides with the one of the (N)-systems for monatomic gases, i.e.,

$$\lambda^{E,\max}_{(N,M)} = \lambda^{E,\max}_{(N)}.$$

Proof In the characteristic polynomial $T^E_{(N,M)}$ described as (10.27), all components of $\tilde{J}^{\mathscr{M}}_{p+r',q+s'+1}$ are included in $\tilde{J}^{\mathscr{M}}_{p+r,q+s}$ since $M < N-1$, therefore these are subtracted from the determinant. Similarly, all components of $\tilde{J}^{\mathscr{M}}_{p'+r',q'+s'+2}$ are subtracted by $\tilde{J}^{\mathscr{M}}_{p'+r,q'+s+1}$. Then, $T^E_{(N,M)}$ may be manipulated as follows:

$$\begin{aligned} T^E_{(N,M)} &= \det\begin{pmatrix} \tilde{J}^{\mathscr{M}}_{p+r,q+s} & 0_{p'+r,q'+s+1} \\ 0_{p+r',q+s'+1} & 4c_s^4(1+\alpha)\tilde{J}^{\mathscr{M}}_{p'+r',q'+s'} \end{pmatrix} \\ &= \left(4c_s^4(1+\alpha)\right)^d \det\left(\tilde{J}^{\mathscr{M}}_{p+r,q+s}\right) \det\left(\tilde{J}^{\mathscr{M}}_{p'+r',q'+s'}\right), \end{aligned} \tag{10.33}$$

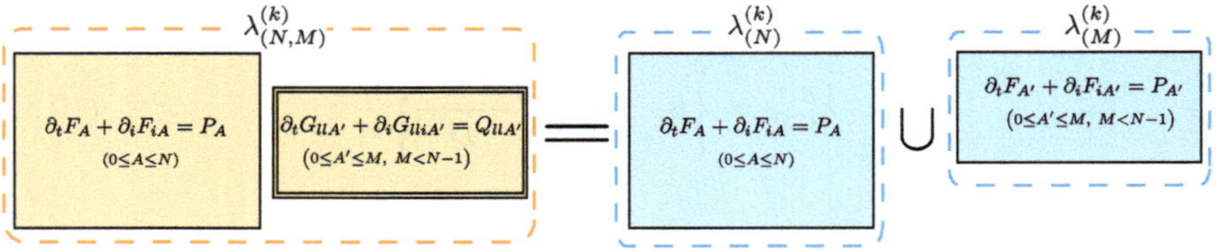

Fig. 10.1 Graphical representation of the collapse of the characteristic velocities derived from (N,M)-system (*left*) into those derived from a (N)-system and a (M)-system (*right*) when $M < N-1$

with

$$d = \frac{M + 1 + \text{mod}(M + 1, 2)}{2} = \begin{cases} \frac{M+1}{2} & \text{if } M \text{ is odd,} \\ \frac{M+2}{2} & \text{if } M \text{ is even,} \end{cases}$$

where 0 with indexes denotes the zero block matrix. From (10.33), it becomes evident that all the characteristic velocities that are obtained as roots of the polynomial $T^E_{(N,M)}$ are independent of the degrees of freedom of a molecule D, and these are identified as the characteristic velocities of (N)-system and of the (M)-system of rarefied monatomic gases.

Finally, the claim that the maximum characteristic velocity coincides with those of the (N)-system for the monatomic gas is a consequence of the subcharacteristic conditions (2.25) (see Sect. 2.4).

It is worth pointing out that Theorem 10.5 is true for the (N^-, M)-system with $M < N - 1$, since the proof comes from the corresponding relationship between $F_{llA'}$ and $G_{llA'}$ $(0 \leqslant A' \leqslant M)$.

As already noticed, the equilibrium characteristic velocities of the 11-moment and of the 6-moment systems do not depend on D. In these cases it is easily seen that the set of the equilibrium characteristic velocities are obtainable according to Theorem 10.5:

The $(2, 0)$-*system (11-moment system):* In this case the equilibrium characteristic velocities of the (2)-system and of the (0)-system of monatomic gases are:

$$\hat{\lambda}^E_{(2)} = \left\{ 0 \text{ (multiplicity 2)}, \ \pm\sqrt{\frac{9}{5}} \right\}, \qquad \hat{\lambda}^E_{(0)} = \{0\},$$

which, compared to (10.30), show that $\lambda^E_{(2,0)} = \lambda^E_{(2)} \cup \lambda^E_{(0)}$ and $\lambda^{E,\max}_{(2,0)} = \lambda^{E,\max}_{(2)}$.

The $(2^-, 0)$-*system (6-moment system):* Also in this case it is easy to see that the equilibrium characteristic velocities (10.31) are those of the (2^-)-system *and* those of the (0)-system of monatomic gases, i.e., $\lambda^E_{(2^-,0)} = \lambda^E_{(2^-)} \cup \lambda^E_{(0)}$ with

$$\hat{\lambda}^E_{(2^-)} = \{0, \ \pm 1\}, \qquad \hat{\lambda}^E_{(0)} = \{0\},$$

and $\lambda^{E,\max}_{(2^-,0)} = \lambda^{E,\max}_{(2^-)}$.

Remark Theorem 10.5 is satisfied also for the system of (N, M^-), (N^-, M) and (N^-, M^-) because the proof of it is true when each G-moment has the correspondence in F-moments. From the requirement of the Galilean invariance, the possible (N^-, M)-system has $M < N - 1$, therefore (N^-, M)-system is always independent of D.

Remark When $M < N - 1$ and the trace part of the N-tensorial equation in the F-series, i.e., $F_{lli_1\cdots i_{N-2}}$ is absent, the characteristic velocities can depend on D. The 10-moment system in which F, F_i, $F_{\langle ij\rangle}$ and G_{ll} are retained as field variables is an example.

In the following, according with the physical case of 5 moments (Euler system) and 14 moments (hyperbolic counterpart of NFS system), the case in which the trace part of the N-tensorial equation always exists is considered. Therefore the above-remarked case is excluded, and combining the results of Theorems 10.1 and 10.5, it is possible to draw the following conclusion:

Statement 3 *The (N, M)-system satisfies the relevant features of Galilean invariance and has equilibrium characteristic velocities depending on the degrees of freedom D if and only if*

$$M = N - 1. \tag{10.34}$$

This conclusion is also true for the $(N, (N-1)^-)$ and $(N^-, (N-1)^-)$-systems, while the $(N^-, (N-1))$-system is excluded.

Taking into account (10.5) and (10.6) in this case the number of moments for an $(N, N-1)$-system, is given by

$$n_{(N,N-1)} = \frac{1}{6}(N+1)(N+2)(2N+3), \tag{10.35}$$

and, for an $(N, (N-1)^-)$-system, is given by

$$n_{(N,(N-1)^-)} = \frac{1}{6}(2N^3 + 9N^2 + N + 12), \tag{10.36}$$

and, for an $(N^-, (N-1)^-)$-system, is given by

$$n_{(N^-,(N-1)^-)} = 1 + \frac{1}{6}N(N-1)(11 + 2N). \tag{10.37}$$

As examples, we present some systems with the list of the densities, the number of moments, and the dependence of $\hat{\lambda}^E$ on D in Table 10.1.

Table 10.1 Densities, number of moments, and the dependence of $\hat{\lambda}^E$ on D for some (N, M)-systems

(N, M)	Densities	Number of moments	Dependence of $\hat{\lambda}^E$ on D
$(1, 0)$	(F, F_i, G_{ll})	5	Yes
$(2^-, 0)$	(F, F_i, F_{ll}, G_{ll})	6	No
$(2, 0)$	(F, F_i, F_{ij}, G_{ll})	11	No
(2,1)	$(F, F_i, F_{ij}, G_{ll}, G_{lli})$	14	Yes
$(3^-, 1)$	$(F, F_i, F_{ij}, F_{lli}, G_{ll}, G_{lli})$	17	No
$(3^-, 2^-)$	$(F, F_i, F_{ij}, F_{lli}, G_{ll}, G_{lli}, G_{lljj})$	18	Yes
$(3, 2^-)$	$(F, F_i, F_{ij}, F_{ijk}, G_{ll}, G_{lli}, G_{lljj})$	25	Yes
(3,2)	$(F, F_i, F_{ij}, F_{ijk}, G_{ll}, G_{lli}, G_{llij})$	30	Yes

10.5 Characteristic Velocities of the $(N, N-1)$-System and the Analysis of the Cases: $D \to 3$ and $D \to \infty$

In this section, an analysis of the equilibrium characteristic velocities of the physically relevant systems, i.e., $(N, N-1)$-systems, is presented. In particular, the equilibrium maximum characteristic velocity is discussed.

The maximum equilibrium characteristic velocity of an $(N, N-1)$-system of rarefied polyatomic gases is limited by those of monatomic gases, as shown below:

Theorem 10.6 *For any truncation order N, the maximum equilibrium characteristic velocity of an $(N, N-1)$-system of a rarefied polyatomic gas, $\lambda^{\max}_{(N,N-1)}$, is bounded by the maximum equilibrium characteristic velocity of the $(N+1^-)$-system and of the (N)-system of a rarefied monatomic gas as follows:*

$$\lambda^{\max}_{(N)} \leqslant \lambda^{\max}_{(N,N-1)} \leqslant \lambda^{\max}_{(N+1^-)}. \tag{10.38}$$

In order to prove this theorem, it suffices to notice that both the upper and lower bounds are obtained from Theorem 10.5 and the subcharacteristic conditions:

$$\lambda^{\max}_{(N+1^-)} = \lambda^{\max}_{(N+1^-,M)} = \lambda^{\max}_{(N+1^-,N-1)} \geqslant \lambda^{\max}_{(N,N-1)}, \qquad \forall M < N$$

$$\lambda^{\max}_{(N,N-1)} \geqslant \lambda^{\max}_{(N,M)} = \lambda^{\max}_{(N)}, \qquad \forall M < N-1.$$

Hereafter, the influences of the degrees of freedom of a molecule D on the characteristic velocities are studied.

10.5.1 Limit Case: $D \to 3$

It was proved [16] that, in the limit case $D \to 3$, the solutions of the 14 moments of polyatomic gas converge to those of the 13 moments system of a monoatomic gas (see Sect. 5.6).

A similar proof can be done for a system with generic number of moments. In [17], it was proved that the $(N, N-1)$-system for a rarefied polyatomic gas (which is composed by $n_{(N,N-1)} = \frac{1}{6}(N+1)(N+2)(2N+3)$ equations) converges to the system of order $(N+1)^-$ for a rarefied monatomic gas (which is composed of $n_{(N+1^-)} = \frac{1}{6}(N+1)(N^2+8N+6)$ equations).

In this limit, the differences between the trace part of F-hierarchy and G-hierarchy,

$$\Pi_\alpha \equiv \lim_{D\to 3}(F_{ll\alpha} - G_{ll\alpha}) \qquad (0 \leqslant \alpha \leqslant N-2)$$

are governed by the following balance equations:

$$\partial_t \Pi_\alpha + \partial_i \Pi_{i\alpha} = \lim_{D\to 3}(P_{ll\alpha} - Q_{ll\alpha}), \tag{10.39}$$

(composed of $n_{(N-2)} = \frac{1}{6}(N-1)N(N+1) = n_{(N,N-1)} - n_{(N+1^-)}$ equations). At least for BGK approximation, it was proved in [17] that $\Pi_{i\alpha}$ and the right-hand side terms are linear combination of Π_α and therefore the system (10.39) admits only the zero solution provided that the initial data are compatible with a monatomic gas, i.e., $\Pi_\alpha(\mathbf{x}, 0) = 0$. Therefore the binary-hierarchy can be regarded as the single-hierarchy of rarefied monatomic gases. A graphical representation of the collapse of the $(N, N-1)$-system into the $(N+1^-)$-system is presented in Fig. 10.2.

As a consequence of the convergence of the solutions of polyatomic gases towards those of monatomic gases, the characteristic velocities follow the same rule. It should be noted that when the characteristic polynomial (10.27) is calculated in this limit, we may obtain the characteristic velocities that is composed of not only those of the $(N+1^-)$-system but also those of the $(N-2)$-system:

$$\lim_{D\to 3} \lambda^E_{(N,N-1)} = \lambda^E_{(N+1^-)} \cup \lambda^E_{(N-2)}. \tag{10.40}$$

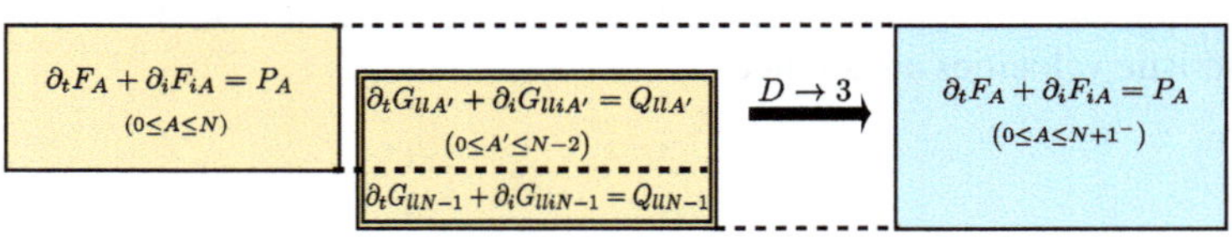

Fig. 10.2 Graphical representation of the collapse of the $(N, N-1)$-system (*left*) into a $(N+1^-)$-system (*right*) when $D \to 3$. The $(N+1^-)$-system is a (N)-system augmented by the trace part of order N, i.e., augmented by the balance laws for the moments $F_{lli_1\cdots i_{N-1}}$

However, the characteristic velocities of the $(N-2)$-system are related to the balance equations (10.39), which vanish in this limit.

Moreover (see Theorem 10.5), it is noticeable that, in the limit case $D \to 3$, the maximum characteristic velocity of the $(N, N-1)$-system coincides with the one of the $(N+1^-)$-system for rarefied monatomic gases, which in turn coincides with that of the $(N+1^-, M)$-system for any $M < N$, i.e.,

$$\lim_{D\to 3} \lambda^{\max}_{(N,N-1)} = \lambda^{\max}_{(N+1^-)} = \lambda^{\max}_{(N+1^-,M)}, \qquad \forall M < N. \tag{10.41}$$

It is easily proved that for the Euler system (10.32) and the 14-moment system (10.29), the relations (10.40) and (10.41) are confirmed:

Euler system: For the Euler system, there is no I_A. Therefore, in the limit case $D \to 3$, it is enough to consider the Euler system for a monatomic gas, i.e., the (2^-)-system. This corresponds to the fact that the Euler system can be applicable for any fluids. The characteristic velocities are given by

$$\lim_{D\to 3} \hat{\lambda}^E_{(1,0)} = \hat{\lambda}^E_{(2^-)} = \{0, \ \pm 1\}.$$

The 14-moment system: For the 14-moment system, in the limit case $D \to 3$, the dynamic pressure $\Pi \propto \lim_{D\to 3}(F_{ll} - G_{ll}) = I_0$ vanishes, then the solutions of 14-moment system coincide with those of the 13-moment rarefied monatomic gas system, i.e., the (3^-)-system. The characteristic velocities, converging to those of the 13-moment system of rarefied monatomic gases, are the following:

$$\lim_{D\to 3} \hat{\lambda}^E_{(2,1)} = \hat{\lambda}^E_{(3^-)} \cup \hat{\lambda}^E_{(0)} = \left\{ 0 \text{ (multiplicity 2)}, \ \pm\sqrt{\frac{3}{5}}\sqrt{\frac{13 \pm \sqrt{94}}{5}} \right\},$$

where

$$\hat{\lambda}^E_{(3^-)} = \left\{ 0, \ \pm\sqrt{\frac{3}{5}}\sqrt{\frac{13 \pm \sqrt{94}}{5}} \right\}$$

are the characteristic velocities of the 13-moment system of a monatomic gas.

10.5.2 Limit Case: $D \to \infty$

On the other hand, for $D \to \infty$, the following theorem holds:

Theorem 10.7 *When $D \to \infty$, the $n_{(N,N-1)} = \frac{1}{6}(N+1)(N+2)(2N+3)$ characteristic velocities of the $(N, N-1)$-system coincide with the $n_{(N)} = \frac{1}{6}(N +$

1)$(N+2)(N+3)$ *characteristic velocities of the* (N)*-system, and with the* $n_{(N-1)} = \frac{1}{6}N(N+1)(N+2)$ *characteristic velocities of the* $(N-1)$*-system.*

$$\lim_{D\to\infty} \lambda^E_{(N,N-1)} = \lambda^E_{(N)} \cup \lambda^E_{(N-1)}.$$

In particular, the maximum characteristic velocity of the $(N, N-1)$*-system coincides with the one of the* (N)*-system for monatomic gases that is in turn coincident with the one of the* (N, M)*-system for any* $M < N - 1$*, i.e.,*

$$\lim_{D\to\infty} \lambda^{\max}_{(N,N-1)} = \lambda^{\max}_{(N)} = \lambda^{\max}_{(N,M)}, \qquad \forall M < N-1. \tag{10.42}$$

Proof The characteristic polynomial $T^E_{(N,N-1)}$, as seen from (10.27), can be written in the form:

$$T^E_{(N,N-1)} = \left(4c_s^4(1+\alpha)\right)^d \det \begin{pmatrix} \tilde{J}^{\mathcal{M}}_{p+r,q+s} & \tilde{J}^{\mathcal{M}}_{p'+r,q'+s+1} \\ \frac{\tilde{J}^{\mathcal{M}}_{p+r',q+s'+1}}{4c_s^4(1+\alpha)} & \frac{\tilde{J}^{\mathcal{M}}_{p'+r',q'+s'+2}}{4c_s^4(1+\alpha)} + \tilde{J}^{\mathcal{M}}_{p'+r',q'+s'} \end{pmatrix} = 0,$$

therefore

$$\begin{aligned} \lim_{\alpha\to\infty} \frac{T^E_{(N,N-1)}}{\left(4c_s^4(1+\alpha)\right)^d} &= \det \begin{pmatrix} \tilde{J}^{\mathcal{M}}_{p+r,q+s} & \tilde{J}^{\mathcal{M}}_{p'+r,q'+s+1} \\ 0_{p+r',q+s'} & \tilde{J}^{\mathcal{M}}_{p'+r',q'+s'} \end{pmatrix} = \\ &= \det\left(\tilde{J}^{\mathcal{M}}_{p+r,q+s}\right) \det\left(\tilde{J}^{\mathcal{M}}_{p'+r',q'+s'}\right) = 0. \end{aligned} \tag{10.43}$$

From (10.43), it is shown that the characteristic velocities of the $(N, N-1)$-system are those of the (N)-system and of the $(N-1)$-system of monatomic gases. Since the $(N-1)$-system is a principal subsystem of the (N)-system, the subcharacteristic conditions guarantee that the maximum characteristic velocity of the $(N, N-1)$-system coincides with the maximum characteristic velocity of the (N)-system. Moreover, from the Theorem 10.5, the maximum characteristic velocity coincides with that of the (N, M)-system for any $M < N - 1$. This means that, in the limit $D \to \infty$, the maximum characteristic velocity is determined only by the order of truncation of the F-hierarchy.

The collapse of the characteristic velocities of the $(N, N-1)$-system into those derived from a (N)-system and a $(N-1)$-system as $D \to \infty$ is graphically shown in Fig. 10.3.

For the Euler system (10.32) and the 14-moment system (10.29), in agreement with Theorem 10.7, the following results are obtained:

The 14-moment system: The characteristic velocities of the 14-moment system for rarefied polyatomic gases (10.29) converge to those of the (2)-system for monatomic gases (i.e., the 10-moment system retaining F, F_i, and F_{ij} as independent fields), and those of the (1)-system (i.e., the 4-moment system retaining F and F_i as

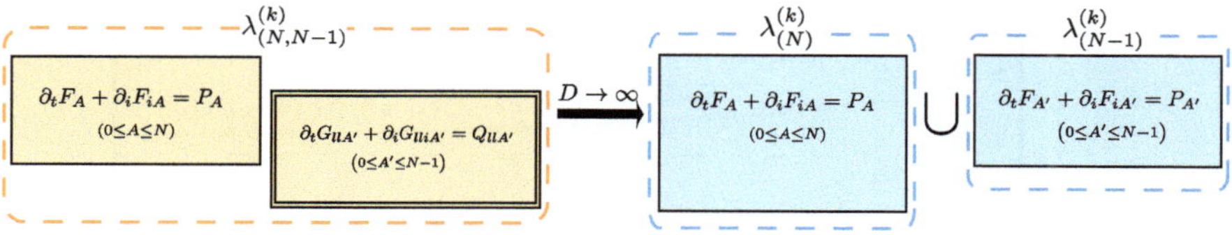

Fig. 10.3 Graphical representation of the collapse of the characteristic velocities derived from $(N, N-1)$-system (*left*) into those derived from a (N)-system and a ($N-1$)-system (*right*) when $D \to \infty$

independent fields):

$$\lim_{D\to\infty} \lambda^E_{(2,1)} = \hat{\lambda}^E_{(2)} \cup \hat{\lambda}^E_{(1)} = \left\{0 \text{ (multiplicity 2)}, \ \pm\sqrt{\frac{9}{5}}\right\} \cup \left\{\pm\sqrt{\frac{3}{5}}\right\} =$$

$$\left\{0 \text{ (multiplicity 2)}, \ \pm\sqrt{\frac{3}{5}}, \ \pm\sqrt{\frac{9}{5}}\right\}.$$

Euler system: The characteristic velocities of the Euler system for rarefied polyatomic gases (10.32) converge to those of the (1)-system (i.e., the 4-moment system retaining F and F_i as independent fields) and that of the (0)-system (i.e., the system retaining only F as independent field):

$$\lim_{D\to\infty} \lambda^E_{(1,0)} = \hat{\lambda}^E_{(1)} \cup \hat{\lambda}^E_{(0)} = \left\{\pm\sqrt{\frac{3}{5}}\right\} \cup \{0\} = \left\{0, \ \pm\sqrt{\frac{3}{5}}\right\}.$$

10.5.3 The Case: 3 < D < ∞

In the case: $3 < D < \infty$, as a corollary of Theorem 10.6, considering (10.41) and (10.42), we obtain the following result:

Theorem 10.8 *For any truncation order N, the maximum equilibrium characteristic velocity of an (N, M)-system of polyatomic gases is bounded as follows:*

$$\lim_{D\to\infty} \lambda^{\max}_{(N,N-1)} \leqslant \lambda^{\max}_{(N,N-1)} \leqslant \lim_{D\to 3} \lambda^{\max}_{(N,N-1)}. \tag{10.44}$$

In Fig. 10.4, dependence of $\lambda^{\max}_{(N,N-1)}$ on D (in the one-dimensional case) for the $(1,0)$-system (5-moment Euler system), for the $(2,1)$-system (14-moment system) and for the $(3,2)$-system (30-moment system) is plotted.

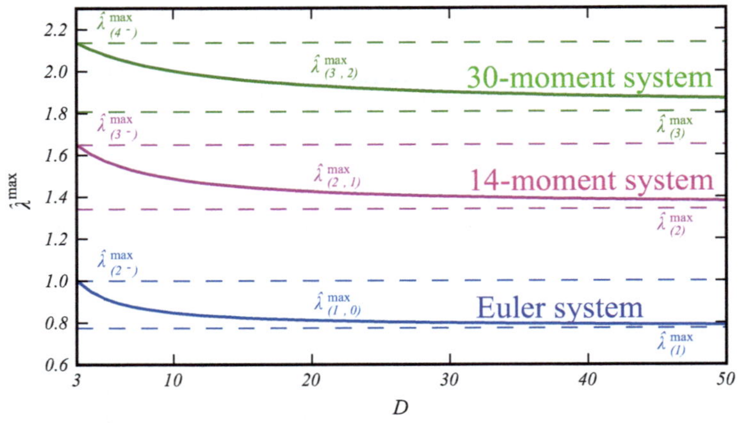

Fig. 10.4 Nondimensional maximum characteristic velocity (at equilibrium) of the $(N, N-1)$-system, $\hat{\lambda}^{\max}_{(N,N-1)} = \lambda^{\max}_{(N,N-1)}/c_0$, for three different values of N ($N = 1$: Euler system; $N = 2$: 14-moment system; $N = 3$: 30-moment system) as a function of the degrees of freedom of a molecule D. The limit value of the non-dimensional characteristic velocity for $D \to 3$ and $D \to \infty$ (respectively, $\hat{\lambda}^{\max}_{(N+1-)} = \lambda^{\max}_{(N+1-)}/c_0$ and $\hat{\lambda}^{\max}_{(N)} = \lambda^{\max}_{(N)}/c_0$) are indicated with *dashed lines* (c_0 being the sound velocity in the monatomic gas)

10.6 Dependence of the Maximum Characteristic Velocity on the Order N

In this section, dependence of the maximum equilibrium characteristic velocity of $(N, N-1)$-system for rarefied polyatomic gases, $\lambda^{\max}_{(N,N-1)}$, on the order of truncation N is analyzed. From Theorem 10.5, the following result can be obtained:

Theorem 10.9 *The maximum equilibrium characteristic velocity of the* $(N, N-1)$*-system has the same lower bound as that for the maximum characteristic velocity of the* (N)*-system of monatomic gases, i.e.,*

$$\hat{\lambda}^{\max}_{(N,N-1)} \geqslant \sqrt{\frac{6}{5}\left(N - \frac{1}{2}\right)}. \tag{10.45}$$

In particular, it can be observed that

$$\lim_{N\to\infty} \lambda^{\max}_{(N,N-1)} = \infty. \tag{10.46}$$

Dependence of the maximum equilibrium characteristic velocity on the order N of the $(N, N-1)$-system has been numerically calculated for various values of the degrees of freedom of a molecule D and the results are shown in Table 10.2.

Table 10.2 Nondimensional maximum characteristic velocity (at equilibrium) of the $(N, N-1)$-system, $\hat{\lambda}^{\max}_{(N,N-1)} = \lambda^{\max}_{(N,N-1)}/c_0$, for different values of the degrees of freedom of a molecule D and for different values of the truncation order N. The lower bound of the maximum characteristic velocity is given in the last column

N	Number of moments	$D = 3.000001$	$D = 5$	$D = 7$	$D = 9$	$D = 10^9$	$\sqrt{\frac{6}{5}\left(N-\frac{1}{2}\right)}$
1	5	1.00000000	0.91651514	0.87831007	0.85634884	0.77459667	0.77459667
2	14	1.65028578	1.57434001	1.52907918	1.49884345	1.34164079	1.34164079
3	30	2.13542008	2.07965378	2.04104554	2.01252658	1.80822948	1.73205081
4	55	2.53553413	2.49590924	2.46579529	2.44196686	2.21299946	2.04939015
5	91	2.88470819	2.85640788	2.83350669	2.81447218	2.57495874	2.32379001
6	140	3.19976009	3.17923587	3.16186390	3.14688733	2.90507811	2.56904652
7	204	3.49004595	3.47488600	3.46161633	3.44985116	3.21035245	2.79284801
8	285	3.76126186	3.74984774	3.73959608	3.73030519	3.49555791	3.00000000
9	385	4.01712781	4.00836824	4.00034138	3.99293937	3.76412372	3.19374388
10	506	4.26021705	4.25336891	4.24699414	4.24103382	4.01860847	3.37638860
11	650	4.49239843	4.48694972	4.48181455	4.47696002	4.26098014	3.54964787
12	819	4.71508666	4.71068005	4.70648638	4.70248678	4.49279023	3.71483512
13	1015	4.92939054	4.92577317	4.92230397	4.91897186	4.71528716	3.87298335
14	1240	5.13620410	5.13319435	5.13029015	5.12748479	4.92949284	4.02492236
15	1496	5.33626485	5.33373027	5.33127253	5.32888746	5.13625617	4.17133072
16	1785	5.53019228	5.52803475	5.52593429	5.52388824	5.33629130	4.31277173
17	2109	5.71851433	5.71666004	5.71484891	5.71307921	5.53020569	4.44971909
18	2470	5.90168619	5.90007879	5.89850458	5.89696241	5.71852112	4.58257569
19	2870	6.08010411	6.07869999	6.07732178	6.07596866	5.90168962	4.71168760
20	3311	6.25411586	6.25288081	6.25166626	6.25047163	6.08010585	4.83735465

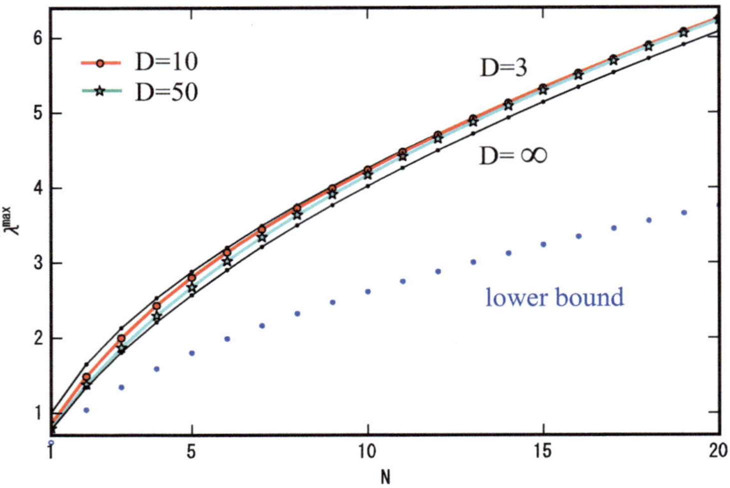

Fig. 10.5 Nondimensional maximum characteristic velocity (at equilibrium) of the $(N, N-1)$-system, $\hat{\lambda}^{\max}_{(N,N-1)} = \lambda^{\max}_{(N,N-1)}/c_0$, for two different values of the degrees of freedom of a molecule D ($D = 10$: *line with circle-shaped markers*; $D = 50$: *line with star-shaped markers*) as a function of the order N. The limit value of the non-dimensional characteristic velocity for $D \to 3$ and $D \to \infty$ (respectively, $\hat{\lambda}^{\max}_{(N+1-)} = \lambda^{\max}_{(N+1-)}/c_0$ and $\hat{\lambda}^{\max}_{(N,M)} = \hat{\lambda}^{\max}_{(N)} = \lambda^{\max}_{(N)}/c_0$ for $M < N-1$) are indicated with the *thin upper* and *lower lines*. The lower bound of the maximum characteristic velocity, given by (10.45), is represented by the *dots* (c_0 being the sound velocity in the monatomic gas)

In Fig. 10.5 this dependence is shown for two representative values: $D = 10$ and $D = 50$, together with the values obtained for the limit cases $D \to 3$ and $D \to \infty$, and the lower bound given by (10.45).

Remark In a special case, the systems characterized by $M < N - 1$ play an interesting role. For example, the 6-moment theory [18–21], which adopts F, F_i, F_{ll}, and G_{ll} as independent fields, can be identified as a good approximation of 14-moment theory when the effect of the dynamic pressure is larger than other dissipative variables, and is the simplest thermodynamic theory of gases with the dissipation. See the next Part IV.

References

1. M. Pavić, T. Ruggeri, S. Simić, Maximum entropy principle for rarefied polyatomic gases. Physica A **392**, 1302 (2013)
2. T. Arima, A. Mentrelli, T. Ruggeri, Molecular extended thermodynamics of rarefied polyatomic gases and wave velocities for increasing number of moments. Ann. Phys. **345**, 111 (2014)
3. T. Arima, A. Mentrelli, T. Ruggeri, Extended thermodynamics of rarefied polyatomic gases and characteristic velocities. Rendiconti Lincei-Matematica e Applicazioni **25**, 275 (2014)

4. G. Boillat, T. Ruggeri, On the evolution law of the weak discontinuities for hyperbolic quasi-linear systems. Wave Motion **1**(2), 149 (1979)
5. T. Ruggeri, Stability and discontinuity waves for symmetric hyperbolic systems, in *Non-linear Wave Motion*, ed. by A. Jeffrey (Longman Press, New York, 1989), pp. 148–161
6. A. Muracchini, T. Ruggeri, L. Seccia, Dispersion relation in the high frequency limit and non linear wave stability for hyperbolic dissipative systems. Wave Motion **15**(2), 143 (1992)
7. Z. Banach, W. Larecki, T. Ruggeri, Dispersion relation in the limit of high frequency for a hyperbolic system with multiple eigenvalues. Wave Motion **51**, 955 (2014)
8. G. Boillat, T. Ruggeri, On the shock structure problem for hyperbolic system of balance laws and convex entropy. Contin. Mech. Thermodyn. **10**, 285 (1998)
9. G. Boillat, T. Ruggeri, Moment equations in the kinetic theory of gases and wave velocities. Contin. Mech. Thermodyn. **9**, 205 (1997)
10. V.M. Zhdanov, The kinetic theory of a polyatomic gas. Sov. Phys. JETP **26**, 1187 (1968)
11. F.J. MacCormack, Kinetic equations for polyatomic gases: the 17-moment approximation. Phys. Fluids **11**, 2533 (1968)
12. T. Ruggeri, Galilean invariance and entropy principle for systems of balance laws. The structure of the extended thermodynamics. Contin. Mech. Thermodyn. **1**, 3 (1989)
13. L. Desvillettes, R. Monaco, F. Salvarani, A kinetic model allowing to obtain the energy law of polytropic gases in the presence of chemical reactions. Eur. J. Mech. B Fluids **24**, 219 (2005)
14. G. Boillat, T. Ruggeri, Hyperbolic principal subsystems: entropy convexity and subcharacteristic conditions. Arch. Ration. Mech. Anal. **137**, 305 (1997)
15. M.C. Carrisi, S. Pennisi, An 18 moments model for dense gases: entropy and galilean relativity principles without expansions. Entropy **17**, 214 (2015)
16. T. Arima, S. Taniguchi, T. Ruggeri, M. Sugiyama, Monatomic rarefied gas as a singular limit of polyatomic gas in extended thermodynamics. Phys. Lett. A **377**, 2136 (2013)
17. T. Arima, T. Ruggeri, M. Sugiyama, S. Taniguchi, A singular limit of rarefied polyatomic gas to monatomic gas in molecular extended thermodynamics (in preparation)
18. T. Arima, S. Taniguchi, T. Ruggeri, M. Sugiyama, A study of linear waves based on extended thermodynamics for rarefied polyatomic gases. Acta Appl. Math. **132**, 15 (2014)
19. S. Taniguchi, T. Arima, T. Ruggeri, M. Sugiyama, Effect of dynamic pressure on the shock wave structure in a rarefied polyatomic gas. Phys. Fluids **26**, 016103 (2014)
20. T. Arima, S. Taniguchi, T. Ruggeri, M. Sugiyama, Extended thermodynamics of real gases with dynamic pressure: an extension of Meixner's theory. Phys. Lett. A **376**, 2799 (2012)
21. T. Arima, T. Ruggeri, M. Sugiyama, S. Taniguchi, On the six-field model of fluids based on extended thermodynamics. Meccanica **49**, 2181 (2014)

Part VI
ET6: A Theory of Far-from-Equilibrium Thermodynamics

Chapter 11
Non-linear ET6 and the Role of the Dynamic Pressure: Phenomenological Approach

Abstract In this chapter, we present ET of real gases with six independent fields, i.e., the mass density, the velocity, the temperature and the dynamic pressure, without adopting the near-equilibrium approximation. We prove its compatibility with the universal principles (the entropy principle, the Galilean invariance and the stability), and obtain the symmetric hyperbolic system with respect to the main field. The correspondence between the ET 6-field (ET6) theory and the Meixner theory of relaxation processes is discussed. The internal variable and the nonequilibrium temperature in the Meixner theory are expressed in terms of the quantities of the ET6 theory, in particular, the dynamic pressure. As an example, we present the case of rarefied polyatomic gases and study the monatomic-gas limit where the system converges to the Euler system of a perfect fluid.

11.1 Introduction

As discussed in Sect. 1.6, one of the limitations on the previous RET was that the gas to which the theory can be applied is only the rarefied monatomic gas. With this limitation in mind, in Part III, we have proved that this can be overcome by adopting the binary hierarchy that takes into account the internal degrees of freedom of a polyatomic molecule. The new RET is, therefore, applicable to the polyatomic gas, and furthermore as seen from the discussion in Sect. 5.5, it may be applied at least to the moderately dense gases.

However, there is another big limitation on RET. In fact, although RET goes beyond the local equilibrium assumption and therefore beyond the NSF theory, the RET theory has been developed only for processes not so far from equilibrium. In the phenomenological RET theory with 14 variables, for example, we have studied only constitutive equations that are linear with respect to the nonequilibrium variables ($\sigma_{\langle ij\rangle}, \Pi, q_i$) because of the enormous difficulties in the analysis. While, in the molecular RET approach, we have seen, in Chap. 6 for the case of 14 moments and in Chap. 10 for the case of generic number of moments, that there exists the problematic point about the convergence of the integrals as we discussed in Sect. 4.5.

As mentioned before modeling of polyatomic gases and of dense gases in nonequilibrium is an active and urgent issue nowadays with several important

T. Ruggeri, M. Sugiyama, *Rational Extended Thermodynamics beyond the Monatomic Gas*, DOI 10.1007/978-3-319-13341-6_11

applications like the study of shock wave structure which is essentially important, for example, for the atmospheric reentry problem of a space vehicle. Therefore it is extremely interesting to have a theory that is valid also far from equilibrium. This was done in [1].

The ET14 theory gives us a complete phenomenological model but its differential system is rather complex. For this reason we have recently constructed a simplified theory with 6 fields (referred to as the ET6 theory) [2, 3]: the mass density ρ, the velocity $\mathbf{v}$, the temperature T, and the dynamic (nonequilibrium) pressure Π. This simplified theory preserves the main physical properties of the more complex theory of 14 variables, in particular, when the bulk viscosity plays more important role than the shear viscosity and the heat conductivity. This situation is observed in many gases such as rarefied hydrogen gases and carbon dioxide gases at some temperature ranges (see Chaps. 7 and 8) [4–6]. ET6 has another advantage to offer us a more affordable hyperbolic partial differential system. In fact, it is the simplest system that takes into account a dissipation mechanism after the Euler system of perfect fluids

In this chapter we consider the ET6 theory in detail. The simplicity of the 6-field model permits us to construct a full non-linear theory, the derivation of which is the main goal of this chapter. From a mathematical point of view, the advantage to have a full non-linear theory is that the hyperbolicity region exists, if an equilibrium state is in it, not only in the neighborhood of an equilibrium state but everywhere provided that the solutions exist and are bounded. For any bounded solutions, we will see below that there exist upper and lower bounds of the dynamic pressure.

We firstly derive the nonlinear system of equations of the ET6 theory by proving its compatibility with the universal requirements: the entropy principle, the Galilean invariance, and the stability, that is, the convexity of the entropy density. As a consequence of this, the differential system is symmetric hyperbolic with respect to the *main field.*

Secondly, through studying real gases, we discuss the connection between the ET6 theory and the Meixner theory more deeply by comparing these nonlinear systems with each other. We prove that the internal variable and the nonequilibrium temperature introduced in the Meixner theory can be expressed explicitly in terms of the macroscopic variables of the ET6 theory, in particular, the dynamic pressure. Meixner's affinity and relaxation time are also evaluated by the macroscopic variables of ET6.

In the last part we specialize the results for the case of an ideal polytropic gas both in the linear and nonlinear approaches. We also clarify the monatomic-gas limit in which the system of ET6 converges to the Euler system of a perfect fluid.

11.2 ET Theory with Six Fields

We consider six independent field-variables (ρ, v_i, T, Π), and we retain the general structure of the binary hierarchy (F-series and G-series) (10.2) but assume only the $(2^-, 0)$-system (see Definition 10.2):

$$\begin{aligned}
&\frac{\partial F}{\partial t}+\frac{\partial F_i}{\partial x_i}=0,\\
&\frac{\partial F_j}{\partial t}+\frac{\partial F_{ji}}{\partial x_i}=0, \qquad\qquad\qquad\qquad\qquad\qquad (11.1)\\
&\frac{\partial F_{ll}}{\partial t}+\frac{\partial F_{lli}}{\partial x_i}=P_{ll}, \qquad \frac{\partial G_{ll}}{\partial t}+\frac{\partial G_{lli}}{\partial x_i}=0,
\end{aligned}$$

where $(11.1)_{1,2,4}$ represent the conservation laws of mass, momentum and energy provided that $F=\rho,\ F_i=\rho v_i,\ F_{ij}=\rho v_i v_j+(p+\Pi)\delta_{ij},\ G_{ll}=\rho v_l v_l+2\rho\varepsilon$, and $G_{lli}=(\rho v_l v_l+2\rho\varepsilon+2p+2\Pi)v_i$. The shear stress $\sigma_{\langle ij\rangle}$ and the heat flux q_i are neglected in this theory. While $(11.1)_3$ is the new balance law corresponding to the dynamic pressure Π.

Therefore undetermined functions at the moment are only F_{lli}, P_{ll}, and the functions h, φ^i and Σ in the entropy law (2.8), (2.46). These functions will be determined by the universal principles in the following analysis.

11.2.1 Galilean Invariance

In the present case, the Galilean invariance (2.36) implies:

$$F_{lli}=\left(5(p+\Pi)+\rho v^2\right)v_i, \quad P_{ll}=\hat{P}_{ll}, \tag{11.2}$$

and

$$\mathbf{X}=\begin{pmatrix}1 & 0 & 0 & 0\\ v_j & \delta_{jk} & 0 & 0\\ v^2 & 2v_j & 1 & 0\\ v^2 & 2v_j & 0 & 1\end{pmatrix}, \tag{11.3}$$

$$\mathbf{A}^r=\left(\frac{\partial \mathbf{X}}{\partial v_r}\right)\bigg|_{v_j=0}=\begin{pmatrix}0 & 0 & 0 & 0\\ \delta_{jr} & 0 & 0 & 0\\ 0 & 2\delta_{jr} & 0 & 0\\ 0 & 2\delta_{jr} & 0 & 0\end{pmatrix}.$$

Then the system (11.1) is written in terms of the physical variables:

$$\begin{aligned}
&\frac{\partial \rho}{\partial t} + \frac{\partial}{\partial x_i}(\rho v_i) = 0, \\
&\frac{\partial \rho v_j}{\partial t} + \frac{\partial}{\partial x_i}\left[(p+\Pi)\delta_{ij} + \rho v_i v_j\right] = 0, \\
&\frac{\partial}{\partial t}(2\rho\varepsilon + \rho v^2) + \frac{\partial}{\partial x_i}\left\{\left[2(p+\Pi) + 2\rho\varepsilon + \rho v^2\right] v_i\right\} = 0, \\
&\frac{\partial}{\partial t}\left[3(p+\Pi) + \rho v^2\right] + \frac{\partial}{\partial x_i}\left\{\left[5(p+\Pi) + \rho v^2\right] v_i\right\} = \hat{P}_{ll}.
\end{aligned} \tag{11.4}$$

We see that, in this exceptional case of ET6, the Galilean invariance fixes completely the structure of the undetermined flux F_{lli} and we notice that the left-hand side of (11.4) is linear in Π without assuming any approximations. Instead the production term $\hat{P}_{ll}$ can be completely nonlinear. Therefore the present theory is valid for any nonequilibrium processes as far as the continuum description is valid.

11.2.2 Entropy Principle

We here study the compatibility with the entropy principle. Let us rewrite the entropy density in the form:

$$h = \rho s + \rho k, \tag{11.5}$$

where $s(\rho, \varepsilon)$ is the specific entropy density in equilibrium and $k(\rho, \varepsilon, \Pi)$ is the nonequilibrium part with the condition:

$$k(\rho, \varepsilon, 0) = 0. \tag{11.6}$$

Let us write the components of the main field $\mathbf{u}'$ as

$$\mathbf{u}' \equiv \left(\lambda, \lambda_j, \mu, \zeta\right), \tag{11.7}$$

then we have, from (2.44),

$$\hat{\lambda}_j = 0, \qquad \hat{h}'^0 = -2(\hat{\mu} + \hat{\zeta})(p + \Pi). \tag{11.8}$$

From $(2.43)_1$, we obtain

$$\hat{\lambda} d\rho + 2\hat{\mu} d(\rho\varepsilon) + 3\hat{\zeta} d(p + \Pi) = (s + k)d\rho + \rho ds + \rho dk. \tag{11.9}$$

Taking into account the Gibbs equation for the equilibrium specific entropy density (1.7) and choosing (ρ, ε, Π) as independents variables, we obtain, from (11.9),

$$\begin{aligned} \hat{\lambda} &= -\frac{g}{T} + k + \rho k_\rho + \varepsilon\left(p_\varepsilon k_\Pi - k_\varepsilon\right) - p_\rho \rho k_\Pi, \\ \hat{\mu} &= \frac{1}{2T} + \frac{1}{2}\left(k_\varepsilon - p_\varepsilon k_\Pi\right), \\ \hat{\zeta} &= \frac{1}{3}\rho k_\Pi, \end{aligned} \tag{11.10}$$

where the suffix indicates a partial differentiation with respect to the corresponding variables, and $g = \varepsilon + p/\rho - Ts$ is the chemical potential. From (2.14) and (11.8) we obtain

$$h = \hat{\lambda}\,\rho + 2\hat{\mu}(\rho\,\varepsilon + p + \Pi) + 5\hat{\zeta}(p + \Pi) = \rho(s + k). \tag{11.11}$$

Substituting (11.10) into (11.11), we obtain a linear first-order partial differential equation for the function $k \equiv k(\rho, \varepsilon, \Pi)$:

$$\rho^2 k_\rho + (p + \Pi)\,k_\varepsilon + \rho\left\{(p + \Pi)\left(\frac{5}{3} - \frac{p_\varepsilon}{\rho}\right) - \rho p_\rho\right\} k_\Pi + \frac{\Pi}{T} = 0, \tag{11.12}$$

with initial data given by (11.6).

The solution exists and is unique provided that the Eq. (11.12) is hyperbolic in the Π-direction. This implies that the coefficient of k_Π in (11.12) must be different from zero:

$$\Gamma = (p + \Pi)\left(\frac{5}{3} - \frac{p_\varepsilon}{\rho}\right) - \rho p_\rho \neq 0. \tag{11.13}$$

As this condition must be satisfied also for $\Pi = 0$, we obtain a particular condition for an admissible thermal equation of state $p \equiv p(\rho, \varepsilon)$:

$$\Gamma^0 = p\left(\frac{5}{3} - \frac{p_\varepsilon}{\rho}\right) - \rho p_\rho \neq 0. \tag{11.14}$$

Changing the pairs of variables from (ρ, ε) to (ρ, s) and taking into account the well-known relations:

$$f_\rho = \left(\frac{\partial f}{\partial \rho}\right)_s - \frac{p}{\rho^2 T}\left(\frac{\partial f}{\partial s}\right)_\rho, \qquad f_\varepsilon = \frac{1}{T}\left(\frac{\partial f}{\partial s}\right)_\rho,$$

we obtain

$$\Gamma = \frac{5}{3}(p + \Pi) - \rho \left(\frac{\partial p}{\partial \rho} \right)_s - \frac{\Pi}{\rho T} \left(\frac{\partial p}{\partial s} \right)_\rho \neq 0, \tag{11.15}$$

and

$$\Gamma^0 = \frac{5}{3}p - \rho \left(\frac{\partial p}{\partial \rho} \right)_s \neq 0. \tag{11.16}$$

We will see in the next subsection that Γ and Γ^0 must be positive.

Moreover, from (2.43), (2.14) and $(2.13)_2$, we have

$$\varphi^i = 0, \qquad \Sigma = \frac{1}{3}\rho k_\Pi \hat{P}_{ll} > 0. \tag{11.17}$$

11.2.3 Convexity Condition and Stability

The convexity condition (2.47) becomes:

$$\begin{aligned} \hat{Q} =& \hat{Q}_5 + (\delta\rho)^2 \left(2k_\rho + \rho k_{\rho\rho} - \rho p_{\rho\rho} k_\Pi\right) + \rho(\delta\varepsilon)^2 \left(k_{\varepsilon\varepsilon} - p_{\varepsilon\varepsilon} k_\Pi\right) \\ &+ \rho(\delta\Pi)^2 k_{\Pi\Pi} + 2\rho\delta\varepsilon\,\delta\Pi\, k_{\varepsilon\Pi} + 2\delta\rho\delta\Pi \left(k_\Pi + \rho k_{\rho\Pi}\right) \\ &+ 2\delta\varepsilon\delta\rho \left\{\rho k_{\rho\varepsilon} - \rho p_{\rho\varepsilon} k_\Pi + p_\varepsilon k_\Pi\right\} < 0, \end{aligned} \tag{11.18}$$

where $\hat{Q}_5$ denotes the quadratic form for the Euler system. We want to prove now a necessary condition for the convexity:

Statement 4 *A necessary condition for the convexity of entropy is that the thermal and caloric equations of state must obey the condition:*

$$\Gamma^0 = \rho \left\{ \frac{5}{3}\frac{p}{\rho} - \left(\frac{\partial p}{\partial \rho} \right)_s \right\} > 0. \tag{11.19}$$

Moreover, in nonequilibrium, the dynamical pressure must satisfy the inequality:

$$\Gamma = \Gamma^0 + \Pi \left(\frac{5}{3} - \frac{p_\varepsilon}{\rho} \right) > 0. \tag{11.20}$$

Proof From (11.6) we have:

$$k_\rho(\rho, \varepsilon, 0) = k_\varepsilon(\rho, \varepsilon, 0) = 0, \tag{11.21}$$

then from (11.12) follows

$$k_{\Pi}(\rho, \varepsilon, 0) = 0. \tag{11.22}$$

Taking into account that all successive derivatives of (11.21) and (11.22) with respect to ρ or ε are identically zero, differentiating (11.12) with respect to Π, and taking the limit $\Pi \to 0$, we obtain;

$$k^0_{\Pi\Pi} = k_{\Pi\Pi}(\rho, \varepsilon, 0) = -\frac{1}{\rho T \Gamma^0}. \tag{11.23}$$

On the other hand, from the convexity condition (11.18) evaluated at $\Pi = 0$, we have:

$$Q^0 = \hat{Q}_5 + \rho(\delta\Pi)^2 k^0_{\Pi\Pi} < 0, \tag{11.24}$$

then $k^0_{\Pi\Pi} < 0$ and from (11.23) follows $\Gamma^0 > 0$. As Γ must be different from zero and, in equilibrium, reduces to Γ^0, we conclude that also Γ must be positive. (Q.E.D.)

11.2.4 Residual Inequality and Production Term

From the residual inequality $(11.17)_2$, the simplest form of the production with only one positive function α is expressed by

$$\hat{P}_{ll} = \alpha k_{\Pi}, \qquad \alpha > 0, \tag{11.25}$$

where α is independent of Π: $\alpha \equiv \alpha(\rho, \varepsilon)$. Hereafter, we will confine our study within this simple, but physically interesting case. We can prove easily that the function α can be related to the relaxation time and to the bulk viscosity ν. In fact, as we assume that α is independent of Π, the function remains the same even if we evaluate $\hat{P}_{ll}$ near equilibrium. For small Π, the production term was evaluated in the linear model in terms of the relaxation time τ as follows [2]:

$$\hat{P}_{ll} = -3\frac{\Pi}{\tau}, \tag{11.26}$$

and, by using the Maxwellian iteration, the Eq. $(11.4)_4$ becomes the Navier-Stokes constitutive equation for the dynamic pressure [3, 7]:

$$\Pi = -\nu \operatorname{div} \mathbf{v} \tag{11.27}$$

with the bulk viscosity ν expressed by the relaxation time τ [2]:

$$\nu = \tau \Gamma^0. \tag{11.28}$$

When we expand (11.25) in Taylor series by taking into account (11.22) and (11.23), we obtain

$$\alpha = -\frac{3}{\tau k^0_{\Pi\Pi}} = \frac{3\rho T \Gamma^0}{\tau} = \frac{3\rho T \Gamma^{02}}{\nu}. \tag{11.29}$$

Therefore we know explicitly the production term in nonlinear case in terms of the relaxation time:

$$\hat{P}_{ll} = 3\frac{\rho T \Gamma^0}{\tau} k_\Pi, \tag{11.30}$$

or in terms of the bulk viscosity:

$$\hat{P}_{ll} = 3\frac{\rho T \Gamma^{02}}{\nu} k_\Pi. \tag{11.31}$$

As the bulk viscosity and the relaxation time can be evaluated by experimental data such as dispersion relation of sound, we have the explicit expression of the production term $\hat{P}_{ll}$.

11.2.5 Main Field and Symmetric Form

The main field (11.7) can be evaluated by (2.42), (11.10) and (2.51):

$$\begin{aligned}
\lambda =& -\frac{g}{T} + \frac{v^2}{2T} + k + \rho k_\rho + \left(\varepsilon - \frac{v^2}{2}\right)(p_\varepsilon k_\Pi - k_\varepsilon) + \frac{1}{3}\rho(v^2 - 3p_\rho)k_\Pi, \\
\lambda_i =& - v_i \left\{\frac{1}{T} + (k_\varepsilon - p_\varepsilon k_\Pi) + \frac{2}{3}\rho k_\Pi\right\}, \\
\mu =& \frac{1}{2T} + \frac{1}{2}(k_\varepsilon - p_\varepsilon k_\Pi), \\
\zeta =& \frac{1}{3}\rho k_\Pi.
\end{aligned} \tag{11.32}$$

The potentials (2.14) become

$$h'^0 = -\frac{p}{T} + \rho\left\{\rho k_\rho + (p - \rho p_\rho + \Pi)k_\Pi\right\}, \quad h'^i = h'^0 v_i. \tag{11.33}$$

According to the general theory, the system of field equations assumes the symmetric form (2.11) if we use these variables.

We can summarize the result obtained as:

Theorem 11.1 *The system* (11.4) *satisfies the entropy principle* (2.8) *if and only if the nonequilibrium entropy density k is a solution of the differential equation* (11.12) *with boundary data* (11.6)*, and the flux of entropy and the production of entropy are given by* $(11.17)_1$ *and* (11.30) *or* (11.31)*. Under the convexity condition* (11.18) *the system* (11.4) *becomes symmetric hyperbolic with respect to the main field with the components given by* (11.32) *and potentials given by (11.33).*

Therefore all the universal principles are satisfied, and the closure is full nonlinear and is valid, in principle, far from equilibrium.

To sum up, we notice that, in this nonlinear approach, the theory is explicitly fixed by the knowledge of the function α in (11.25) and of the thermal and caloric equations of state that must satisfy the inequality (11.19). Inserting the equations of state into (11.12) and taking into account (11.6), we can obtain uniquely $k \equiv k(\rho, \varepsilon, \Pi)$. Then we need to verify whether the quadratic form (11.18) is negative definite. Inserting k into (11.10) and into (11.25), we have the ET theory that fully satisfies the universal principles.

11.2.6 Alternative Form of the Differential System

By subtracting the equation of the energy $(11.4)_3$ from the equation of the dynamic pressure $(11.4)_4$ we can rewrite the system (11.4) in an equivalent form:

$$
\begin{aligned}
&\frac{\partial \rho}{\partial t} + \frac{\partial}{\partial x_i}(\rho v_i) = 0, \\
&\frac{\partial \rho v_j}{\partial t} + \frac{\partial}{\partial x_i}\left[(p + \Pi)\delta_{ij} + \rho v_i v_j\right] = 0, \\
&\frac{\partial}{\partial t}(2\rho\varepsilon + \rho v^2) + \frac{\partial}{\partial x_i}\left\{\left[2(p+\Pi) + 2\rho\varepsilon + \rho v^2\right] v_i\right\} = 0, \\
&\frac{\partial}{\partial t}\left[3(p+\Pi) - 2\rho\varepsilon\right] + \frac{\partial}{\partial x_i}\left\{\left[3(p+\Pi) - 2\rho\varepsilon\right] v_i\right\} = \alpha k_\Pi .
\end{aligned}
\tag{11.34}
$$

It is easy to verify from $(2.13)_1$ that the corresponding main field $\left(\bar{\lambda}, \bar{\lambda}_i, \bar{\mu}, \bar{\zeta}\right)$ that symmetrize the system (11.34) have the same components of the main field of the system (11.4) except for the Lagrange multiplier $\bar{\mu}$ of the energy equation $(11.34)_3$

that is given by $\bar{\mu} = \mu + \zeta$:

$$
\begin{aligned}
\bar{\lambda} &= -\frac{g}{T} + \frac{v^2}{2T} + k + \rho k_\rho + \left(\varepsilon - \frac{v^2}{2}\right)(p_\varepsilon k_\Pi - k_\varepsilon) + \frac{1}{3}\rho(v^2 - 3p_\rho)k_\Pi,\\
\bar{\lambda}_i &= -v_i\left\{\frac{1}{T} + (k_\varepsilon - p_\varepsilon k_\Pi) + \frac{2}{3}\rho k_\Pi\right\},\\
\bar{\mu} &= \frac{1}{2T} + \frac{1}{2}(k_\varepsilon - p_\varepsilon k_\Pi) + \frac{1}{3}\rho k_\Pi,\\
\bar{\zeta} &= \frac{1}{3}\rho k_\Pi.
\end{aligned}
\tag{11.35}
$$

By introducing the material time derivative, the system (11.34) can be rewritten as

$$
\begin{aligned}
&\dot{\rho} + \rho\frac{\partial v_k}{\partial x_k} = 0,\\
&\rho\dot{v}_i + \frac{\partial}{\partial x_i}(p + \Pi) = 0,\\
&\rho\dot{\varepsilon} + (p + \Pi)\frac{\partial v_k}{\partial x_k} = 0,\\
&\left(\frac{p + \Pi}{\rho} - \frac{2}{3}\varepsilon\right)^{\bullet} = \frac{\alpha k_\Pi}{3\rho}.
\end{aligned}
\tag{11.36}
$$

We observe that any differentiable solution of (11.36) satisfies the entropy law (2.8) that can be rewritten, from (11.5) and (11.17), as:

$$
(s + k)^{\bullet} = \frac{\alpha}{3}k_\Pi^2. \tag{11.37}
$$

11.2.7 Euler Fluid as a Principal Subsystem of the ET6 System and Subcharacteristic Conditions

System of an Euler fluid is a principal subsystem of the ET6 theory according to the definition given by Boillat and Ruggeri in Sect. 2.4. In fact, when $\Pi = 0$, k_Π vanishes as shown in (11.22) and the corresponding component $\bar{\zeta}$ of the main field in the last equation of (11.35) vanishes. The remaining five components of the main field in (11.35) converge to the symmetrizable variables, which were deduced first by Godunov [8] for the Euler-fluid system:

$$
\bar{\lambda} \to -\frac{g}{T} + \frac{v^2}{2T}, \quad \bar{\lambda}_i \to -\frac{v_i}{T}, \quad \bar{\mu} \to \frac{1}{2T}. \tag{11.38}
$$

According to the result in Sect. 2.4, the convexity of a system implies automatically the so-called subcharacteristic condition, i.e., the spectrum of the characteristic velocities of the principal subsystem is contained in the spectrum of the full system evaluated in the subspace of the subsystem. This implies, in particular, that the maximum characteristic velocity of the subsystem is not greater than the one of the full system. In the present particular case of ET6, the maximum characteristic velocity is given by (see Sect. 14.1)

$$v_i n_i + \sqrt{\frac{5}{3}\frac{p+\Pi}{\rho}} \tag{11.39}$$

with $\mathbf{n} \equiv (n_i)$ being the unit normal to the wave front. While the maximum characteristic velocity of the Euler fluid is:

$$v_i n_i + \sqrt{\left(\frac{\partial p}{\partial \rho}\right)_s}. \tag{11.40}$$

We notice that the condition (11.19) corresponds exactly to the subcharacteristic condition, i.e., the maximum characteristic velocity (11.39) evaluated in equilibrium ($\Pi = 0$) is greater than the corresponding one of the Euler fluid (11.40).

11.3 Comparison Between Nonlinear Systems of the ET6 Theory and of the Meixner Theory

In order to describe thermodynamic properties of a material with internal structures, theories with internal variables were developed [9–11], and have been applied to various phenomena like chemical reaction, sound wave and shock wave propagation in a polyatomic gas, inelastic behavior of a solid, dynamics of a dielectric material, and so on. See, for example, [12]. The Meixner theory [13, 14] may be regarded as the prototype of the theories of internal variables, but it is still frequently applied to various nonequilibrium phenomena. For a recent study, where dynamic degrees of freedom and internal variables are discussed in a uniform way, see, for example, [15].

In this section, we compare the ET6 theory with the Meixner theory, and discuss the physical implications involved.

We assume the simplest version of the Meixner theory composed of the Euler equations and only one relaxation equation for an internal variable ξ. In this case the system of field equations is summarized as follows [9, 13, 14]:

$$
\begin{aligned}
&\dot{\rho} + \rho \frac{\partial v_k}{\partial x_k} = 0, \\
&\rho \dot{v}_i + \frac{\partial \mathscr{P}}{\partial x_i} = 0, \\
&\rho \dot{\mathscr{E}} + \mathscr{P} \frac{\partial v_k}{\partial x_k} = 0, \\
&\dot{\xi} = -\beta \mathscr{A},
\end{aligned}
\tag{11.41}
$$

where $\mathscr{P}$, $\mathscr{E}$ and $\mathscr{A}$ are, respectively, the pressure, the specific internal energy and the affinity of relaxation processes, and β is a positive phenomenological coefficient. The generalized Gibbs relation in the Meixner theory is assumed to be

$$
\mathscr{T} \mathrm{d}\mathscr{S} = \mathrm{d}\mathscr{E} - \frac{\mathscr{P}}{\rho^2} \mathrm{d}\rho - \mathscr{A} \mathrm{d}\xi,
\tag{11.42}
$$

where $\mathscr{T}$ is the temperature and $\mathscr{S}$ is the specific entropy. Note that the quantities $\mathscr{T}$, $\mathscr{S}$, $\mathscr{P}$ and $\mathscr{A}$ depend not only on the mass density ρ and the specific internal energy $\mathscr{E}$ but also on the internal variable ξ. From (11.42), taking into account (11.41), we obtain:

$$
\dot{\mathscr{S}} = \beta \frac{\mathscr{A}^2}{\mathscr{T}}.
\tag{11.43}
$$

Comparing the system of the ET6 theory (11.36) and (11.37) with the system of the Meixner theory (11.41) and (11.43), we obtain the perfect correspondence provided that

$$
\begin{aligned}
&\xi = \frac{p + \Pi}{\rho} - \frac{2}{3}\varepsilon, \\
&\mathscr{P} = p + \Pi, \\
&\mathscr{E} = \varepsilon, \\
&\mathscr{A} = -\rho \mathscr{T} k_\Pi, \\
&\mathscr{S} = s + k, \\
&\beta = \frac{\alpha}{3\,\rho^2\,\mathscr{T}},
\end{aligned}
\tag{11.44}
$$

where $\mathscr{T}$ is given by

$$
\frac{1}{\mathscr{T}} = \frac{1}{T} + \left\{ k_\varepsilon + \left(\frac{2}{3}\rho - p_\varepsilon \right) k_\Pi \right\}.
\tag{11.45}
$$

Inserting the correspondence relations (11.44) into (11.35), we obtain the simple expression of the main field in terms of the variables in the Meixner theory:

$$
\begin{aligned}
\bar{\lambda} &= -\frac{\mathscr{G}}{\mathscr{T}} + \frac{v^2}{2\mathscr{T}}, \\
\bar{\lambda}_i &= -\frac{v_i}{\mathscr{T}}, \\
\bar{\mu} &= \frac{1}{2\mathscr{T}}, \\
\bar{\zeta} &= -\frac{1}{3}\frac{\mathscr{A}}{\mathscr{T}},
\end{aligned}
\tag{11.46}
$$

where we define

$$
\mathscr{G} \equiv \mathscr{E} + \frac{\mathscr{P}}{\rho} - \mathscr{T}\mathscr{S} - \mathscr{A}\xi \tag{11.47}
$$

as the chemical potential in the Meixner theory. The main field (11.46) makes the system (11.48) below with Meixner's variables be symmetric:

$$
\begin{aligned}
&\frac{\partial \rho}{\partial t} + \frac{\partial}{\partial x_i}(\rho v_i) = 0, \\
&\frac{\partial \rho v_j}{\partial t} + \frac{\partial}{\partial x_i}\left(\mathscr{P}\delta_{ij} + \rho v_i v_j\right) = 0, \\
&\frac{\partial}{\partial t}(2\rho\mathscr{E} + \rho v^2) + \frac{\partial}{\partial x_i}\left\{\left[2\mathscr{P} + 2\rho\mathscr{E} + \rho v^2\right] v_i\right\} = 0, \\
&\frac{\partial 3\rho\xi}{\partial t} + \frac{\partial}{\partial x_i}(3\rho\xi v_i) = -3\beta\rho\mathscr{A}.
\end{aligned}
\tag{11.48}
$$

This system is equivalent to the system (11.34). It is interesting to compare the expression of the main field (11.46) with the one for the Euler fluid (11.38).

To sum up, we have proved the following statement:

Statement 5 *The system of ET6 is completely equivalent to the system of the Meixner theory that is composed of the Euler system and a relaxation equation for an internal variable, provided that the change of variables is given by* (11.44). *This correspondence gives us the possibility to write explicitly the relationship between the two kinds of the coldness (inverse of the temperature) as shown in* (11.45). *Therefore a natural nonequilibrium temperature* $\mathscr{T}$ *appears. The Meixner system, rewritten as a system of balance laws* (11.48), *becomes to be symmetric hyperbolic by adopting the main field given by* (11.46).

Physical implications of this correspondence may be summarized as follows:

As stated above, the nonlinear system of the Meixner theory with one internal variable has the same *mathematical* structure as the nonlinear system of the ET6 theory. However, *physical* contents of the ET6 theory are more precise than those of the Meixner theory as explained (i)–(iii) below, although it is true that the Meixner theory is more general than the ET6 theory of real gases in the following sense: The internal variables may describe not only relaxation processes of the internal degrees of freedom of molecules in a polyatomic gas but also many other kinds of phenomena like chemical reactions [9, 16]. But, as is pointed out in the paper [2], the theoretical basis of the Meixner theory has not been clear enough, in particular, for phenomena out of local equilibrium. We will also discuss this point from the view point of the ET theory with many fields in the concluding remarks.

(i) By using the correspondence relations (11.44), the affinity $\mathscr{A}$, the entropy $\mathscr{S}$ and the phenomenological coefficient β in the Meixner theory can be explicitly expressed in terms of the macroscopic variables of the ET6 theory. For an example, the case of ideal polytropic gases will be shown below.

(ii) In the Meixner theory, it is usually a difficult task to identify a proper internal variable. We notice, however, that the explicit dependence of the internal variable ξ on the dynamic pressure Π is given by $(11.44)_1$. The dynamic pressure is identified as a suitable quantity that indicates the energy transfer between the translational modes and the internal modes, such as molecular rotation and vibration, in a polyatomic molecule.

(iii) The nonequilibrium temperature $\mathscr{T}$ of the Meixner theory is introduced through the generalized Gibbs relation (11.42). This can be interpreted as the temperature of a state *in equilibrium* with a constraint that the system is kept at a given ξ by applying a suitable external field [17], which may be or may not be virtual. We sometimes call such a field thermodynamic external field or effective field. This technique has been frequently used, for example, in the linear response theory [18, 19]. However, in applications, it is usually difficult for the Meixner theory to specify the nonequilibrium temperature $\mathscr{T}$ in an explicit and operational way.

On the other hand the temperature T of ET6 is defined in Chap. 5 as the temperature of an equilibrium state with the same values of the mass density ρ and the specific internal energy ε in a nonequilibrium state under consideration. This definition has been usually adopted, for example, in the kinetic theory [4, 20], in nonequilibrium statistical mechanics [18]. We may call this, in the case of rarefied monatomic gases, *kinetic temperature* because it is proportional to the thermal average of the kinetic (or translational) energy of a molecule. In this respect, the analysis based on molecular ET from a view point of the kinetic theory in Chap. 6 is helpful for understanding this definition. Hereafter we call the nonequilibrium temperature of ET simply *temperature*.

In the present analysis, we have discovered the relation (11.45) that discloses the new link between the nonequilibrium temperature $\mathscr{T}$ in the

Meixner theory and the temperature T in ET6. We will discuss, in a general way, the nonequilibrium temperature (and also chemical potential) in Chap. 15.

A survey on nonequilibrium temperatures is given in [21, 22]. See also [23] for the analysis of the thermodynamic temperature in the ET theory.

11.4 ET6 Theory Near Equilibrium

In this section we study the system near equilibrium, where we can adopt linear constitutive equations with respect to the dynamic pressure. The entropy density in this case must be in the form:

$$h = \rho s - \Psi(\rho, \varepsilon)\Pi^2, \tag{11.49}$$

then we have

$$k = -\frac{1}{\rho}\Psi\Pi^2. \tag{11.50}$$

From (11.23) it follows that

$$\Psi = \frac{1}{2T\Gamma^0}. \tag{11.51}$$

The production term has the form expressed by (11.26), and all universal principles are satisfied provided that the inequalities

$$\Gamma^0 > 0, \qquad \tau > 0 \tag{11.52}$$

hold. The main field in this case becomes

$$\begin{aligned}
\bar{\lambda} &= -\frac{g}{T} + \frac{v^2}{2T} + \left\{\frac{2}{\rho}\left(p_\rho\,\rho - p_\varepsilon\,\varepsilon\right) - v^2\left(\frac{2}{3} - \frac{p_\varepsilon}{\rho}\right)\right\}\Psi\Pi, \\
\bar{\lambda}_i &= -\frac{v_i}{T} - 2v_i\left(\frac{p_\varepsilon}{\rho} - \frac{2}{3}\right)\Psi\Pi, \\
\bar{\mu} &= \frac{1}{2T} + \left(\frac{p_\varepsilon}{\rho} - \frac{2}{3}\right)\Psi\Pi, \\
\bar{\zeta} &= -\frac{2}{3}\Psi\Pi,
\end{aligned} \tag{11.53}$$

where we have neglected the second-order terms with respect to Π.

The system of field equations is the same as (11.36) except that the last equation is replaced by

$$\left(\frac{p+\Pi}{\rho}-\frac{2}{3}\varepsilon\right)^{\bullet}=-\frac{\Pi}{\rho\tau}. \tag{11.54}$$

And the entropy law (11.37) in this case is given by

$$\left(s-\frac{\Psi}{\rho}\Pi^2\right)^{\bullet}=\frac{2\Psi}{\rho\tau}\Pi^2. \tag{11.55}$$

11.4.1 Comparison Between the ET Theory Near Equilibrium and the Meixner Theory

The correspondence between the ET6 theory and the Meixner theory is the same as (11.44) except that the last three equations are replaced by

$$\begin{aligned}\mathscr{A}&=2\mathscr{T}\Psi\Pi,\\ \mathscr{S}&=s-\frac{\Psi}{\rho}\Pi^2,\\ \beta&=\frac{1}{2\rho\mathscr{T}\Psi\tau},\end{aligned} \tag{11.56}$$

where only the leading terms with respect to Π are retained. The temperature $\mathscr{T}$ in the Meixner theory, within the present approximation, is given by

$$\frac{1}{\mathscr{T}}=\frac{1}{T}-2\left(\frac{2}{3}-\frac{p_\varepsilon}{\rho}\right)\Psi\Pi. \tag{11.57}$$

Noteworthy points are summarized as follows:

1. We can find explicit expressions of the quantities $\mathscr{P}, \mathscr{E}, \mathscr{A}$, and $\mathscr{S}$ in terms of the internal variable ξ. In fact, eliminating Π in (11.56), we obtain

$$\begin{aligned}\mathscr{P}&=p+\rho(\xi-\xi_{eq}),\\ \mathscr{E}&=\varepsilon,\\ \mathscr{A}&=2\rho\mathscr{T}\Psi(\xi-\xi_{eq}),\\ \mathscr{S}&=s-\rho\Psi(\xi-\xi_{eq})^2,\end{aligned} \tag{11.58}$$

where ξ_{eq} is given by

$$\xi_{eq} \equiv \frac{p}{\rho} - \frac{2}{3}\varepsilon. \tag{11.59}$$

We notice that, from (11.51) and (11.59), there exists a relation between Ψ and ξ_{eq}:

$$\Psi = -\frac{1}{2\rho^2 T \left(\frac{\partial \xi_{eq}}{\partial \rho}\right)_s}.$$

2. The relaxation time in the Meixner theory is defined by de Groot and Mazur [9]

$$\tau_M \equiv \frac{1}{\beta}\left(\frac{\partial \xi}{\partial \mathscr{A}}\right)_{\rho,\mathscr{E}}. \tag{11.60}$$

From $(11.56)_3$, $(11.58)_{2,3}$ and (11.60), we can prove that this relaxation time coincides with the relaxation time τ in the ET6 theory:

$$\tau_M = \tau. \tag{11.61}$$

In the paper [2], the equality was confirmed only within the linear system of field equations through studying the dispersion relation of the ultrasonic sounds. Now we have the relation (11.61) in the nonlinear case.

3. As we have observed before, by using the Maxwellian iteration, the Eq. (11.54) becomes the Navier-Stokes constitutive equation for the dynamic pressure (11.27). The equivalence between the relaxation times (11.61) and the relation (11.28) enables us to evaluate the value of the bulk viscosity from the data on the relaxation time τ_M for the internal degrees of freedom, which are accumulated in literature [24], in many kinds of polyatomic gases. See also Sect. 11.5.2.

   We observe that the shear viscosity and the heat conductivity come from the evolution equations for the heat flux and the shear stress in the 14-field theory with the use of the Maxwellian iteration. Therefore, in this simplified model ET6 in which we consider only the dynamic pressure, we cannot describe the cross effects between the fundamental irreversible processes.

   If we apply the Maxwellian iteration to $(11.36)_4$ instead of (11.54), we have a nonlinear constitutive equation for Π with respect to the velocity gradient, that is, the so-called non-Newtonian constitutive equation. Details are omitted here for simplicity.

   Lastly it is interesting to point out that there exist works [25–27] that, in the parabolic Navier-Stokes framework, neglect the shear viscosity but take into account the bulk viscosity expressed by the Eq. (11.27). And the qualitative analysis of the system was made.

11.5 Examples of an Ideal Polytropic Gas

In this section we present the case of ideal polytropic gases, i.e., ideal gases with the constant heat capacity, where the equilibrium equations of state are expressed by

$$p = \frac{k_B}{m}\rho T \quad \text{and} \quad \varepsilon = \frac{D}{2}\frac{k_B}{m}T. \tag{11.62}$$

It should be noted that, although the application of the ET6 theory to polytropic gases is limited, the present example gives us theoretical features of ET6 explicitly.

11.5.1 *Far-from-Equilibrium Case*

After cumbersome calculation using the method of characteristics, we obtain as a solution of (11.12) with (11.6):

$$k = \frac{k_B}{m}\log\left\{\left(1+\frac{\Pi}{p}\right)^{\frac{3}{2}}\left(1-\frac{3}{D-3}\frac{\Pi}{p}\right)^{\frac{D-3}{2}}\right\}. \tag{11.63}$$

We notice that k depends on (ρ, ε, Π) only through a single variable $Z = \Pi/p$. For $D > 3$, k exists and is bounded in the domain that contains an equilibrium state:

$$-1 < Z < \frac{D-3}{3}, \tag{11.64}$$

in which $k(Z) < 0, \forall Z \neq 0$ and k has a global maximum $k(0) = 0$ at the equilibrium state (see Fig. 11.1).

Therefore the convexity condition is satisfied in the range (11.64) and, according with a theorem proved in [28], the entropy h has, as is expected, the maximum value at the equilibrium state where $h = \rho s$.

The production term (11.30) becomes

$$\hat{P}_{ll} = -\frac{3(D-3)p^2\Pi}{\{(D-3)p - 3\Pi\}(p+\Pi)\tau}, \tag{11.65}$$

or equivalently in terms of the bulk viscosity (11.31):

$$\hat{P}_{ll} = -\frac{2}{\nu}\frac{(D-3)^2p^3\Pi}{D(p+\Pi)((D-3)p-3\Pi)}. \tag{11.66}$$

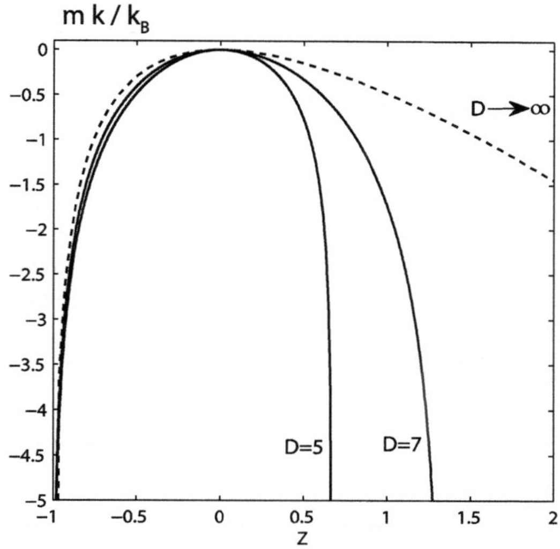

Fig. 11.1 Nonequilibrium entropy density k as the function of Z for different values of D

The main field (11.35) is given by

$$\begin{aligned}
\bar{\lambda} &= -\frac{g}{T} + k + \frac{v^2}{2T}\left(1 + \frac{\Pi}{p}\right)^{-1}, \\
\bar{\lambda}_i &= -\frac{v_i}{T}\left(1 + \frac{\Pi}{p}\right)^{-1}, \\
\bar{\mu} &= \frac{1}{2T}\left(1 + \frac{\Pi}{p}\right)^{-1}, \\
\bar{\xi} &= -\frac{1}{2T}\frac{D}{D-3}\frac{\Pi}{p}\left(1 + \frac{\Pi}{p}\right)^{-1}\left(1 - \frac{3}{D-3}\frac{\Pi}{p}\right)^{-1},
\end{aligned} \tag{11.67}$$

while the potentials (11.33) become

$$h'^0 = -\frac{p}{T}, \quad h'^i = -\frac{p}{T}v_i. \tag{11.68}$$

The Meixner temperature (11.45) becomes

$$\mathscr{T} = T\frac{p + \Pi}{p}. \tag{11.69}$$

It is interesting to observe that, from (11.69), the ratio of the two temperatures is equal to the ratio between the pressure in equilibrium and the total pressure:

$$\frac{T}{\mathcal{T}} = \frac{p}{\mathcal{P}}. \tag{11.70}$$

Furthermore, taking (11.62) into account, we obtain:

$$\mathcal{T} = T + \frac{m}{k_B}\frac{\Pi}{\rho}. \tag{11.71}$$

It is evident that the dynamical pressure (or the internal variable of the Meixner theory) is responsible for the difference between the two temperatures. Similar situation occurs for the nonequilibrium chemical potential. In fact, from $(11.67)_1$, $(11.44)_5$ and $(11.46)_1$, we have

$$\mathcal{G} = \left(\varepsilon + \frac{p}{\rho} - T\mathcal{S}\right)\left(1 + \frac{\Pi}{p}\right). \tag{11.72}$$

In the present particular case, the previous general results are rephrased as follows:

Statement 6 *In the case of ideal polytropic gases, the nonequilibrium entropy density is given by the expression* (11.63)*. Bounded entropic solutions satisfy the following inequalities:*

$$D > 3, \quad -p < \Pi < p\frac{D-3}{3}, \quad \tau > 0. \tag{11.73}$$

The entropy is convex and maximal in equilibrium. The system is symmetric hyperbolic and is thermodynamically stable.

In fact, the first inequality is a consequence of (11.19) where Γ^0 and Γ are given by

$$\Gamma^0 = \frac{2}{3}\frac{D-3}{D}p, \tag{11.74}$$

$$\Gamma = \Gamma^0\left\{1 + \frac{5D-6}{2(D-3)}\frac{\Pi}{p}\right\}. \tag{11.75}$$

The second inequalities are (11.64) and the last inequality is a consequence of the residual inequality (11.17) that requires $\alpha > 0$ (see (11.29)).

11.5.2 Near-Equilibrium Case

In the case of ideal polytropic gases (11.62), the internal variable is given by

$$\xi = \frac{\Pi}{\rho} - \frac{D-3}{3}\frac{p}{\rho}. \tag{11.76}$$

The expressions of the other quantities are easily expressed by using the expression of Ψ:

$$\Psi = \frac{3D}{4(D-3)}\frac{1}{pT}. \tag{11.77}$$

The main field (11.67) becomes

$$\begin{aligned} \bar{\lambda} &= -\frac{g}{T} + \frac{v^2}{2T}\left(1 - \frac{\Pi}{p}\right), \\ \bar{\lambda}_i &= -\frac{v_i}{T}\left(1 - \frac{\Pi}{p}\right), \\ \bar{\mu} &= \frac{1}{2T}\left(1 - \frac{\Pi}{p}\right), \\ \bar{\zeta} &= -\frac{1}{2T}\frac{D}{D-3}\frac{\Pi}{p}. \end{aligned} \tag{11.78}$$

If we know τ from experimental data, we know, from (11.28) and (11.74), the bulk viscosity:

$$\nu = \frac{2}{3}\tau\frac{D-3}{D}p, \tag{11.79}$$

and also we know the production term: (11.26).

By comparing (11.57) with (11.71), we notice that Meixner's temperature far from equilibrium has the same expression with the one near equilibrium for a ideal polytropic gas.

11.5.3 Rarefied Monatomic-Gas Limit

As the convexity condition in this case is expressed by $D > 3$, the above relation (11.77) is not valid in the case of monatomic gases where $D = 3$. Nevertheless we have shown in [3] (see also Sect. 5.6 for the case of ET14 and Sect. 10.5.1 for molecular ET) that the limit $D \to 3$ is a singular limit and that the solutions of the system of field equations converge to the solutions of the Euler fluid provided that initial data are consistent with monatomic gases, i.e., $\Pi(\mathbf{x}, 0) = 0$. In fact equation (11.54) becomes, for $D \to 3$, a homogeneous equation

$$\left(\frac{\Pi}{\rho}\right)^{\bullet} = -\frac{\Pi}{\rho\tau} \tag{11.80}$$

that admits only the solution $\Pi(\mathbf{x}, t) = 0, \forall t \geqslant 0$, while the system of the remaining field equations becomes the Euler system. The first five components of the main field (11.78) converge to the one of the Euler fluid (11.38) while the last component of the main field, the Lagrange multiplier $\bar{\zeta}$ of the vanishing equation in the system of ET6, remains undetermined. Then the convergence of the system of ET6 to the Euler system when $D \to 3$ is fully verified.

Remark We observe that, although, for $D > 3$, the two differential symmetric systems (11.4) and (11.34) are equivalent to each other, only the main field (11.78) of (11.34) converges to the main field of the Euler fluid. In fact, as we have noticed, the difference between the main fields (11.32) and (11.35) of the two equivalent systems consists only in the component of the energy equation, i.e., μ and $\bar{\mu}$ with the relation $\mu = \bar{\mu} - \bar{\zeta}$. Now $\bar{\mu}$ converges to the corresponding value of the Euler fluids $1/(2T)$, instead the limit of μ is undetermined due to the indetermination of $\bar{\zeta}$. This indicates that, for the convergence in the polyatomic to monatomic limit, it is preferable to use the system (11.34) instead of the system (11.4).

11.6 Conclusion

We have established the nonlinear ET6 theory valid also far from equilibrium of real gases based on extended thermodynamics in a general and complete way. We have obtained the main field explicitly, by using which the differential system becomes to be symmetric hyperbolic. The correspondence between the fully nonlinear systems of the ET6 theory and of the Meixner theory is discovered. The Meixner theory is rewritten into the closed system of balance equations that satisfies the Galilean invariance [29]. The internal variable and the nonequilibrium temperature in the Meixner theory are related to the quantities of the ET6 theory, in particular, the dynamic pressure and the temperature. The nonlinear theory includes, as a special case, the results near equilibrium. An interesting example of rarefied polytropic gases was presented. In this case the nonequilibrium entropy is explicitly derived and all of the calculations are given in the closed form. The monatomic-gas limit in which the system converges to the Euler system is also discussed.

Lastly we make four remarks:

(i) In our knowledge the ET6 theory is the only one that can address irreversible processes far from equilibrium in RET.
(ii) We have found the relationship between the ET theory and the Meixner theory within the simplest cases, namely, the ET6 theory and the Meixner

theory composed of the Euler equations with only one relaxation equation. The following question may naturally arise: Is there a possibility that, by establishing some correspondence relations like (11.56), the ET theory with more fields (for example, 14 fields studied in Chaps. 5 and 6, many fields in Chap. 10) has a counterpart in the Meixner theory with many internal variables? This question is interesting because several authors have tried to present models with internal variables to describe nonequilibrium processes and shown its applicability (see, for example, [10, 15] and the review paper [30]). In our opinion the answer is negative. In fact, only in the simplest ET theory with 6 fields, that is, the ET6 theory, a combination of the balance laws $(11.4)_3$ and $(11.4)_4$ gives us the possibility to have the equation in the form $(11.36)_4$ in which the spatial derivative exists only within the material time derivative, which is typical in the internal variable equation $(11.41)_4$. When the number of independent fields is more than 6, however, we have a system of balance laws (2.1) for the fields that is perfectly consistent with the kinetic theory [31, 32]. In our opinion, this is the reason why the ET theory is much more powerful than the other theories using internal variables in a similar way to the Meixner theory.

(iii) In the case of rarefied polyatomic gases, there exists large literature on kinetic models to explain relaxation processes and chemical reacting flows. See for example [33].

(iv) One of the promising applications of the present nonlinear analysis is the study of shock wave phenomena in a *non-polytropic* gas. This study is highly expected to enrich the previous studies [6, 34] by using the concept of the internal variable and the nonequilibrium temperature of the Meixner theory, because some authors studied this subject by using these quantities [35]. We will discuss this subject in Chap. 13 briefly.

References

1. T. Arima, S. Taniguchi, T. Ruggeri, M. Sugiyama, Nonlinear extended thermodynamics of real gases with 6 fields. Int. J. Non-Linear Mech. **72**, 6 (2015)
2. T. Arima, S. Taniguchi, T. Ruggeri, M. Sugiyama, Extended thermodynamics of real gases with dynamic pressure: an extension of Meixner's theory. Phys. Lett. A **376**, 2799 (2012)
3. T. Arima, T. Ruggeri, M. Sugiyama, S. Taniguchi, On the six-field model of fluids based on extended thermodynamics. Meccanica **49**, 2181 (2014)
4. S. Chapman, T.G. Cowling, *The Mathematical Theory of Non-uniform Gases* (Cambridge University Press, Cambridge, 1991)
5. T. Arima, S. Taniguchi, T. Ruggeri, M. Sugiyama, Dispersion relation for sound in rarefied polyatomic gases based on extended thermodynamics. Contin. Mech. Thermodyn. **25**, 727 (2013)

6. S. Taniguchi, T. Arima, T. Ruggeri, M. Sugiyama, Thermodynamic theory of the shock wave structure in a rarefied polyatomic gas: beyond the Bethe-Teller theory. Phys. Rev. E **89**, 013025 (2014)
7. T. Arima, S. Taniguchi, T. Ruggeri, M. Sugiyama, Monatomic rarefied gas as a singular limit of polyatomic gas in extended thermodynamics. Phys. Lett. A **377**, 2136 (2013)
8. S.K. Godunov, An interesting class of quasilinear systems. Sov. Math. **2**, 947 (1961)
9. S.R. de Groot, P. Mazur, *Non-equilibrium Thermodynamics* (North-Holland, Amsterdam, 1963)
10. B.D. Coleman, M.E. Gurtin, Thermodynamics with internal state variables. J. Chem. Phys. **47**, 597 (1967)
11. G.A. Maugin, W. Muschik, Thermodynamics with internal variables. Part I. General concepts. J. Non-Equilib. Thermodyn. **19**, 217 (1994)
12. G.A. Maugin, W. Muschik, Thermodynamics with internal variables. Part II. Applications. J. Non-Equilib. Thermodyn. **19**, 250 (1994)
13. J. Meixner, Absorption und dispersion des schalles in gasen mit chemisch reagierenden und anregbaren komponenten. I. Teil. Ann. Phys. **43**, 470 (1943)
14. J. Meixner, Allgemeine theorie der schallabsorption in gasen und flussigkeiten unter beruck-sichtigung der transporterscheinungen. Acoustica **2**, 101 (1952)
15. P. Ván, D. Berezovski, J. Engelbrecht, Internal variables and dynamic degrees of freedom. J. Non-Equilib. Thermodyn. **33**, 235 (2008)
16. W.G. Vincenti, C.H. Kruger Jr., *Introduction to Physical Gas Dynamics* (Wiley, New York/London/Sydney, 1965)
17. M.A. Leontovich, *Introduction to Thermodynamics*. Statistical Physics (Nauka, Moscow, 1983)
18. D.N. Zubarev, *Nonequilibrium Statistical Thermodynamics* (Consultants Bureau, New York/London, 1974)
19. N. Pottier, *Nonequilibrium Statistical Physics* (Oxford University Press, Oxford, 2010)
20. H. Grad, On the kinetic theory of rarefied gases. Commun. Pure Appl. Math. **2**(4), 331 (1949)
21. J. Casas-Vázquez, D. Jou, Nonequilibrium temperature versus local-equilibrium temperature. Phys. Rev. E **49**, 1040 (1994)
22. J. Casas-Vázquez, D. Jou, Temperature in non-equilibrium states: a review of open problems and current proposals. Rep. Prog. Phys. **66**, 1937 (2003)
23. E. Barbera, I. Müller, M. Sugiyama, On the temperature of a rarefied gas in non-equilibrium. Meccanica **34**, 103 (1999)
24. R.G. Gordon, W. Klemperer, J.I. Steinfeld, Vibrational and rotational relaxation. Ann. Rev. Phys. Chem. **19**, 215 (1968)
25. P. Secchi, Existence theorems for compressible viscous fluid having zero shear viscosity. Rend. Sem. Padova **70**, 73 (1983)
26. V. Shelukhin, Vanishing shear viscosity in a free-boundary problem for the equations of compressible fluids. J. Differ. Equ. **167**, 73 (2000)
27. H. Frid, V. Shelukhin, Vanishing shear viscosity in the equations of compressible fluids for the flows with the cylinder symmetry. SIAM J. Math. Anal. **31**, 1144 (2000)
28. T. Ruggeri, D. Serre, Stability of constant equilibrium state for dissipative balance laws system with a convex entropy. Quart. Appl. Math. **62**, 163 (2004)
29. T. Ruggeri, Galilean invariance and entropy principle for systems of balance laws. The structure of the extended thermodynamics. Contin. Mech. Thermodyn. **1**, 3 (1989)
30. V.A. Cimmelli, D. Jou, T. Ruggeri, P. Ván, Entropy principle and recent results in non-equilibrium theories. Entropy **16**, 1756 (2014)
31. T. Ruggeri, M. Sugiyama, Recent developments in extended thermodynamics of dense and rarefied polyatomic gases. Acta Appl. Math. **132**, 527 (2014)
32. T. Arima, A. Mentrelli, T. Ruggeri, Molecular extended thermodynamics of rarefied poly-atomic gases and wave velocities for increasing number of moments. Ann. Phys. **345**, 111 (2014)
33. E. Nagnibeda, E. Kustova, *Non-equilibrium Reacting Gas Flow* (Springer, Berlin/Heidelberg, 2009)

34. S. Taniguchi, T. Arima, T. Ruggeri, M. Sugiyama, Effect of dynamic pressure on the shock wave structure in a rarefied polyatomic gas. Phys. Fluids **26**, 016103 (2014)
35. Y.B. Zel'dovich, Y.P. Raizer, *Physics of Shock Waves and High-Temperature Hydrodynamic Phenomena* (Dover Publications, New York, 2002)

Chapter 12
Molecular Non-linear ET6 for Rarefied Polyatomic Gas

Abstract We establish extended thermodynamics of rarefied polyatomic gases with six independent fields via the maximum entropy principle. The distribution function is not necessarily near equilibrium. The result is in perfect agreement with the phenomenological ET theory explained in the previous Chap. 11. This is the first example of molecular extended thermodynamics with a non-linear closure. The integrability condition of the moments requires that the dynamical pressure should be bounded from below and from above.

12.1 Introduction

We discussed the MEP procedure in Sect. 1.5.5, and in particular we put in evidence the problem of the convergence of moments in a far-from-equilibrium case in Sect. 4.5. All closures by the MEP procedure studied up to here are valid only near equilibrium, including the 14-moment case of polyatomic gases discussed in Chap. 6. One of the worse circumstances is that hyperbolicity exists only in some small domain of the configuration space near equilibrium (see [1, 2] for monatomic non-degenerate gas case and [3] for Fermi and Bose gases).

The aim of this chapter is to prove that, in the case of ET6 of rarefied polyatomic gases, a theory can be established with the closure that is valid even far from equilibrium in accord with the phenomenological theory presented in the previous Chap. 11. We will show that this non-linear closure matches completely the previous result obtained by using only the macroscopic method [4]. The result of this chapter is presented firstly in the paper [5].

T. Ruggeri, M. Sugiyama, *Rational Extended Thermodynamics beyond the Monatomic Gas*, DOI 10.1007/978-3-319-13341-6_12

12.2 Nonequilibrium Distribution Function

We have deduced the equilibrium distribution function of any polyatomic gas in Theorem 6.1 (see (6.9)). Now we consider 6 moments (11.1):

$$\begin{pmatrix} F \\ F_i \\ F_{ll} \end{pmatrix} = \begin{pmatrix} \rho \\ \rho v_i \\ \rho v^2 + 3(p + \Pi) \end{pmatrix} = \int_{R^3} \int_0^\infty m \begin{pmatrix} 1 \\ c_i \\ c^2 \end{pmatrix} f I^\alpha \, dI \, d\mathbf{c} \qquad (12.1)$$

and

$$G_{ll} = \rho v^2 + 2\rho\varepsilon = \int_{R^3} \int_0^\infty m(c^2 + 2I/m) f I^\alpha \, dI \, d\mathbf{c}. \qquad (12.2)$$

We want to prove the following theorem [5]:

Theorem 12.1 *The distribution function that maximizes the entropy (6.5) under the constraints (12.1) and (12.2) has the form:*

$$f = \frac{\rho}{m\,(k_B T)^{1+\alpha} \Gamma(1+\alpha)} \left(\frac{m}{2\pi k_B T} \frac{1}{1 + \frac{\Pi}{p}} \right)^{3/2} \left(\frac{1}{1 - \frac{3}{2(1+\alpha)} \frac{\Pi}{p}} \right)^{1+\alpha}$$
$$\exp\left\{ -\frac{1}{k_B T} \left(\frac{1}{2} m C^2 \left(\frac{1}{1 + \frac{\Pi}{p}} \right) + I \left(\frac{1}{1 - \frac{3}{2(1+\alpha)} \frac{\Pi}{p}} \right) \right) \right\}. \qquad (12.3)$$

All the moments are convergent provided that

$$-1 < \frac{\Pi}{p} < \frac{2}{3}(1+\alpha), \qquad \alpha > -1. \qquad (12.4)$$

Proof The proof of the theorem is accomplished with the use of the Lagrange multiplier method. Introducing the vector of the multipliers $(\lambda, \lambda_i, \zeta, \mu)$, we define the functional:

$$\begin{aligned} \mathscr{L} = & -\int_{R^3} \int_0^\infty k_B f \log f \, I^\alpha \, dI \, d\mathbf{c} + \lambda \left(\rho - \int_{R^3} \int_0^\infty m f \, I^\alpha \, dI \, d\mathbf{c} \right) \\ & + \lambda_i \left(\rho v_i - \int_{R^3} \int_0^\infty m f c_i \, I^\alpha \, dI \, d\mathbf{c} \right) \\ & + \zeta \left(\rho v^2 + 3(p + \Pi) - \int_{R^3} \int_0^\infty m c^2 f \, I^\alpha \, dI \, d\mathbf{c} \right) \\ & + \mu \left(\rho v^2 + 2\rho\varepsilon - \int_{R^3} \int_0^\infty m \left(c^2 + 2\frac{I}{m} \right) f \, I^\alpha \, dI \, d\mathbf{c} \right). \end{aligned}$$

As this is a functional of the distribution function f and we want to maximize it with respect to f with the given macroscopic quantities, this functional can be substituted by the following one:

$$\mathscr{L} = -\int_{R^3}\int_0^\infty k_B f \log f\, I^\alpha\, dI\, d\mathbf{c} - \lambda \int_{R^3}\int_0^\infty m f\, I^\alpha\, dI\, d\mathbf{c}$$
$$-\lambda_i \int_{R^3}\int_0^\infty m f c_i\, I^\alpha\, dI\, d\mathbf{c} - \zeta \int_{R^3}\int_0^\infty m c^2 f\, I^\alpha\, dI\, d\mathbf{c} \tag{12.5}$$
$$-\mu \int_{R^3}\int_0^\infty m\left(c^2 + 2\frac{I}{m}\right) f\, I^\alpha\, dI\, d\mathbf{c}.$$

Since $\mathscr{L}$ is a scalar, it must retain the same value in the case of zero hydrodynamic velocity $\mathbf{v} = \mathbf{0}$ due to the Galilean invariance. Therefore, with $c_i = C_i + v_i$, we have

$$\mathscr{L} = -\int_{R^3}\int_0^\infty k_B f \log f\, I^\alpha\, dI\, d\mathbf{C} - \hat{\lambda} \int_{R^3}\int_0^\infty m f\, I^\alpha\, dI\, d\mathbf{C}$$
$$\hat{\lambda}_i \int_{R^3}\int_0^\infty m f C_i\, I^\alpha\, dI\, d\mathbf{C} - \hat{\zeta} \int_{R^3}\int_0^\infty m C^2 f\, I^\alpha\, dI\, d\mathbf{C} \tag{12.6}$$
$$\hat{\mu} \int_{R^3}\int_0^\infty m\left(C^2 + 2\frac{I}{m}\right) f\, I^\alpha\, dI\, d\mathbf{C}.$$

Comparison between (12.5) and (12.6) yields the relations between the Lagrange multipliers and the corresponding zero-velocity (or intrinsic) Lagrange multipliers indicated by hat:

$$\lambda = \hat{\lambda} - \hat{\lambda}_i v_i + (\hat{\zeta} + \hat{\mu})v^2; \quad \lambda_i = \hat{\lambda}_i - 2(\hat{\zeta} + \hat{\mu})v_i; \quad \zeta = \hat{\zeta} \quad \mu = \hat{\mu}, \tag{12.7}$$

which dictate the velocity dependence of the Lagrange multipliers. We notice that these relations are in accordance with the general results of the Galilean invariance [6] (see (2.42)). The Euler-Lagrange equation $\delta\mathscr{L}/\delta f = 0$ leads to the following form of the distribution function in agreement with the general Theorem 10.2 valid for any (N, M) system (see (10.10)) :

$$f = \exp^{-1-\frac{m}{k_B}\chi}, \tag{12.8}$$

where

$$\chi = \hat{\lambda} + \hat{\lambda}_i C_i + \hat{\zeta} C^2 + \hat{\mu}\left(C^2 + 2\frac{I}{m}\right).$$

By introducing the following variables:

$$\xi = \frac{m}{k_B}(\hat{\zeta} + \hat{\mu}), \quad \eta_i = \frac{m}{k_B}\hat{\lambda}_i, \quad \omega = \frac{2}{k_B}\hat{\mu}, \quad \Omega = \exp\left(-1 - \frac{m}{k_B}\hat{\lambda}\right), \tag{12.9}$$

the distribution function can be rewritten as

$$f = \Omega e^{-\omega I} e^{-\xi C^2 - \eta_i C_i}. \tag{12.10}$$

Inserting (12.10) into the second equation of (12.1) evaluated at the zero velocity, we obtain immediately $\eta_i = 0$. Then the remaining equations of (12.1) and (12.2) evaluated at $\mathbf{v} = 0$ become

$$\begin{aligned}
\rho &= \int_{R^3}\int_0^\infty mfI^\alpha \, dI \, d\mathbf{C} = m\pi^{3/2}\Gamma(1+\alpha)\frac{\Omega}{\xi^{3/2}\omega^{1+\alpha}}, \\
p + \Pi &= \frac{1}{3}\int_{R^3}\int_0^\infty mfC^2 I^\alpha \, dI \, d\mathbf{C} = m\pi^{3/2}\Gamma(1+\alpha)\frac{\Omega}{2\xi^{5/2}\omega^{1+\alpha}}, \\
\rho\varepsilon &= \int_{R^3}\int_0^\infty mf\left(\frac{C^2}{2} + \frac{I}{m}\right) I^\alpha \, dI \, d\mathbf{C} \\
&= m\pi^{3/2}\Gamma(1+\alpha)\frac{\Omega}{4\xi^{5/2}\omega^{1+\alpha}}\left(3 + \frac{4}{m}(1+\alpha)\frac{\xi}{\omega}\right).
\end{aligned} \tag{12.11}$$

From the integrability condition, we have

$$\omega > 0, \qquad \xi > 0, \qquad \alpha > -1. \tag{12.12}$$

From (12.11) and (11.62), we obtain

$$\begin{aligned}
\varepsilon &= \frac{1}{4\xi}\left\{3 + \frac{2}{m}(D-3)\frac{\xi}{\omega}\right\}, \\
p &= \frac{m}{2D}\pi^{3/2}\Gamma\left(\frac{D-3}{2}\right)\frac{\Omega}{\xi^{5/2}\omega^{\frac{D-3}{2}}}\left\{3 + \frac{2}{m}(D-3)\frac{\xi}{\omega}\right\}, \\
\Pi &= \frac{m}{2}\pi^{3/2}\Gamma\left(\frac{D-3}{2}\right)\frac{D-3}{D}\frac{1 - \frac{2}{m}\frac{\xi}{\omega}}{\xi^{5/2}\omega^{\frac{D-3}{2}}}\Omega.
\end{aligned} \tag{12.13}$$

We can invert these relations as follows:

$$\xi = \frac{\rho}{2p}\frac{1}{1+\frac{\Pi}{p}},$$
$$\omega = \frac{\rho}{m}\frac{(D-3)}{2\rho\varepsilon - 3(p+\Pi)} = \frac{\rho}{mp}\frac{1}{1-\frac{3}{D-3}\frac{\Pi}{p}}, \tag{12.14}$$
$$\Omega = \frac{\rho}{m\pi^{3/2}\Gamma\left(\frac{D-3}{2}\right)}\left(\frac{\rho}{2p}\frac{1}{1+\frac{\Pi}{p}}\right)^{\frac{3}{2}}\left(\frac{\rho}{mp}\frac{1}{1-\frac{3}{D-3}\frac{\Pi}{p}}\right)^{\frac{D-3}{2}}.$$

The integrability conditions (12.12) imply that, for a bounded solution, the ratio Π/p must satisfy the same inequalities in the macroscopic approach (11.64). Inserting (12.14) into the distribution function (12.10), we obtain (12.3) and the proof is completed. When $\Pi \to 0$ (12.3) becomes the equilibrium distribution function (6.9).

12.3 Closure and the Field Equations

Substituting (12.3) into the fluxes F_{llk}, G_{llk} and the production term P_{ll}, we obtain after some calculations

$$\begin{aligned}
F_{ik} &= \int_{R^3}\int_0^\infty mc_ic_kfI^\alpha\, dI\, d\mathbf{c} = \rho v_iv_k + (p+\Pi)\delta_{ik},\\
F_{llk} &= \int_{R^3}\int_0^\infty mc^2c_kfI^\alpha\, dI\, d\mathbf{c} = \left(5(p+\Pi)+\rho v^2\right)v_k,\\
G_{llk} &= \int_{R^3}\int_0^\infty m\left(c^2+\frac{2I}{m}\right)c_kfI^\alpha\, dI\, d\mathbf{c} = (\rho v^2 + 2\rho\varepsilon + 2p + 2\Pi)v_k,\\
P_{ll} &= \hat{P}_{ll} = \int_{R^3}\int_0^\infty mC^2Q(f)\, I^\alpha\, dI\, d\mathbf{C}.
\end{aligned} \tag{12.15}$$

From the balance equations of momentum and of energy in continuum mechanics, we know that

$$F_{ik} = \rho v_iv_k - t_{ik}, \qquad G_{llk} = (\rho v^2 + 2\rho\varepsilon)v_k - 2t_{ik}v_i + 2q_k,$$

where the stress tensor t_{ik} is given by (1.5). Comparing with the closure $(12.15)_{(1,3)}$, we conclude that the closure gives the result that, in the 6-moment theory, $\sigma_{<ik>} = 0$ and $q_k = 0$. This is the expected result that there exist no shear viscosity and no heat conductivity in the 6-moment theory. For what concerns $(12.15)_2$, taking into account the Galilean invariance, we obtain the zero-velocity part of F_{llk}:

$$\hat{F}_{llk} = \int_{R^3} \int_0^\infty mC^2 C_k f I^\alpha \, dI \, d\mathbf{C} = 0.$$

Concerning the production term $(12.15)_4$, the main problem is that, in order to have explicit expression of the production, we need a specific model for the collision term, which is, in general, not easy to obtain in the case of polyatomic gases.

With (12.15) we obtain the same differential system of 6 moments as the one obtained via macroscopic approach (11.4). This system (11.4) with the thermal and caloric equations of state (11.62) is a closed system for the 6 unknowns (ρ, v_i, T, Π) provided that we know the collision term in $(12.15)_4$.

We conclude that the results obtained via MPE is in perfect agreement with the results derived from the phenomenological theory in Chap. 11 [4]. A possible expression for the production term $\hat{P}_{ll}$ is given in (6.34) for 14 moments putting $\sigma_{<ij>} = 0$ or, via phenomenological considerations, in (11.65) and (11.66) in terms of the relaxation time of the bulk viscosity. In the case of the BGK approach we obtain $\hat{P}_{ll} = -3\Pi/\tau$.

12.3.1 Entropy Density

Let us study the entropy density h with non-linear distribution function:

$$h = -k_B \int \int f \log f I^\alpha \mathrm{d}I \, \mathrm{d}\mathbf{C} = \frac{k_B}{m} \rho \left(\frac{D}{2} - \log \Omega \right),$$

with Ω being given by $(12.14)_3$. The equilibrium part of the entropy density h_E is expressed as

$$h_E = \rho s = \frac{k_B}{m} \rho \left(\frac{D}{2} - \log \Omega_E \right). \tag{12.16}$$

Moreover we notice that the chemical potential $g = \varepsilon + \frac{p}{\rho} - Ts$ is expressed as

$$\frac{g}{T} = \frac{k_B}{m} \left(1 + \log \Omega_E \right). \tag{12.17}$$

On the other hand, the nonequilibrium part of the entropy becomes

$$k = \frac{1}{\rho} (h - h_E) = -\frac{k_B}{m} \log \frac{\Omega}{\Omega_E}. \tag{12.18}$$

Since

$$\frac{\Omega}{\Omega_E} = \left(1 + \frac{\Pi}{p}\right)^{-\frac{3}{2}} \left(1 - \frac{3}{D-3}\frac{\Pi}{p}\right)^{-\frac{D-3}{2}}, \tag{12.19}$$

we find that k coincides with the one obtained by the phenomenological ET approach (11.63). As we have seen in Sect. 11.5, k depends on a single variable $Z = \Pi/p$ and has a global maximum at an equilibrium state.

12.4 Main Field and Symmetric System

From (12.14), (12.9) and (12.7), we obtain the full expression of the Lagrange multipliers after some cumbersome calculations and they coincide exactly with the one obtained by macroscopic approach (11.67). Notice that, in equilibrium where $\Pi = 0$, the first five components of the main field (11.67) coincide with those obtained by Godunov for the Euler fluid [7]:

$$\lambda|_E = -\frac{1}{T}\left(g - \frac{v^2}{2}\right), \quad \lambda_i|_E = -\frac{v_i}{T}, \quad \mu|_E = \frac{1}{2T},$$

while $\zeta|_E = 0$ according to the fact that the Euler fluid is a *principal subsystem* of the 6-moment system [8].

We observe that, if we apply the Maxwellian iteration, the last equation of (11.4) with the production given by the BGK approximation reduces to the Navier-Stokes law in the case of no shear stress [4, 9, 10]:

$$\Pi = -\nu \ \mathrm{div}\mathbf{v}, \quad \text{with} \quad \nu = \frac{2}{3}\frac{D-3}{D}p\tau, \tag{12.20}$$

where ν is the bulk viscosity. The system (11.4) in which the last equation is replaced by (12.20) was studied by Secchi [11] and by Frid and Shelukhin [12, 13].

References

1. I. Müller, T. Ruggeri, *Rational Extended Thermodynamics*, 2nd edn. (Springer, New York, 1998)
2. F. Brini, Hyperbolicity region in extended thermodynamics with 14 moments. Contin. Mech. Thermodyn. **13**, 1 (2001)
3. T. Ruggeri, M. Trovato, Hyperbolicity in extended thermodynamics of Fermi and Bose gases. Contin. Mech. Thermodyn. **16**(6), 551 (2004)

4. T. Arima, S. Taniguchi, T. Ruggeri, M. Sugiyama, Nonlinear extended thermodynamics of real gases with 6 fields. Int. J. Non-Linear Mech. **72**, 6 (2015)
5. T. Ruggeri, Non-linear maximum entropy principle for a polyatomic gas subject to the dynamic pressure. Bull. Inst. Math. Acad. Sin. Special Issue in honor of Tai-Ping Liu 70 birthday **11** (2016)
6. T. Ruggeri, Galilean invariance and entropy principle for systems of balance laws. The structure of the extended thermodynamics. Contin. Mech. Thermodyn. **1**, 3 (1989)
7. S.K. Godunov, An interesting class of quasilinear systems. Sov. Math. **2**, 947 (1961)
8. G. Boillat, T. Ruggeri, Hyperbolic principal subsystems: entropy convexity and subcharacteristic conditions. Arch. Ration. Mech. Anal. **137**, 305 (1997)
9. T. Arima, S. Taniguchi, T. Ruggeri, M. Sugiyama, Extended thermodynamics of real gases with dynamic pressure: an extension of Meixner's theory. Phys. Lett. A **376**, 2799 (2012)
10. T. Arima, T. Ruggeri, M. Sugiyama, S. Taniguchi, On the six-field model of fluids based on extended thermodynamics. Meccanica **49**, 2181 (2014)
11. P. Secchi, Existence theorems for compressible viscous fluid having zero shear viscosity. Rend. Sem. Padova **70**, 73 (1983)
12. V. Shelukhin, Vanishing shear viscosity in a free-boundary problem for the equations of compressible fluids. J. Differ. Equ. **167**, 73 (2000)
13. H. Frid, V. Shelukhin, Vanishing shear viscosity in the equations of compressible fluids for the flows with the cylinder symmetry. SIAM J. Math. Anal. **31**, 1144 (2000)

Chapter 13
Application of ET6: Shock Wave and Sub-shock Formation

Abstract In this chapter, we show the usefulness of the ET6 theory for the analysis of the shock wave structure in a rarefied polyatomic gas. We compare the theoretical prediction derived from the ET6 theory with that from the ET14 theory. We see, in particular, that the thin layer in Type C with finite thickness described by the ET14 theory is replaced by a discontinuous jump, the sub-shock. The strength and the stability of a sub-shock is also discussed. Lastly the temperature overshoot at a sub-shock in terms of Meixner's temperature is shown.

13.1 Introduction

In Chap. 8, the shock wave structure in a rarefied polyatomic gas has been studied on the basis of the ET14 theory. We found that the ET14 theory can describe three types of the shock wave structure, Types A, B and C, in a rarefied polyatomic gas in a unified and consistent way.

In the analysis of Chap. 8, we also found that the dynamic pressure Π plays an essential role in the formation of the shock wave structure but the shear stress and the heat flux are negligibly small everywhere except for the inside of a thin layer. This is due to the fact that the relaxation time of the dynamic pressure is much larger than the other relaxation times of the shear stress and the heat flux in a rarefied polyatomic gas, such as H_2 gas, CO_2 gas [2, 3]. Therefore, it is natural to expect that we may study the shock wave structure properly on the basis of the ET6 theory.

In this chapter, we study the same problem of Chap. 8, but with the use of the ET6 theory [1].

13.2 Basis of the Present Analysis

13.2.1 Characteristic Velocities

Let us consider the problem of one space dimension. The characteristic velocities in equilibrium λ of the ET6 system have the following expression for any gas (see for

T. Ruggeri, M. Sugiyama, *Rational Extended Thermodynamics beyond the Monatomic Gas*, DOI 10.1007/978-3-319-13341-6_13

the proof Sect. 14.1):

$$\lambda = 0, \ 0, \ \pm\sqrt{\frac{5}{3}\frac{k_B}{m}T}. \tag{13.1}$$

Let us consider a non-polytropic gas. In this case it is convenient to write the ratio with the sound velocity c given in (7.9) :

$$\frac{\lambda}{c} = 0, \ 0, \ \pm\sqrt{\frac{5}{3}\frac{1}{\gamma(T)}}, \tag{13.2}$$

where $\gamma(T)$ is the ratio of specific heat given in (7.10). As in Chap. 8, we study the shock wave structure in a CO_2 gas at $T = 295$ [K] and $p = 69$ [mmHg] in the unperturbed state. In this case, the maximum characteristic velocity is estimated as $\lambda_{max}/c \approx 1.137$. Therefore, from Theorem 3.1, the continuous shock wave structure may exist in the Mach number region: $1 < M_0 < 1.137$. It is interesting to notice that, in the case of the ET14 system studied in Chap. 8, $\lambda_{max}/c \approx 1.74$.

13.2.2 Parameters

Figure 8.1 shows the dependence of the specific heat of a rarefied CO_2 gas on the temperature. The temperature and mass density dependence of the relaxation time of the dynamic pressure is expressed by (see $(8.10)_1$)

$$\frac{\tau_\Pi(\rho, T)}{\tau_\Pi(\rho_0, T_0)} = \frac{5 - 3\gamma_0}{5 - 3\gamma(T)}\frac{\rho_0}{\rho}\left(\frac{T_0}{T}\right)^{1-n}, \tag{13.3}$$

where T_0 is the temperature in the unperturbed state and $\gamma_0 \equiv \gamma(T_0)$. The exponent n is, as before, given by $n = 0.935$ [4]. The values of the specific heat, the sound velocity, and the relaxation time of the dynamic pressure are summarized in Tables 8.1 and 8.2.

Remark The ET6 theory is also applicable to some other gases like rarefied hydrogen and deuterium gases. In fact, the analysis of ultrasonic sounds in Chap. 7 reveals that τ_Π of these gases is much larger than τ_S and τ_q. However, as there is no suitable experimental datum on the shock wave structure in these gases, we take only a CO_2 gas as a typical example in the present chapter.

13.2.3 Dimensionless Form of the Balance Equations

We hereafter study the structure of a stationary plane shock wave. For convenience we again introduce the dimensionless quantities (8.4). Then the balance equations (11.4) are rewritten as

$$\begin{aligned}
&\frac{\mathrm{d}}{\mathrm{d}\hat{x}}(\hat{\rho}\hat{v}) = 0, \\
&\frac{\mathrm{d}}{\mathrm{d}\hat{x}}\left\{\frac{1}{\gamma_0}\left(\hat{\rho}\hat{T} + \hat{\Pi}\right) + \hat{\rho}\hat{v}^2\right\} = 0, \\
&\frac{\mathrm{d}}{\mathrm{d}\hat{x}}\left\{\frac{2}{\gamma_0}\left((\hat{\rho}\hat{T} + \hat{\Pi})\hat{v} + \hat{\rho}\hat{v}\frac{1}{T_0}\int_{T_R}^{T}\hat{c}_v(\xi)\,d\xi\right) + \hat{\rho}\hat{v}^3\right\} = 0, \\
&\frac{\mathrm{d}}{\mathrm{d}\hat{x}}\left\{\frac{5}{\gamma_0}(\hat{\rho}\hat{T} + \hat{\Pi})\hat{v} + \hat{\rho}\hat{v}^3\right\} = -\frac{3}{\gamma_0}\frac{\hat{\Pi}}{\hat{\tau}_\Pi}.
\end{aligned} \tag{13.4}$$

Here we have assumed the linear relation (11.26) for $\hat{P}_{ll}$. Compare these equations with (8.5).

By integrating the conservation laws $(13.4)_{1-3}$, the balance equations (13.4) become as follows:

$$\begin{aligned}
&\hat{\rho} = \frac{M_0}{\hat{v}}, \\
&\frac{1}{\gamma_0}\left(\frac{M_0\hat{T}}{\hat{v}} + \hat{\Pi}\right) + M_0\hat{v} = \frac{1}{\gamma_0} + M_0^2, \\
&\frac{2}{\gamma_0}\left(M_0\hat{T} + \hat{\Pi}\hat{v} + \frac{M_0}{T_0}\int_{T_0}^{T}\hat{c}_v(\xi)\,d\xi\right) + M_0\hat{v}^2 = \frac{2}{\gamma_0}M_0 + M_0^3, \\
&\frac{\mathrm{d}}{\mathrm{d}\hat{x}}\left\{\frac{5}{\gamma_0}\left(M_0\hat{T} + \hat{\Pi}\hat{v}\right) + M_0\hat{v}^2\right\} = -\frac{3}{\gamma_0}\frac{\hat{\Pi}}{\hat{\tau}_\Pi}.
\end{aligned} \tag{13.5}$$

13.2.4 Boundary Conditions

The quantities in the perturbed state are derived from the Rankine-Hugoniot (RH) conditions (8.8).

13.2.5 RH Conditions for a Sub-shock in Type C

From the general conditions (3.34), we obtain the RH conditions for a sub-shock in Type C:

$$
\begin{aligned}
&\hat{\rho}_* = \frac{M_0}{\hat{v}_*}, \\
&\frac{1}{\gamma_0}\left(\frac{M_0 \hat{T}_*}{\hat{v}_*} - 1 + \hat{\Pi}_*\right) + M_0\left(\hat{v}_* - M_0\right) = 0, \\
&\frac{2}{\gamma_0}\left(M_0 \hat{T}_* + \hat{\Pi}_* \hat{v}_* + \frac{M_0}{T_0}\int_{T_0}^{T_*} \hat{c}_v(\xi)\, d\xi - M_0\right) + M_0\left(\hat{v}_*^2 - M_0^2\right) = 0, \\
&\frac{5}{\gamma_0}\left(M_0 \hat{T}_* + \hat{\Pi}_* \hat{v}_* - M_0\right) + M_0\left(\hat{v}_*^2 - M_0^2\right) = 0.
\end{aligned}
\tag{13.6}
$$

The quantities with the subscript $*$ mean the quantities in the state just after a sub-shock. Note that the state $*$ in Type C is different from the perturbed state 1. See the shock profile of Type C in Fig. 1.1.

If we subtract $(13.6)_3$ from $(13.6)_4$, we obtain:

$$
\frac{\hat{\Pi}_*}{\hat{\rho}_*} = \frac{2}{3}\frac{1}{T_0}\int_{T_0}^{T_*}\left(\hat{c}_v(\xi) - \frac{3}{2}\right) d\xi
$$

that gives the value of $\hat{\Pi}_*$. If the gas is monatomic, we notice that $\hat{c}_v = 3/2$ and then Π vanishes everywhere. Furthermore, from $(13.6)_2$ and $(13.6)_4$, we have

$$
4M_0\hat{v}_*^2 - 5\left(M_0 + \frac{1}{\gamma_0}\right)\hat{v}_* + \frac{5}{\gamma_0}M_0 + M_0^3 = 0. \tag{13.7}
$$

Then we have

$$
\hat{v}_* = M_0 \quad \text{and} \quad \frac{1}{4M_0}\left(M_0^2 + \frac{5}{\gamma_0}\right). \tag{13.8}
$$

The first solution is a trivial one, and the second one corresponds to the solution for a sub-shock.

13.2.6 Numerical Methods

We solve numerically the system (13.5) under the boundary conditions by adopting the numerical methods proposed by Weiss [5, 6]. Because the mass density is already expressed by the other variables $(13.5)_1$, we need to solve the system $(13.5)_{2-4}$ for $\mathbf{u} = (\hat{v}, \hat{T}, \hat{\Pi})$.

13.2.7 Case 1: $M_0 < \lambda_{max}/c_0$

In this case, no sub-shock appears. Therefore the numerical method explained in Sect. 8.3.4 can be adopted also in this case.

13.2.8 Case 2: $M_0 > \lambda_{max}/c_0$

As a sub-shock appears in this case, which is assumed to be at $x = 0$, the boundary condition is replaced by the RH conditions for a sub-shock (13.6). We introduce the $N + 1$ grid points such that the range $[0, L]$ in the x-axis is discretized with constant intervals $\Delta x = L/N$ as follows:

$$x^i = \frac{L}{N} i \quad \text{for} \quad i = 0, 1, \cdots, N. \tag{13.9}$$

The boundary conditions are given by

$$\begin{aligned} \mathbf{u}^0 &= \mathbf{u}_*, \\ \mathbf{u}^N &= \mathbf{u}_1, \end{aligned} \tag{13.10}$$

where $\mathbf{u}^i$ represents $\mathbf{u}|_{x=x^i}$, $\mathbf{u}_0 = (M_0, T_0, 0)$ and $\mathbf{u}_* = (\hat{v}_*, T_*, \hat{\Pi}_*)$. We solve numerically the basic equations in the same way as above.

13.3 Shock Wave Structure with and without a Sub-shock

13.3.1 Shock Wave Structure without a Sub-shock

Let us analyze the continuous shock wave structure in the Mach number region $1 < M_0 < \lambda_{max}/c_0$. The profiles of the mass density, the velocity, the temperature and the dynamic pressure at $M_0 = 1.04$ is shown in Fig. 13.1. This is a typical example of the shock wave structure of Type A. We also depict the shock wave structure at $M_0 = 1.12$ as an example of Type B in Fig. 13.2.

From these figures, we understand clearly that the ET6 theory can reproduce nearly the same shock waves structures of Types A and B as those by the ET14

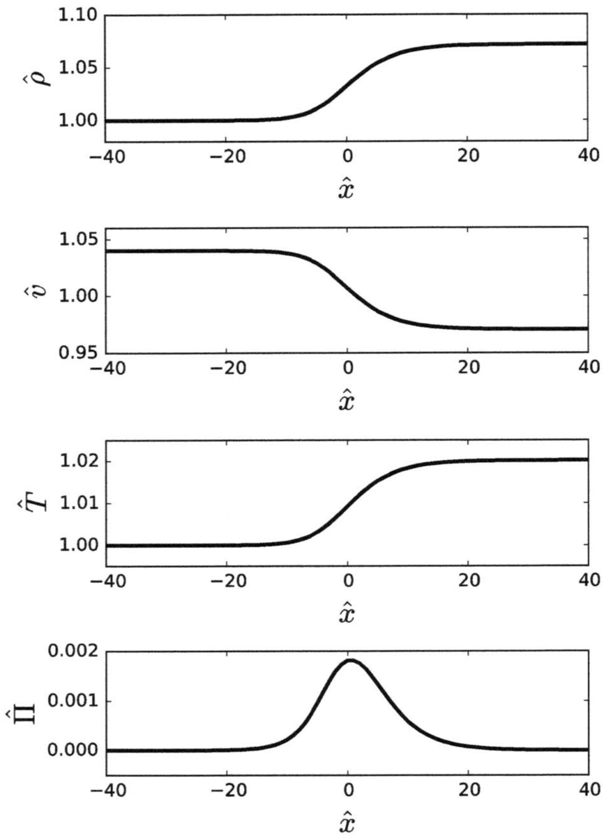

Fig. 13.1 Type A: Profiles of the dimensionless mass density, the velocity, the temperature and the dynamic pressure predicted by the ET6 theory (*black thick curves*) and by the ET14 theory (*red thin curves*). $M_0 = 1.04$. The *black* and *red curves* coincide with each other within the thickness of the *black curve*

theory. Note that this conclusion is true only when the relaxation time of the dynamic pressure is much larger than the other two relaxation times.

13.3.2 Shock Wave Structure with a Sub-shock

Figures 13.3 and 13.4 show the predictions of the shock wave structure at $M_0 = 1.15$ and $M_0 = 1.47$ by the ET6 and ET14 theories. These shock wave structures belong to Type C.

We notice from Figs. 13.3 and 13.4 that the ET6 theory can describe the shock wave structure quite well. Only the difference between ET6 and ET14 can be

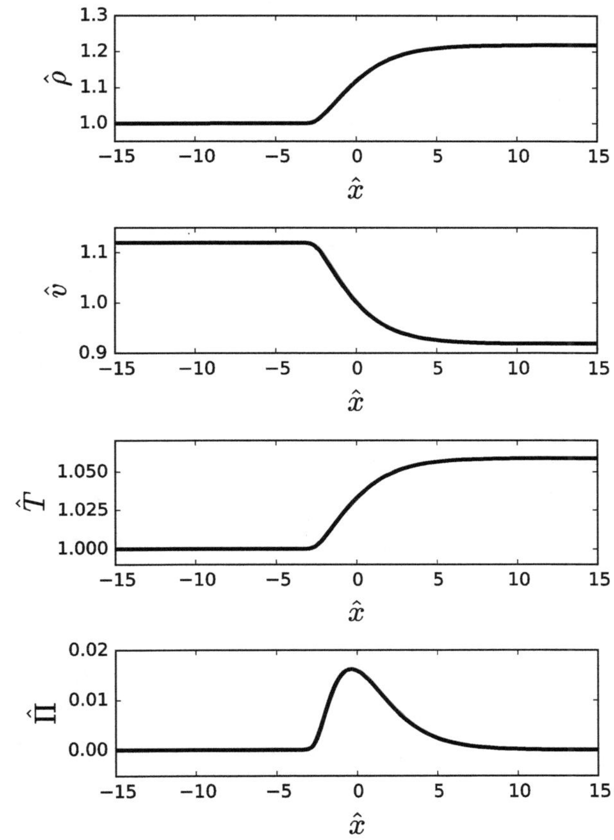

Fig. 13.2 Type B: Shock wave structures predicted by the ET6 theory (*black thick curves*) and by the ET14 theory (*red thin curves*). $M_0 = 1.12$. The *black* and *red curves* coincide with each other within the thickness of the *black curve*

detected in the thin layer. See the magnifications shown in the sub-figures of Figs. 13.3 and 13.4. The thin layer in the ET6 theory is represented by a sub-shock, but, in the ET14 theory, it is represented by a thin layer with finite thickness. In other words, the ET6 theory can describe the thin layer only as a discontinuous surface because of the limited resolution inherent in the theory. On the other hand, owing to its finer resolution, the ET14 theory can describe the fine structure of the thin layer. In this respect, it is also interesting to note that the Bethe-Teller theory [7] also regards the thin layer as a discontinuous surface.

In conclusion, the ET6 theory can describe also the shock wave structure of Type C very well if we are not interested in the fine structure of the thin layer.

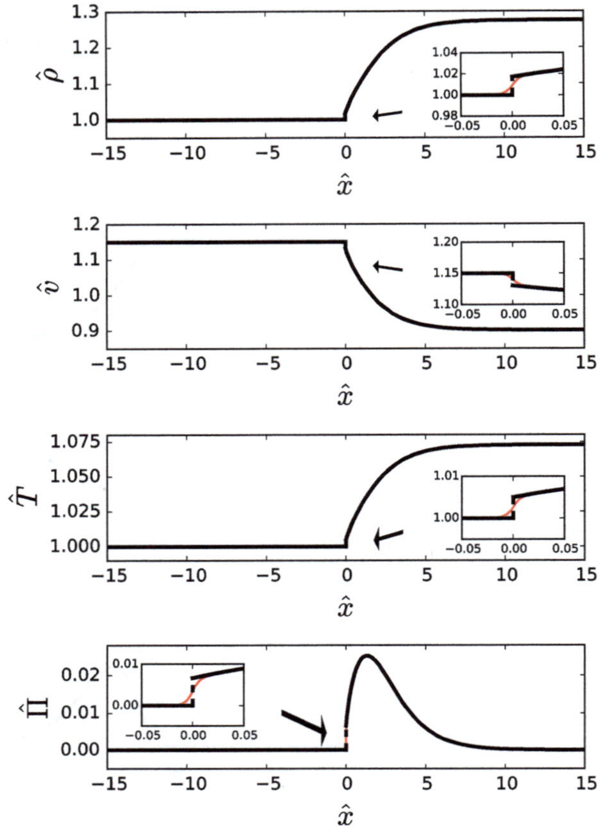

Fig. 13.3 Type C: Shock wave structures predicted by the ET6 theory (*black thick curves*) and by the ET14 theory (*red thin curves*). $M_0 = 1.15$

13.3.3 Discussions

(1) We have seen above that the ET6 theory can describe the shock wave structures of all Types A, B, and C in a unified way within its resolution. We may use the ET6 theory for analyzing the shock wave structure even at the Mach number larger than λ_{max}/c_0.

(2) The interpretation in (1) is consistent with the fact that the maximum characteristic velocity monotonically increases with the increase of the number of the independent variables in hyperbolic systems (see Sect. 4.4). The ET theory can describe more and more fine shock wave structure as the number of the independent variables increases.

(3) As pointed out before, in the ET14 theory, the continuous shock wave structure is obtained up to the Mach number $M_0 \approx 1.74$. Above this Mach number, the ET14 theory describes the thin layer as the layer with a sub-shock. If we want to analyze the continuous shock wave structure at a larger Mach number than 1.74, we need to adopt more independent variables.

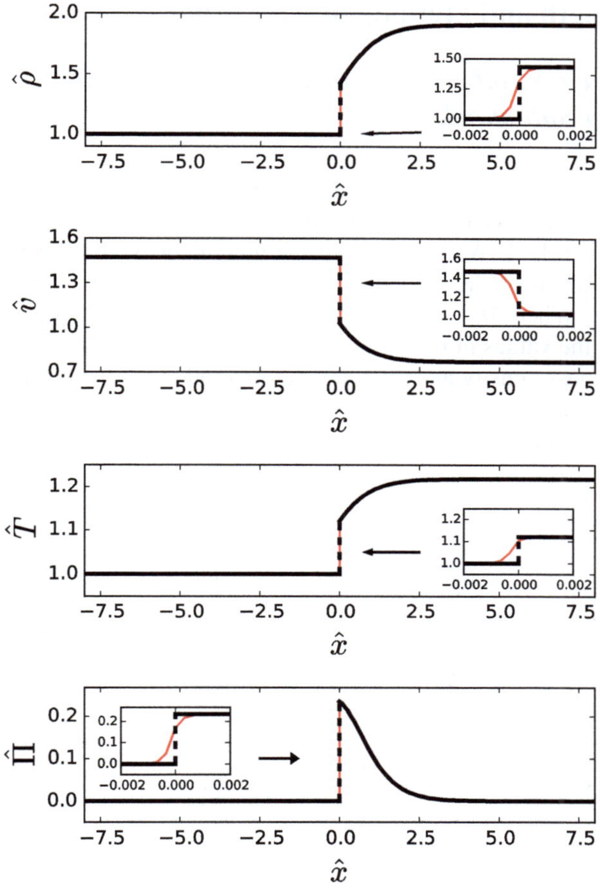

Fig. 13.4 Type C: Shock wave structures predicted by the ET6 theory (*black thick curves*) and by the ET14 theory (*red thin curves*). $M_0 = 1.47$

(4) We can define, without ambiguity, the critical Mach number at which the transition between Types B and C occurs by using the characteristic velocity of ET6 theory. We obtain $M_0 \approx 1.137$, which is consistent with the experimental data [8–13] although, at present, the experimental data are too few to determine the definite value of the critical Mach number.

13.4 Strength and Stability of a Sub-shock

13.4.1 *Mach Number Dependence of the Strength of a Sub-shock*

We depict the Mach number dependence of the mass density, the velocity, the temperature and the dynamic pressure in the state $*$, that is, the state just after a sub-shock in Fig. 13.5. We can see that the mass density, the temperature and the dynamic pressure (and the velocity) increase (decreases) with the increase of the Mach number. When $M_0 = \lambda_{max}/c_0$, the values of the quantities in the state $*$ coincide with the values in the unperturbed state 0. It is interesting to note that the values of the mass density and the temperature (and the velocity) in the state $*$ are always smaller (larger) than the values of those in the perturbed state 1. This means that there is no overshoot of these quantities in a sub-shock.

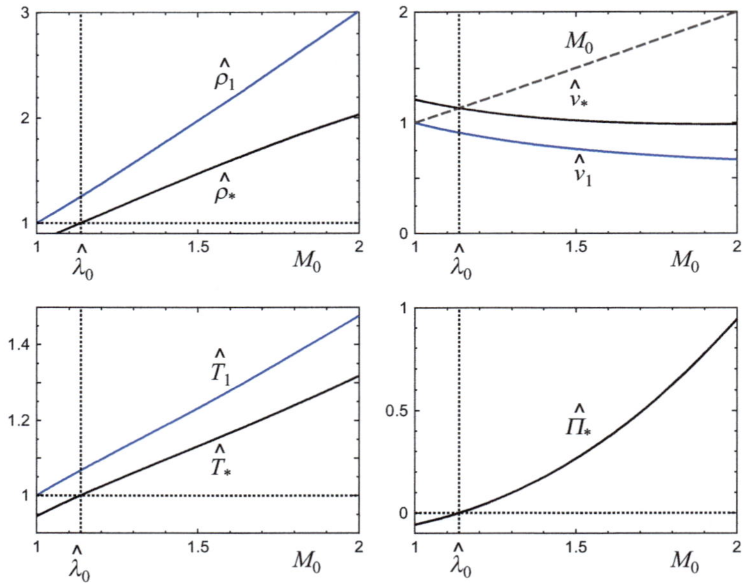

Fig. 13.5 Mach number dependence of the mass density, the velocity, the temperature and the dynamic pressure in the state $*$ just after a sub-shock (*black curves*). Similar dependence of the quantities in the perturbed state 1 (*blue curves*) is also shown. Here $\hat{\lambda}_0$ is defined by $\hat{\lambda}_0 \equiv \lambda_{max}/c_0$

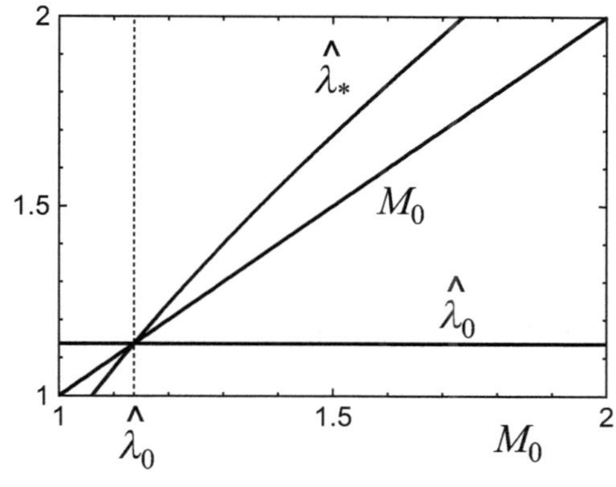

Fig. 13.6 The dependence of the characteristic velocities $\hat{\lambda}_0$ and $\hat{\lambda}_*$ on the Mach number for a sub-shock

13.4.2 Stability of a Sub-shock

All solutions of the RH conditions for a sub-shock (13.6) are not necessarily stable (admissible). In order to select a stable solution, we can utilize the Lax condition (see Sect. 3.3.2.1), which , in the present case, is written by

$$\hat{\lambda}_0 < M_0 < \hat{\lambda}_* \tag{13.11}$$

with $\hat{\lambda}_0$ and $\hat{\lambda}_*$ being

$$\hat{\lambda}_0 = \sqrt{\frac{5}{3}\frac{1}{\gamma_0}} \quad \text{and} \quad \hat{\lambda}_* = \sqrt{\frac{5}{3}\frac{1}{\gamma(T_*)}\frac{c_*}{c_0}} + M_0 - \hat{v}_*. \tag{13.12}$$

The dependence of $\hat{\lambda}_0$ and $\hat{\lambda}_*$ on the Mach number is shown in Fig. 13.6. We can see that the Lax condition is satisfied when the Mach number is larger than the dimensionless maximum characteristic velocity in the unperturbed state, that is, $M_0 > \hat{\lambda}_0 \equiv \lambda_{max}/c_0$. This is consistent with the theorem about the formation of a sub-shock explained in Sect. 3.4. We conclude that all and only *compressive* sub-shocks are admissible.

13.5 Meixner's Temperature and the Temperature Overshoot

We discussed in Chap. 11 that we can also describe nonequilibrium phenomena in terms of the quantities of the Meixner theory. Therefore, by using the correspondence relation (11.44), the shock wave structure described by the ET6 theory can be converted into the structure from the viewpoint of the Meixner theory.

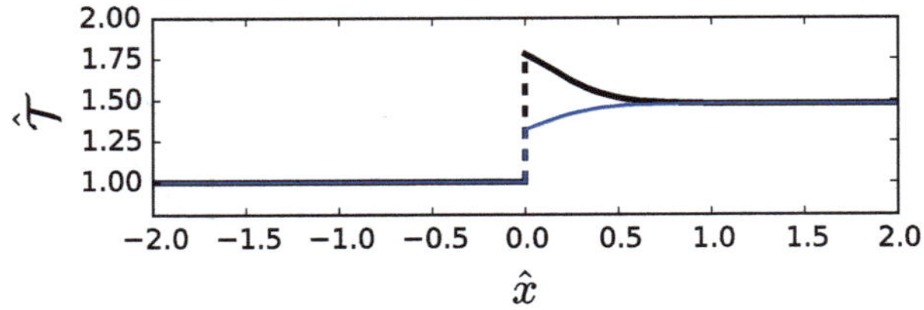

Fig. 13.7 Shock wave structure in the Meixner theory (*black thick curves*). The temperature profile in the ET6 theory (*blue thin curve*) is also shown. $M_0 = 2.0$

Comparison of the nonequilibrium temperatures in both theories is especially interesting. Numerical analysis [14] reveals the following facts:

(i) When the Mach number M_0 is small both temperatures behave in nearly the same way.
(ii) When M_0 increases, only the temperature of the Meixner theory overshoots. One typical example is shown in Fig. 13.7.

These facts clearly show that the existence of the temperature overshoot in a shock wave depends on the definition of the nonequilibrium temperature adopted.

In Chap. 15, we will propose a well-defined nonequilibrium temperature and chemical potential, which, within the ET6 theory, coincide with the corresponding quantities in the Meixner theory.

13.6 Conclusion

In this chapter, we have studied the shock wave structure by using the ET6 theory with the linear constitutive equation (5.60) and we have restricted our analysis within the cases with small Mach numbers. We can also analyze the shock wave structure on the basis of the nonlinear ET6 theory (11.34), if we do not care about the complications encountered in the analysis.

Numerical analysis with the use of the nonlinear ET6 theory [14] reveals that difference in the shock wave structure between nonlinear and linear cases are not so remarkable at least up to $M_0 \sim 5$. This is not surprising because of the following facts: As the left-hand side of the system of field equations is the same in both linear and nonlinear cases, the strength of a sub-shock is the same in both cases. And, if the Mach number is not so large, the relaxation after the sub-shock is not so large. This finding also means that the reliability of the results derived from the linear theory in the present chapter is quite high.

References

1. S. Taniguchi, T. Arima, T. Ruggeri, M. Sugiyama, Effect of dynamic pressure on the shock wave structure in a rarefied polyatomic gas. Phys. Fluids **26**, 016103 (2014)
2. S. Taniguchi, T. Arima, T. Ruggeri, M. Sugiyama, Thermodynamic theory of the shock wave structure in a rarefied polyatomic gas: beyond the Bethe-Teller theory. Phys. Rev. E **89**, 013025 (2014)
3. T. Arima, S. Taniguchi, T. Ruggeri, M. Sugiyama, Dispersion relation for sound in rarefied polyatomic gases based on extended thermodynamics. Contin. Mech. Thermodyn. **25**, 727 (2013)
4. D. Gilbarg, D. Paolucci, The structure of shock waves in the continuum theory of fluids. J. Ration. Mech. Anal. **2**, 617 (1953)
5. W. Weiss, Continuous shock structure in extended thermodynamics. Phys. Rev. E **52**, R5760 (1995)
6. I. Müller, T. Ruggeri, *Rational Extended Thermodynamics*, 2nd edn. (Springer, New York, 1998)
7. H.A. Bethe, E. Teller, Deviations from Thermal Equilibrium in Shock Waves. Reprinted by Engineering Research Institute. University of Michigan
8. E.F. Smiley, E.H. Winkler, Z.I. Slawsky, Measurement of the vibrational relaxation effect in CO_2 by means of shock tube interferograms. J. Chem. Phys. **20**, 923 (1952)
9. E.F. Smiley, E.H. Winkler, Shock-tube measurements of vibrational relaxation. J. Chem. Phys. **22**, 2018 (1954)
10. W.C. Griffith, W. Bleakney, Shock waves in gases. Am. J. Phys. **22**, 597 (1954)
11. W. Griffith, D. Brickl, V. Blackman, Structure of shock waves in polyatomic gases. Phys. Rev. **102**, 1209 (1956)
12. N.H. Johannesen, H.K. Zienkiewicz, P.A. Blythe, J.H. Gerrard, Experimental and theoretical analysis of vibrational relaxation regions in carbon dioxide. J. Fluid Mech. **13**, 213 (1962)
13. W.C. Griffith, A. Kenny, On fully-dispersed shock waves in carbon dioxide. J. Fluid Mech. **3**, 286 (1957)
14. S. Taniguchi, T. Arima, T. Ruggeri, M. Sugiyama, Overshoot of the nonequilibrium temperature in the shock wave structure of a rarefied polyatomic gas subject to the dynamic pressure (submitted).

Chapter 14
Acceleration Wave, K-Condition, and Global Existence in ET6

Abstract We verify the K-condition for the non-linear ET6 model and show for any gas the existence of global smooth solutions provided that initial data are sufficiently small. As an example, in the case of polyatomic gases, we study acceleration waves. We evaluate the Bernoulli equation for the amplitude of the wave. If the initial amplitude of an acceleration wave is sufficiently small compared with the critical amplitude, the acceleration wave exists for all time and decays to zero as the time t becomes large. Vice versa, for large initial amplitude, there exists a critical time at which we have the blow up of the solution and the formation of a shock wave. We show the peculiarity of this model, that is, the velocity of a disturbance and the critical time are universal: these are independent of the degrees of freedom of the constituent molecule D.

14.1 Characteristic Velocities and the K-Condition

We discussed, in Sects. 2.6.2 and 2.6.3, the role of the K-condition (2.35) [1]. Together with the entropy convexity, it is a sufficient condition for the existence of global smooth solutions. Lou and Ruggeri [2] noticed a connection between the K-condition and the global existence of acceleration waves (see Sect. 3.2) and they rewrote the condition (2.35) in the form (3.33). In this chapter, we study acceleration waves and prove that the K-condition is satisfied for the ET6 model for any gas with a convex entropy.

The system (11.36) in the BGK approximation is given by

$$\begin{aligned}
&\dot{\rho} + \rho \frac{\partial v_k}{\partial x_k} = 0, \\
&\rho \dot{v}_i + \frac{\partial}{\partial x_i}(p + \Pi) = 0, \\
&\rho \dot{\varepsilon} + (p + \Pi)\frac{\partial v_k}{\partial x_k} = 0, \\
&\left(\frac{p + \Pi}{\rho} - \frac{2}{3}\varepsilon\right)^{\bullet} = -\frac{\Pi}{\rho \tau}.
\end{aligned} \qquad (14.1)$$

T. Ruggeri, M. Sugiyama, *Rational Extended Thermodynamics beyond the Monatomic Gas*, DOI 10.1007/978-3-319-13341-6_14

Using the chain rule (3.32) we obtain, from the differential system, characteristic eigenvalues and right eigenvectors. In particular, for the material time derivative, the chain rule is expressed by

$$\bullet \to -V\delta, \quad V = \lambda - v_n, \quad v_n = v_i n_i.$$

In the present case, from the system (14.1), we obtain

$$\begin{aligned}
&- V\delta\rho + \rho\delta v_n = 0, \\
&- \rho V\delta \mathbf{v} + \mathbf{n}\delta(p + \Pi) = 0, \\
&- \rho V\delta\varepsilon + (p + \Pi)\delta v_n = 0, \\
&- V\delta\left(\frac{p + \Pi}{\rho} - \frac{2}{3}\varepsilon\right) = 0.
\end{aligned} \tag{14.2}$$

Taking into account the equations of state (11.62), we have:

1) $$V = 0 \quad \longleftrightarrow \quad \lambda = v_n, \qquad \text{contact waves,} \tag{14.3}$$

with $\delta\rho$, $\delta\mathbf{v}_T$, δp arbitrary (multiplicity 4), and $\delta v_n = 0$, $\delta\Pi = -\delta p$ ($\mathbf{v}_T$ denotes the tangential velocity).

2) $$V = \pm\sqrt{\frac{5}{3}\frac{p+\Pi}{\rho}} \quad \longleftrightarrow \quad \lambda = v_n \pm \sqrt{\frac{5}{3}\frac{p+\Pi}{\rho}}, \qquad \text{sound waves,} \tag{14.4}$$

with $\delta\rho$ arbitrary,

$$\delta\mathbf{v} = \mathbf{n}V\frac{\delta\rho}{\rho}, \quad \delta\varepsilon = \frac{\delta\rho}{\rho}\left(\frac{p+\Pi}{\rho}\right), \quad \delta\Pi = \frac{\delta\rho}{\rho}\Gamma, \tag{14.5}$$

where Γ is given by (11.13). As only the last component of the production term $\mathbf{f}$ of the generic system (2.1) is non-zero (see (14.1)), the K-condition (2.35) is satisfied if $\delta\Pi \neq 0$ in equilibrium ($\Pi = 0$). This is true for both contact waves and for sound waves because of (14.5) and the inequality $\Gamma_0 > 0$ [see (11.19)] due to the convexity condition. Therefore the K-condition is satisfied for any gas and, together with the convexity of the entropy, we can conclude that, according to the general theorems of Sect. 2.6.3, the 6-moment system has global smooth solutions for all time, and the solution converges to the equilibrium one provided that the initial data are sufficiently smooth. This proof was given in [3].

We notice that the sound velocity in (14.4) is independent of the degrees of freedom of a molecule and that, in equilibrium, it coincides with the sound velocity of a monatomic gas. In the case of polytropic polyatomic fluids this result is in agreement with the Theorem 10.5. In fact for polytropic polyatomic gases, the

6-moment system is a $(2^-, 0)$ system and belongs to the class for which the characteristic velocities are independent of D, see Sect. 10.4.2.

14.2 Time-Evolution of the Amplitude and the Critical Time

Let us consider the system (14.1) in the one-dimensional case for the field $\mathbf{u} \equiv (\rho, v, \varepsilon, \Pi)^T$ with $\mathbf{v} \equiv (v, 0, 0)$. We consider for simplicity a polytropic fluid. We study time-evolution of the amplitude of an acceleration wave along the characteristic line of the transport equation. We focus on the fastest wave propagating in an equilibrium state at rest: $\mathbf{u}_0 \equiv (\rho_0, 0, \varepsilon_0, 0)$.

In the present case, the amplitude evolves according to (3.27) with a and b given by (3.26). The system (14.1) can be rewritten as

$$\dot{\mathbf{u}} + \mathbf{A}\mathbf{u}_x = \mathbf{f} \tag{14.6}$$

with

$$\mathbf{u} \equiv \begin{pmatrix} \rho \\ v \\ \varepsilon \\ \Pi \end{pmatrix}, \quad \mathbf{A} \equiv \begin{pmatrix} 0 & \rho & 0 & 0 \\ \frac{3c^2}{5\rho} & 0 & \frac{2}{D} & \frac{1}{\rho} \\ 0 & \frac{3c^2}{5} + \frac{\Pi}{\rho} & 0 & 0 \\ 0 & \frac{2c^2(D-3)\rho}{5D} + \frac{5D-6}{3D}\Pi & 0 & 0 \end{pmatrix}, \quad \mathbf{f} \equiv \begin{pmatrix} 0 \\ 0 \\ 0 \\ -\frac{\Pi}{\tau} \end{pmatrix}, \tag{14.7}$$

where we have put

$$c = \sqrt{\frac{5}{3}\frac{p}{\rho}}. \tag{14.8}$$

The eigenvalues of $\mathbf{A}$ are given by

$$V = \left\{0, 0, -\sqrt{\frac{5}{3}\frac{p+\Pi}{\rho}}, \sqrt{\frac{5}{3}\frac{p+\Pi}{\rho}}\right\}.$$

Therefore the fastest velocity is expressed in terms of the field $\mathbf{u}$ as

$$\lambda = v + \sqrt{10\epsilon/(3D) + 5\Pi/(3\rho)}. \tag{14.9}$$

The gradient evaluated in equilibrium (we omit the index E) is given by

$$\operatorname{grad}\lambda = \left(\partial_\rho\lambda, \partial_v\lambda, \partial_\epsilon\lambda, \partial_\Pi\lambda\right) = \left(0, 1, \frac{5}{3cD}, \frac{5}{6c\rho}\right). \tag{14.10}$$

The orthonormal right and left eigenvectors in equilibrium are obtained as

$$\mathbf{d}_1 = \left(-\frac{5}{3c^2}, 0, 0, 1\right)^T, \quad \mathbf{d}_2 \equiv \left(-\frac{10\rho}{3c^2 D}, 0, 1, 0\right)^T,$$

$$\mathbf{d}_3 \equiv \left(\frac{5D}{2c^2(D-3)}, -\frac{5D}{2c\rho(D-3)}, \frac{3D}{2(D-3)\rho}, 1\right)^T,$$

$$\mathbf{d} = \mathbf{d}_4 \equiv \left(\frac{5D}{2c^2(D-3)}, \frac{5D}{2c\rho(D-3)}, \frac{3D}{2(D-3)\rho}, 1\right)^T, \tag{14.11}$$

$$\mathbf{l}_1 \equiv \left(-\frac{6c^2(D-3)}{25D}, 0, -\frac{4(D-3)\rho}{5D^2}, \frac{3(2+D)}{5D}\right),$$

$$\mathbf{l}_2 \equiv \left(\frac{3c^2(6-5D)}{25D\rho} + \frac{6c^2(D-3)}{25D\rho}, 0, -\frac{6-5D}{5D}, -\frac{3}{5\rho}\right),$$

$$\mathbf{l}_3 \equiv \left(\frac{3c^2(D-3)}{25D}, -\frac{c(D-3)\rho}{5D}, \frac{2(D-3)\rho}{5D^2}, \frac{-3+D}{5D}\right),$$

$$\mathbf{l} = \mathbf{l}_4 \equiv \left(\frac{3c^2(-3+d)}{25D}, \frac{c(D-3)\rho}{5D}, \frac{2(D-3)\rho}{5D^2}, \frac{-3+D}{5D}\right).$$

Therefore from (3.26) we have

$$a = \frac{10D}{3c(D-3)\rho}, \quad b = \frac{1}{\tau}. \tag{14.12}$$

The condition (3.12) with (3.13) becomes, in the present case,

$$\begin{pmatrix} \left[\frac{[\partial\rho]}{\partial x}\right] \\ \left[\frac{[\partial v]}{\partial x}\right] \\ \left[\frac{[\partial\varepsilon]}{\partial x}\right] \\ \left[\frac{[\partial\Pi]}{\partial x}\right] \end{pmatrix} = \mathscr{A} \begin{pmatrix} \frac{5D}{2c^2(D-3)} \\ \frac{5D}{2c\rho(D-3)} \\ \frac{3D}{2(D-3)\rho} \\ 1 \end{pmatrix}. \tag{14.13}$$

Introducing the acceleration jump

$$G = \left[\frac{\partial v}{\partial t}\right]$$

and taking into account both the Hadamard condition and the second component of (14.13), we obtain

$$\mathscr{A} = -\frac{2\rho(D-3)}{5D}G.$$

Therefore the transport Bernoulli equation (3.14) becomes very simple:

$$\frac{dG}{dt} - \frac{4}{3c}G^2 + \frac{G}{\tau} = 0. \tag{14.14}$$

An interesting remark is that this equation is independent of D as is the sound velocity (14.8)! The solution of (14.14) (see (3.27)) is

$$G(t) = \frac{G(0)e^{-\frac{t}{\tau}}}{1 + G(0)\frac{4\tau}{3c}\left(e^{-\frac{t}{\tau}} - 1\right)}. \tag{14.15}$$

Therefore if the initial amplitude of the acceleration jump satisfies the condition:

$$G(0) < G_{cr} = \frac{3}{4}\frac{c}{\tau}, \tag{14.16}$$

we have no critical time. The acceleration jump decays in agreement with the global existence and the K-condition. Instead, if $G(0)$ is greater than the critical value G_{cr}, we have the critical time (see (3.29)):

$$t_{cr} = -\tau \log\left(1 - \frac{3c}{4G(0)\tau}\right) \tag{14.17}$$

and blow-up arises with the formation of a shock wave.

It is interesting to observe that the Bernoulli equation for Euler monatomic fluids is the same as (14.14) if the last term is absent. In the case of Euler fluids, any compressive wave $G(0) > 0$ has the critical time that, in the case of monatomic gases, is given by [4]

$$t_{cr}^{5M} = \frac{3c}{4G(0)}. \tag{14.18}$$

We can rewrite (14.17) by using (14.18):

$$\frac{t_{cr}}{\tau} = -\log\left(1 - \frac{t_{cr}^{5M}}{\tau}\right). \tag{14.19}$$

If $w = t_{cr}^{5M}/\tau \ll 1$, then

$$\frac{t_{cr}}{\tau} = \frac{t_{cr}^{5M}}{\tau} + O(w^2). \tag{14.20}$$

14.3 Conclusion

We have evaluated the time-evolution of an acceleration wave for the ET6 model. And we have seen such peculiarity of this simple model that the velocity of disturbance and the critical time are universal in the sense that these are independent of the degrees of freedom of a constituent molecule D. This is probably due to the fact that the model is oversimplified and more realistic model is the 14-moment model.

Nevertheless, even though the model may be oversimplified, it has the advantage that the system is the simplest dissipative system that corrects the classical Euler fluid. The dissipation is enough to produce a competition with hyperbolicity. Therefore, as the K-condition is satisfied and the entropy is convex, smooth solutions exist for all time and they converge to the equilibrium state.

The acceleration wave is a good example to understand this situation. In fact, if the initial acceleration amplitude is sufficiently small compared with the critical amplitude (14.16), the acceleration wave exists for all time and decays to zero as the time t becomes large as seen in (14.15).

References

1. Y. Shizuta, S. Kawashima, Systems of equations of hyperbolic-parabolic type with applications to the discrete Boltzmann equation. Hokkaido Math. J. **14**, 249–275 (1985)
2. J. Lou, T. Ruggeri, Acceleration waves and weak Shizuta-Kawashima Condition. Suppl. Rend. Circ. Mat. Palermo **78**, 187–200 (2006) [*Non Linear Hyperbolic Fields and Waves. A Tribute to Guy Boillat*]
3. T. Ruggeri, Non-linear maximum entropy principle for a polyatomic gas subject to the dynamic pressure. Bull. Inst. Math. Acad. Sin. **11** (2016) [Special issue in honor of Tai-Ping Liu 70th birthday]
4. T. Ruggeri, Interaction between a discontinuity wave and a shock wave: critical time for the fastest transmitted wave, example of the polytropic fluid. Appl. Anal. **11**, 103–112 (1980)

Chapter 15
Nonequilibrium Temperature and Chemical Potential

Abstract In this chapter, we propose a natural definition of nonequilibrium temperature and chemical potential. The main field, with which the generalized Gibbs equation is expressed in a differential form, is the key quantity in the definition. In the ET6 theory, in particular, the nonequilibrium quantities coincide exactly with those in the Meixner theory explained in Chap. 11.

15.1 Generalized Gibbs Equation, Nonequilibrium Temperature and Chemical Potential

As mentioned in Sect. 1.6.7, one of the most delicate and controversial questions in nonequilibrium thermodynamics is the following one: What is the most appropriate definition of the nonequilibrium temperature? This question has been considered by many authors, but here is not a suitable place to discuss its history and many tentatives to this intriguing problem. Some general discussions are summarized in the paper [1]. See also a previous tentative in ET [2, 3]. Instead, we here want to reconsider the idea that was proposed in [4] in a relativistic framework.

For this aim, let us firstly observe the structure of ET given in the case of monatomic gases, by (2.13). For this structure, the temporal part of the differential conditions $(2.43)_1$ derived from the requirement of the entropy principle reads

$$
\begin{aligned}
dh^0 &= \mathbf{u}' \cdot d\mathbf{F}^0 = \hat{\mathbf{u}}' \cdot d\hat{\mathbf{F}}^0 = \\
&= \hat{u}' d\hat{F} + \hat{u}'_{k_1} d\hat{F}_{k_1} + \hat{u}'_{k_1 k_2} d\hat{F}_{k_1 k_2} + \cdots + \hat{u}'_{k_1 k_2 \ldots k_n} d\hat{F}_{k_1 k_2 \ldots k_n} = \\
&= \hat{u}' d\hat{F} + \hat{u}'_{ll} d\hat{F}_{kk} + \hat{u}'_{<k_1 k_2>} d\hat{F}_{<k_1 k_2>} + \cdots + \hat{u}'_{k_1 k_2 \ldots k_n} d\hat{F}_{k_1 k_2 \ldots k_n},
\end{aligned}
\tag{15.1}
$$

where we write explicitly the components of the main field:

$$
\mathbf{u}' \equiv (u', u'_{k_1}, u'_{k_1 k_2}, \ldots, u'_{k_1 k_2 \ldots k_n}). \tag{15.2}
$$

By taking into account the general properties that, in equilibrium, all the components of the main field corresponding to the balance laws vanish $(2.29)_2$ and that the first five fields F, F_i, F_{kk} are equilibrium quantities (mass density,

T. Ruggeri, M. Sugiyama, *Rational Extended Thermodynamics beyond the Monatomic Gas*, DOI 10.1007/978-3-319-13341-6_15

momentum, 2 times energy), the relation (15.1) reduces to

$$dh_E^0 = \hat{\mathbf{u}}'_E \cdot d\hat{\mathbf{F}}_E^0 = \hat{u}'_E d\hat{F} + +\hat{u}'_{Ell} dF_{kk}. \tag{15.3}$$

This is nothing else the Gibbs equation (1.7) with non-vanishing components of the main field. These components coincide with those of the main field of the Euler equilibrium principal subsystem:

$$\hat{u}'_E = -\frac{g}{T}, \qquad \hat{u}'_{Ell} = \frac{1}{2T}. \tag{15.4}$$

Therefore, firstly we may conclude that (15.1) represents the generalized Gibbs equation in nonequilibrium. In this respect, we emphasize the fact that, in ET, we do not adopt the local equilibrium assumption characterized by the equilibrium Gibbs equation (15.3).

Secondly we can understand the physical meaning of the intrinsic Lagrange multipliers in equilibrium corresponding to the mass conservation and the energy conservation owing to the relation (15.4). They are, respectively, the ratios between the chemical potential and the absolute temperature except for the sign, and the coldness (the inverse of the temperature) except for the factor $1/2$. It is therefore natural that, also for a nonequilibrium state, we assume the following relations:

$$\hat{u}' = -\frac{\mathscr{G}}{\mathscr{T}}, \qquad \hat{u}'_{ll} = \frac{1}{2\mathscr{T}}, \tag{15.5}$$

where $\mathscr{G}$ and $\mathscr{T}$ are interpreted as the nonequilibrium chemical potential and the nonequilibrium temperature. In other words, these nonequilibrium quantities are defined through the relations (15.5).

Finally, an important remark is the following: in the old ET [5] or in the EIT [6], the Gibbs equation (15.1) is the starting point and therefore is assumed as a hypothesis, while, in RET, (15.1) is not assumed but is deduced. This means that, as we saw and will see in several chapters, we are able to evaluate all the main field components *without any ambiguity*. Therefore we are able to measure the nonequilibrium temperature and the nonequilibrium chemical potential indirectly through the relations:

$$\mathscr{T} = \frac{1}{2\hat{\mu}}, \qquad \mathscr{G} = -\frac{\hat{\lambda}}{2\hat{\mu}} \tag{15.6}$$

where we have put $\hat{\lambda} = \hat{u}'$ and $\hat{\mu} = \hat{u}'_{ll}$ for simplicity and also for uniformity with previous chapters.

In the case of rarefied monatomic gases, among other ET theories, the ET theory with 13 fields is the most interesting one in the sense that all fields have concrete physical meanings. For such a theory, the definition (15.6) gives a simple but non-trivial answer. In fact, from the expression of the main field of ET with

13 fields (4.1), the definition (15.6) gives the relations:

$$\mathscr{T} = T, \qquad \mathscr{G} = g. \tag{15.7}$$

In this case, two kinds of the nonequilibrium quantities coincide with each other.

We have to recall that the ET theory with 13 fields is valid only near equilibrium, where we assume that all quantities are linear with respect to the nonequilibrium quantities. Because of the two facts that (1) the scalar quantities $\mathscr{T}$ and $\mathscr{G}$ can be influenced only by nonequilibrium scalars obtained from the tensor F's and (2) we do not have any nonequilibrium scalar in the first order, we have obtained the relation (15.7). Here, the fact that the dynamical pressure vanishes identically in a rarefied monatomic gas is essential.

In the case of ET with more fields than 13, however, other nonequilibrium scalars may play a role. For example, in the ET theory of monatomic gases with 14 fields, we adopt one more balance equation for the new independent variable F_{llkk}. Then we have [7]

$$\hat{\lambda} = -\frac{g}{T} - \frac{\Delta}{8pT}, \qquad \hat{\mu} = \frac{1}{2T}\left(1 + \frac{\rho}{6p^2}\Delta\right),$$

where Δ is the nonequilibrium part of F_{llkk}.

In the case of polyatomic gases and dense gases, the situation is quite different. The nonequilibrium temperature $\mathscr{T}$ and chemical potential $\mathscr{G}$ are different from the temperature T and chemical potential g due to the existence of the dynamic pressure Π. We will establish, in the following sections, an explicit relationship between ($\mathscr{T}$, $\mathscr{G}$) and (T, g) in the cases of ET6, ET14, and ET with arbitrary number of fields.

15.2 Nonequilibrium Temperature and Chemical Potential in ET with the Binary Hierarchy

In the case of ET of polyatomic gases and dense gases with the binary hierarchy (10.1), one important point concerning the Lagrange multiplier of the energy equation should be made clear in order to define the nonequilibrium temperature and chemical potential properly. We start to consider this point in the case of ET6.

In ET6, we already observed that we can write the system of field equations in two different ways. The one is written in the form of F-series and G-series (1.27), which corresponds to (11.4) and to the Lagrange multipliers $(\lambda, \lambda_i, \mu, \zeta)$ given by (11.32). The other one is written in the form (1.27) except that, instead of the equation for F_{ll}, we adopt the equation:

$$\partial_t (F_{ll} - G_{ll}) + \partial_i (F_{ill} - G_{ill}) = P_{ll}, \tag{15.8}$$

which is obtained by subtracting the energy equation from the equation of the dynamical pressure (11.34). In this case, the Lagrange multipliers (11.35), indicated by a bar, are the same of the former ones except that the Lagrange multiplier of the energy equation is now given by

$$\bar{\mu} = \mu + \zeta. \tag{15.9}$$

As explained in the Remark in Sect. 11.5.3, the two possible forms of the system of ET6 are completely equivalent to each other, but only the second form gives the possibility to obtain the singular limit $D \to 3$ of a monatomic gas. In fact, Eq. (15.8) coincides with the (11.80) and, in the limit, admits the solution $\Pi = 0$. The Lagrange multipliers that converge to those of Euler fluids are the ones of the second form.

Therefore, from the existence condition of the limit of the monoatomic case, the correct intrinsic Lagrange multiplier of the energy equation must be $\hat{\bar{\mu}}$. And the previous definition (15.6) for $\mathscr{T}$ and $\mathscr{G}$ should be expressed as

$$\mathscr{T} = \frac{1}{2\hat{\bar{\mu}}}, \qquad \mathscr{G} = -\frac{\hat{\bar{\lambda}}}{2\hat{\bar{\mu}}}. \tag{15.10}$$

Or equivalently, in terms of the intrinsic Lagrange multipliers of the F- and G-series, we have

$$\mathscr{T} = \frac{1}{2(\hat{\mu} + \hat{\zeta})}, \qquad \mathscr{G} = -\frac{\hat{\lambda}}{2(\hat{\mu} + \hat{\zeta})}, \tag{15.11}$$

where $\hat{\lambda}, \hat{\zeta}, \hat{\mu}$, are, respectively, the Lagrange multipliers with respect to the conservation equations for F, F_{ll}, G_{ll}. Above considerations are valid for ET with any number of fields.

15.3 Nonequilibrium Temperature and Chemical Potential in ET6 and ET14

In the case of ET6, from the main field (11.35) and (15.11), we obtain, as the nonequilibrium temperature:

$$\mathscr{T} = \frac{T}{1 + T(k_\varepsilon - p_\varepsilon k_\Pi) + \frac{2}{3}\rho T k_\Pi}, \tag{15.12}$$

and as the nonequilibrium chemical potential:

$$\mathcal{G} = \frac{g - T(k + (-k_\varepsilon + k_\Pi p_\varepsilon)\varepsilon + k_\rho \rho - k_\Pi p_\rho \rho)}{1 + (k_\varepsilon - k_\Pi p_\varepsilon)T + \frac{2k_\Pi T\rho}{3}}. \tag{15.13}$$

Taking into account the condition for k (11.12), after some cumbersome calculations, we prove that the above expressions are completely equivalent to the temperature and the chemical potential of Meixner: (11.45) and (11.47)! In Sect. 13.5, we have seen that the behavior of the nonequilibrium temperature $\mathcal{T}$ can be quite different from the behavior of the temperature T in the shock wave structure in a polyatomic gas.

In the case of polytropic gases, we can simplify above expressions a lot and obtain:

$$\mathcal{T} = T\frac{p + \Pi}{p}, \qquad \mathcal{G} = (g - Tk)\left(1 + \frac{\Pi}{p}\right), \tag{15.14}$$

which are again equal to the ones for the Meixner theory (11.69) and (11.72). We can see that the dynamic pressure plays a dominant role to cause the difference between the kinetic temperature T and the nonequilibrium temperature $\mathcal{T}$, and the same is true for the chemical potential.

If we consider a theory near equilibrium, by using the expression of k given by (11.50) and retaining only the linear part with respect to Π, we have

$$\mathcal{T} = T\left(1 - \frac{p_\varepsilon - \frac{2}{3}\rho}{\rho \Gamma_0}\Pi\right), \tag{15.15}$$

$$\mathcal{G} = g + \left(\frac{3(\varepsilon p_\varepsilon - \rho p_\rho) + g(2\rho - 3p_\varepsilon)}{3\rho \Gamma_0}\right)\Pi. \tag{15.16}$$

In the case of nonpolytropic ideal gases, these expressions become

$$\mathcal{T} = T\frac{p + \Pi}{p}, \qquad \mathcal{G} = g\left(1 + \frac{\Pi}{p}\right) + 3\frac{\varepsilon - c_v T}{2\hat{c}_v - 3}\frac{\Pi}{p}. \tag{15.17}$$

Finally, in the case of polytropic ideal gases where $\varepsilon = c_v T$, we have

$$\mathcal{T} = T\frac{p + \Pi}{p}, \qquad \mathcal{G} = g\left(1 + \frac{\Pi}{p}\right). \tag{15.18}$$

In the case of ET14 within a linear approximation, as the nonequilibrium scalar is only Π, we have no difference of the nonequilibrium temperature and the nonequilibrium chemical potential from those of the ET6. Therefore, the same expressions are valid: for generic gases, (15.15) and (15.16), for nonpolytropic ideal gases (15.17), and for polytropic ideal gases (15.18).

For a general N-moment system, the definition of the nonequilibrium quantities are given by the general one, i.e., (15.11). As $\mathscr{T}$ and $\mathscr{G}$ are scalars, in order to have a new nonequilibrium contribution to these quantities, it is necessary to consider more moments such that there appear other scalars of the densities. For example, we expect a new contribution in a $(3^-, 2^-)$-system.

15.4 Conclusion

We have proposed a natural definition of the nonequilibrium temperature and chemical potential by comparing the generalized Gibbs equation and the classical one. The component of the main field are the coefficients of the generalized Gibbs equation in a differential form. In equilibrium, only the components that are related to the mass density and the energy density remain to be non-zero. The first one is strictly related to the chemical potential and the second to the coldness. Therefore we can define the nonequilibrium coldness (or temperature) and chemical potential by assuming that the coefficients do not change their meanings even in nonequilibrium.

We have proved that, in the most interesting cases, that is, ET6 and ET14 theories where all fields have concrete physical meanings, the dynamical pressure is responsible for the difference between nonequilibrium quantities and the corresponding kinetic ones. In the limit of monatomic gas, the difference disappears. This means that the internal degrees of freedom in a polyatomic molecule play a crucial role to cause the difference. Appropriateness of the definition is supported, in the case of ET6, by the perfect coincidence of the nonequilibrium quantities with those in the Meixner theory. See also the discussions made in Sect. 11.3.

Finally, in the case of ET with many moments, all nonequilibrium scalars constructed by the tensors of the density-fields play roles in the nonequilibrium quantities.

References

1. J. Casas-Vázquez, D. Jou, Temperature in non-equilibrium states: a review of open problems and current proposals. Rep. Prog. Phys. **66**, 1937 (2003)
2. E. Barbera, I. Müller, M. Sugiyama, On the temperature of a rarefied gas in non-equilibrium. Meccanica **34**, 103 (1999)
3. J. Au, I. Müller, T. Ruggeri, Temperature jumps at the boundary of a rarefied gas. Contin. Mech. Thermodyn. **12**(1), 19 (2000)
4. T. Ruggeri, Relativistic extended thermodynamics: general assumptions and mathematical procedure, in *Corso CIME Noto (Giugno 1987)*. Lecture Notes in Mathematics, vol. 1385, ed. by A. Anile, Y. Choquet-Bruhat (Springer, New York, 1989), pp. 269–277

5. I. Müller, Zum Paradoxon der Wermeleitungstheorie. Zeitschrift für Physik **198**, 329 (1967)
6. D. Jou, J. Casas-Vázquez, G. Lebon, *Extended Irreversible Thermodynamics* (Springer, Berlin, 1996)
7. F. Brini, T. Ruggeri, Entropy principle for the moment systems of degree α associated to the Boltzmann equation. Critical derivatives and non controllable boundary data. Contin. Mech. Thermodyn. **14**, 165 (2002)

Part VII
Mixture of Gases with Multi-Temperature

Chapter 16
Multi-Temperature Mixture of Fluids

Abstract We present a survey on recent results concerning some different models of a mixture of compressible fluids. In particular, we discuss the most realistic case of a mixture where each constituent has its own temperature (*MT*). We first compare the solutions of this model with the one with unique common temperature (*ST*). In the case of Eulerian fluids, it will be shown that the corresponding *ST* differential system is a *principal subsystem* of the *MT* system. Global behavior of smooth solutions for large time for both systems will also be discussed through the application of the Shizuta-Kawashima K-condition.

Then we introduce the concept of the average temperature of a mixture based on the consideration that the internal energy of the mixture is the same as that in the case of a single-temperature mixture. As a consequence, it is shown that the entropy of the mixture reaches a local maximum in equilibrium. Through the procedure of the Maxwellian iteration, a new constitutive equation for nonequilibrium temperatures of constituents is obtained in a classical limit, together with the Fick law for the diffusion flux.

Finally, in order to justify the Maxwellian iteration, we present, for dissipative fluids, a possible approach to a classical theory of mixtures with the multi-temperature. We prove that the differences of temperatures between the constituents imply the existence of a new *dynamic pressure* even if fluids have zero bulk viscosities.

16.1 Introduction

Modeling and analysis of mixtures is a challenging and stimulating problem. In the case of gaseous mixtures, it can be successfully studied by using not only the method of the kinetic theory of gases but also the method of the continuum theory of fluids. In either case, appropriate macroscopic equations can be derived in order to explain irreversible phenomena like diffusion, heat transfer and chemical reactions. However, since there still remain many open problems, the study of mixtures is one of the fields of active research.

In the classical theory of diffusion, although different concentrations of the components in a mixture are taken into account, but only one common global

T. Ruggeri, M. Sugiyama, *Rational Extended Thermodynamics beyond the Monatomic Gas*, DOI 10.1007/978-3-319-13341-6_16

velocity and one temperature are considered; the velocities of each components are obtained through the constitutive equation, that is, the classical Fick law.

Then this classical model is followed by more sophisticated models that are constructed on the basis of either the continuum theory or the kinetic theory of gases. Among these, there appeared two main approaches with two constitutive theories. They can be classified by the answer to the question: Should the constituents of a mixture have a common temperature or not? Both of these approaches, i.e., single-temperature (ST) one and multi-temperature (MT) one, gained considerable successes in modeling a behavior of the mixture.

The MT approach is naturally embedded into Maxwell's kinetic theory of mixtures [1, 2]. This theory comes on its own especially in gases where atomic masses of the constituents are different, for example, in plasmas where the constituents are electrons, ions and neutral atoms. The relevance of the MT model is thus put in evidence and can be further supported by the analysis of plasmas at high temperatures. See for details the references by Kannappan and Bose [3, 4] and Bose and Seeniraj [5]. Influence of electron and phonon temperature on the efficiency of thermoelectric conversion was the subject of a recent paper of Sellitto et al. [6].

The idea of multiple temperatures thus reflects the physically justified intention to get a deeper insight into nonequilibrium processes in mixtures, but this concept seems to be mostly overlooked in the context of macroscopic theories. Nevertheless, it was appreciated and naturally embedded in the kinetic theory of gases, which is perfectly designed to monitor the processes far from equilibrium. For example, it appeared as an efficient tool in nonequilibrium flow computations [7].

Apart from the physical reasons, these two theories are completely different from each other from a mathematical point of view: In general, the MT system does not admit the solution with single temperature $T_1 = T_2 = \ldots = T$ even if we pose this condition initially.

The theory of homogeneous mixtures was developed within the framework of rational thermodynamics by Truesdell [8] under the assumption that each constituent obeys the same balance laws as a single fluid. A huge amount of literature appeared after that in the context of continuum approach, see, for example, [9–15].

The aim of this chapter is to present recent results of the theory of multi-temperature at the continuum level [16–18]. A survey on these results can be seen also in review papers [19–21].

16.2 Mixtures in Rational Thermodynamics

In the context of rational thermodynamics, the description of a homogeneous mixture of n constituents is based on the postulate that each constituent obeys to the same balance laws as those to which a single fluid obeys [8–10]. The laws express the balance equations of masses, momenta and energies:

$$\begin{cases} \dfrac{\partial \rho_\alpha}{\partial t} + \operatorname{div}(\rho_\alpha \mathbf{v}_\alpha) = \tau_\alpha, \\ \dfrac{\partial(\rho_\alpha \mathbf{v}_\alpha)}{\partial t} + \operatorname{div}(\rho_\alpha \mathbf{v}_\alpha \otimes \mathbf{v}_\alpha - \mathbf{t}_\alpha) = \mathbf{m}_\alpha, \qquad (\alpha = 1, 2, \ldots, n) \\ \dfrac{\partial\left(\frac{1}{2}\rho_\alpha v_\alpha^2 + \rho_\alpha \varepsilon_\alpha\right)}{\partial t} + \operatorname{div}\left\{\left(\dfrac{1}{2}\rho_\alpha v_\alpha^2 + \rho_\alpha \varepsilon_\alpha\right)\mathbf{v}_\alpha - \mathbf{t}_\alpha \mathbf{v}_\alpha + \mathbf{q}_\alpha\right\} = e_\alpha. \end{cases} \tag{16.1}$$

On the left-hand side, ρ_α is the density, $\mathbf{v}_\alpha$ is the velocity, ε_α is the specific internal energy, $\mathbf{q}_\alpha$ is the heat flux and $\mathbf{t}_\alpha$ is the stress tensor of the constituent α. The stress tensor $\mathbf{t}_\alpha$ can be decomposed into a pressure part $-p_\alpha \mathbf{I}$ and a viscous part $\boldsymbol{\sigma}_\alpha$ as

$$\mathbf{t}_\alpha = -p_\alpha \mathbf{I} + \boldsymbol{\sigma}_\alpha.$$

On the right-hand sides τ_α, $\mathbf{m}_\alpha$ and e_α represent the production terms related to the interactions between constituents. Due to the total conservation of mass, momentum and energy of the mixture, the sum of the production terms over all constituents must vanish

$$\sum_{\alpha=1}^{n} \tau_\alpha = 0, \quad \sum_{\alpha=1}^{n} \mathbf{m}_\alpha = \mathbf{0}, \quad \sum_{\alpha=1}^{n} e_\alpha = 0. \tag{16.2}$$

Global mixture quantities ρ, $\mathbf{v}$, ε, $\mathbf{t}$ and $\mathbf{q}$ are defined as

$$\begin{aligned} &\rho = \sum_{\alpha=1}^{n} \rho_\alpha && \text{total mass density,} \\ &\mathbf{v} = \frac{1}{\rho}\sum_{\alpha=1}^{n} \rho_\alpha \mathbf{v}_\alpha && \text{mixture velocity,} \\ &\varepsilon = \varepsilon_I + \frac{1}{2\rho}\sum_{\alpha=1}^{n} \rho_\alpha u_\alpha^2 && \text{internal energy,} \\ &\mathbf{t} = -p\mathbf{I} + \boldsymbol{\sigma}_I - \sum_{\alpha=1}^{n} (\rho_\alpha \mathbf{u}_\alpha \otimes \mathbf{u}_\alpha) && \text{stress tensor,} \\ &\mathbf{q} = \mathbf{q}_I + \sum_{\alpha=1}^{n} \rho_\alpha \left(\varepsilon_\alpha + \frac{p_\alpha}{\rho_\alpha} + \frac{1}{2}u_\alpha^2\right)\mathbf{u}_\alpha && \text{flux of internal energy,} \end{aligned} \tag{16.3}$$

where

$$\mathbf{u}_\alpha = \mathbf{v}_\alpha - \mathbf{v}, \qquad \left(\sum_{\alpha=1}^{n} \rho_\alpha \mathbf{u}_\alpha = 0\right) \tag{16.4}$$

is the diffusion velocity of the component α,

$$p = \sum_{\alpha=1}^{n} p_\alpha$$

is the total pressure, and

$$\varepsilon_I = \frac{1}{\rho}\sum_{\alpha=1}^{n} \rho_\alpha \varepsilon_\alpha, \quad \mathbf{q}_I = \sum_{\alpha=1}^{n} \mathbf{q}_\alpha, \quad \boldsymbol{\sigma}_I = \sum_{\alpha=1}^{n} \boldsymbol{\sigma}_\alpha$$

are, respectively, the total intrinsic internal energy, heat flux and shear stress.

Summing up all the equations (16.1) and taking (16.2) into account, we obtain

$$\begin{cases} \dfrac{\partial \rho}{\partial t} + \operatorname{div}(\rho \mathbf{v}) = 0, \\[2ex] \dfrac{\partial (\rho \mathbf{v})}{\partial t} + \operatorname{div}(\rho \mathbf{v} \otimes \mathbf{v} - \mathbf{t}) = \mathbf{0}, \\[2ex] \dfrac{\partial \left(\frac{1}{2}\rho v^2 + \rho\varepsilon\right)}{\partial t} + \operatorname{div}\left\{\left(\dfrac{1}{2}\rho v^2 + \rho\varepsilon\right)\mathbf{v} - \mathbf{t}\mathbf{v} + \mathbf{q}\right\} = 0, \end{cases} \tag{16.5}$$

which are the conservation laws of mass, momentum and energy of the mixture. They are in the same form as for a single fluid.

In order to compare the balance equations of mixture and single fluid, we rewrite (16.1), by taking into account that the production terms are not independent (see (16.2)), in the following equivalent form:

$$\begin{cases} \dfrac{\partial \rho}{\partial t} + \operatorname{div}(\rho \mathbf{v}) = 0, \\ \dfrac{\partial (\rho \mathbf{v})}{\partial t} + \operatorname{div}(\rho \mathbf{v} \otimes \mathbf{v} - \mathbf{t}) = \mathbf{0}, \\ \dfrac{\partial \left(\frac{1}{2}\rho v^2 + \rho\varepsilon\right)}{\partial t} + \operatorname{div}\left\{\left(\dfrac{1}{2}\rho v^2 + \rho\varepsilon\right)\mathbf{v} - \mathbf{t}\mathbf{v} + \mathbf{q}\right\} = 0, \\ \dfrac{\partial \rho_b}{\partial t} + \operatorname{div}(\rho_b \mathbf{v}_b) = \tau_b, \qquad (b = 1, \dots, n-1) \\ \dfrac{\partial (\rho_b \mathbf{v}_b)}{\partial t} + \operatorname{div}(\rho_b \mathbf{v}_b \otimes \mathbf{v}_b - \mathbf{t}_b) = \mathbf{m}_b, \\ \dfrac{\partial \left(\frac{1}{2}\rho_b v_b^2 + \rho_b\varepsilon_b\right)}{\partial t} + \operatorname{div}\left\{\left(\dfrac{1}{2}\rho_b v_b^2 + \rho_b\varepsilon_b\right)\mathbf{v}_b - \mathbf{t}_b\mathbf{v}_b + \mathbf{q}_b\right\} = e_b, \end{cases} \tag{16.6}$$

where the index b runs from 1 to $n-1$.

In this multi-temperature model (*MT*), used in particular in plasma physics [22], we have $5n$ independent field variables ρ_α, $\mathbf{v}_\alpha$ and T_α $(\alpha = 1, 2, \dots, n)$, where T_α is the temperature of the constituent α. To close the system (16.6) of the field equations of the mixture, we must write the constitutive equations for the quantities $p_\alpha, \varepsilon_\alpha, \mathbf{q}_\alpha, \boldsymbol{\sigma}_\alpha$ $(\alpha = 1, 2, \dots, n)$ and τ_b, $\mathbf{m}_b$, e_b $(b = 1, \dots, n-1)$ in terms of the field variables $\rho_\alpha, \mathbf{v}_\alpha$ and T_α $(\alpha = 1, 2, \dots, n)$.

16.2.1 Galilean Invariance of Field Equations

The system (16.6) is a particular case of the balance laws (2.1) with (2.38):

$$\mathbf{F}^0 = \begin{pmatrix} \rho \\ \rho v^j \\ \frac{1}{2}\rho v^2 + \rho\varepsilon \\ \rho c_b \\ \rho c_b (u_b^j + v^j) \\ \frac{1}{2}\rho c_b (\mathbf{u}_b + \mathbf{v})^2 + \rho c_b \varepsilon_b \end{pmatrix}, \quad \hat{\mathbf{F}}^0 = \begin{pmatrix} \rho \\ 0^k \\ \rho\varepsilon \\ \rho c_b \\ \rho c_b u_b^k \\ \frac{1}{2}\rho c_b u_b^2 + \rho c_b \varepsilon_b \end{pmatrix}, \tag{16.7}$$

$$\boldsymbol{\Phi}^i = \begin{pmatrix} 0^i \\ -t^{ij} \\ -t^{ik} v_k + q^i \\ \rho c_b u_b^i \\ \rho c_b u_b^i (u_b^j + v^j) - t_b^{ij} \\ \left\{ \begin{array}{c} \left(\frac{1}{2}\rho c_b (\mathbf{u}_b + \mathbf{v})^2 + \rho c_b \varepsilon_b\right) u_b^i \\ -t_b^{ik}(u_b^k + v_k) + q_b^i \end{array} \right. \end{pmatrix}, \quad \hat{\boldsymbol{\Phi}}^i = \begin{pmatrix} 0^i \\ -t^{ik} \\ q^i \\ \rho c_b u_b^i \\ \rho c_b u_b^i u_b^k - t_b^{ik} \\ \left\{ \begin{array}{c} \left(\frac{1}{2}\rho c_b u_b^2 + \rho c_b \varepsilon_b\right) u_b^i \\ -t_b^{ik} u_b^k + q_b^i \end{array} \right. \end{pmatrix}, \tag{16.8}$$

$$\mathbf{f} = \begin{pmatrix} 0 \\ 0^j \\ 0 \\ \tau_b \\ m_b^j \\ e_b \end{pmatrix}, \quad \hat{\mathbf{f}} = \begin{pmatrix} 0 \\ 0^k \\ 0 \\ \hat{\tau}_b \\ \hat{m}_b^k \\ \hat{e}_b \end{pmatrix}. \tag{16.9}$$

Theorem 16.1 *The linear operator* $\mathbf{X}(\mathbf{v})$ *(see (2.51) and (2.36)), which assures the Galilean invariance of the field equation (16.6), has the form:*

$$\mathbf{X}(\mathbf{v}) = \begin{pmatrix} 1 & 0_k & 0 & 0 & 0_k & 0 \\ v^j & \delta^{jk} & 0^j & 0^j & 0^{jk} & 0^j \\ \frac{v^2}{2} & v_k & 1 & 0 & 0_k & 0 \\ 0 & 0_k & 0 & 1 & 0_k & 0 \\ 0^j & 0^{jk} & 0^j & v^j & \delta^{jk} & 0^j \\ 0 & 0_k & 0 & \frac{v^2}{2} & v_k & 1 \end{pmatrix}. \tag{16.10}$$

As a consequence, the following relations between the production terms and their internal (intrinsic) counterparts are obtained:

$$\begin{aligned} \tau_b &= \hat{\tau}_b, \\ m_b^j &= \hat{\tau}_b v^j + \hat{m}_b^j, \\ e_b &= \hat{\tau}_b \frac{v^2}{2} + \hat{m}_b^k v_k + \hat{e}_b. \end{aligned} \tag{16.11}$$

In (16.10), 0^i, 0_i and 0^{ik} indicate, respectively, the zero column vector (3×1), the zero row vector (1×3), and the (3×3) null matrix, δ^{jk} is the Kronecker delta and only in this case we have difference between v_k and v^j that indicates the velocity components in row or in column respectively.

The first part of the theorem can be proved by the direct application of relations $(2.36)_{1,2}$. Once the operator $\mathbf{X}(\mathbf{v})$ is determined, the second part of the statement can be derived from $(2.36)_3$.

16.3 Coarse-Grained Theories: Single Temperature Model and Classical Mixture

Due to the difficulty to measure the temperature of each component, a common practice among engineers and physicists is to consider only one temperature for the mixture. When we use a single temperature (*ST*), $(16.6)_6$ disappears and we get a unique global conservation law of the total energy in the form $(16.6)_3$ (see for example [10]):

$$\begin{cases} \dfrac{\partial \rho}{\partial t} + \operatorname{div}(\rho \mathbf{v}) = 0, \\[2ex] \dfrac{\partial (\rho \mathbf{v})}{\partial t} + \operatorname{div}(\rho \mathbf{v} \otimes \mathbf{v} - \mathbf{t}) = \mathbf{0}, \\[2ex] \dfrac{\partial \left(\frac{1}{2}\rho v^2 + \rho \varepsilon\right)}{\partial t} + \operatorname{div}\left\{\left(\dfrac{1}{2}\rho v^2 + \rho\varepsilon\right)\mathbf{v} - \mathbf{t}\mathbf{v} + \mathbf{q}\right\} = 0, \\[2ex] \dfrac{\partial \rho_b}{\partial t} + \operatorname{div}(\rho_b \mathbf{v}_b) = \tau_b, \qquad\qquad (b = 1, \dots, n-1) \\[2ex] \dfrac{\partial (\rho_b \mathbf{v}_b)}{\partial t} + \operatorname{div}(\rho_b \mathbf{v}_b \otimes \mathbf{v}_b - \mathbf{t}_b) = \mathbf{m}_b. \end{cases} \tag{16.12}$$

A further step to a coarse-grained theory is the classical approach of mixtures (*CT*), in which the independent field variables are the density, the mixture velocity, the single temperature of the mixture and the concentrations of constituents. In this case, the last equation in (16.12) also disappears and the system reduces to the system of equations:

$$\begin{cases} \dfrac{d\rho}{dt} + \rho \operatorname{div} \mathbf{v} = 0, \\[2ex] \rho \dfrac{d\mathbf{v}}{dt} - \operatorname{div}\mathbf{t} = 0, \\[2ex] \rho \dfrac{d\varepsilon}{dt} - \mathbf{t} \operatorname{grad} \mathbf{v} + \operatorname{div} \mathbf{q} = 0, \\[2ex] \rho \dfrac{dc_b}{dt} + \operatorname{div} \mathbf{J}_b = 0, \quad (b = 1, \cdots n-1), \end{cases} \tag{16.13}$$

where

$$\frac{d}{dt} = \frac{\partial}{\partial t} + \mathbf{v} \cdot \frac{\partial}{\partial \mathbf{x}}$$

represents the material derivative of the mixture motion,

$$c_\alpha = \frac{\rho_\alpha}{\rho}, \qquad \left(\sum_{\alpha=1}^{n} c_\alpha = 1\right)$$

are the components concentrations, and

$$\mathbf{J}_\alpha = \rho_\alpha \mathbf{u}_\alpha = \rho_\alpha \left(\mathbf{v}_\alpha - \mathbf{v}\right) \qquad \left(\sum_{\alpha=1}^{n} \mathbf{J}_\alpha = 0\right) \tag{16.14}$$

are the diffusion fluxes of the components.

In the classical approach, the stress tensor—as in a single fluid—splits into the pressure (isotropic part) and the viscous stress tensor (for Stokesian fluids this is a deviatoric tensor)

$$\mathbf{t} = -p\mathbf{I} + \boldsymbol{\sigma}.$$

The system (16.13) determines the field variables $\rho, T, \mathbf{v}$ and c_b $(b = 1, \cdots n-1)$. Consequently, we need constitutive relations for $\varepsilon, \boldsymbol{\sigma}, \mathbf{q}$ and $\mathbf{J}_b$ $(b = 1, \cdots n-1)$.

The pressure $p(\rho, T, c_b)$, the internal energy $\varepsilon(\rho, T, c_b)$, and the chemical potentials $g_\alpha(\rho, T, c_b)$ $(\alpha = 1, \cdots, n)$ are assumed to be given by the equilibrium equations of state. The Gibbs equation in the case of mixtures is given by

$$T ds = d\varepsilon - \frac{p}{\rho^2}\, d\rho - \sum_{b=1}^{n-1} \left(g_b - g_n\right)\, dc_b, \tag{16.15}$$

where s is the entropy density of a mixture [9, 23]. Compare this relation to (1.7) for a single-component fluid.

The entropy balance law is a consequence of equation (16.15) and system (16.13). For dissipative fluids, by using similar arguments of the thermodynamics of irreversible processes *(TIP)* presented in Chap. 1 (see also [10, 24]), we obtain the classical constitutive equations of mixtures:

$$\begin{aligned}
\boldsymbol{\sigma} &= \nu\, (\operatorname{div} \mathbf{v})\, \mathbf{I} + 2\, \mu\, \mathbf{D}^D, \\
\mathbf{q} &= L \operatorname{grad}\left(\frac{1}{T}\right) + \sum_{b=1}^{n-1} L_b \operatorname{grad}\left(\frac{g_b - g_n}{T}\right), \\
\mathbf{J}_a &= \tilde{L}_a \operatorname{grad}\left(\frac{1}{T}\right) - \sum_{b=1}^{n-1} L_{ab} \operatorname{grad}\left(\frac{g_b - g_n}{T}\right),
\end{aligned} \tag{16.16}$$

where $\mathbf{D}^D$ denotes the deviatoric part of the strain velocity tensor $\mathbf{D} = \frac{1}{2}\left(\nabla\mathbf{v} + (\nabla\mathbf{v})^T\right)$. The *phenomenological coefficients* $L, L_b, \tilde{L}_a$ and $L_{ab}\,(a, b = 1, \cdots, n-1)$ are the transport coefficients of heat conduction and diffusion.

Let us note that relation $(16.16)_1$ is the classical Navier-Stokes equations of a Newtonian fluid $(1.10)_{1,2}$, while $(16.16)_{2,3}$ are generalizations of the original phenomenological laws of Fourier $(1.10)_3$ and Fick according to which the heat flux and the diffusion flux depend on the gradients of temperature and concentrations respectively. While *TIP* permits the temperature gradient to influence the diffusion fluxes and concentration gradients to influence the heat flux; both effects (cross effects) are indeed observed and they are called, respectively, thermo-diffusion effect and diffusion-thermo effect or Soret effect. Furthermore, the Onsager reciprocal theorem yields the following symmetries of coefficients [25]:

$$L_{ab} = L_{ba}\,, \quad \tilde{L}_b = L_b, \quad (a, b = 1, \cdots, n-1)$$

and the following inequalities must be satisfied

$$\begin{bmatrix} L & L_b \\ \tilde{L}_a & L_{ab} \end{bmatrix} \quad \text{is a positive definite matrix}$$

$$\text{and} \quad \mu, \nu \geqslant 0, \tag{16.17}$$

such that the entropy inequality can be satisfied.

16.4 Mixtures of Euler Fluids

We return to the general case of a mixture with multi-temperature (16.6). We observe that, up to now, it was not necessary to introduce the constitutive equations in order to close the system (16.6). Now, let us introduce the assumption that all the constituents of the mixture are Eulerian fluids, i.e., neither viscous nor heat-conducting:

$$\mathbf{t}_\alpha = -p_\alpha \mathbf{I}, \quad \mathbf{q}_\alpha = \mathbf{0} \quad (\alpha = 1, \ldots, n). \tag{16.18}$$

As a consequence of this assumption the stress tensor and the heat flux (16.3) are reduced to

$$\mathbf{t} = -p\mathbf{I} + \sum_{\alpha=1}^{n}(\rho_\alpha \mathbf{u}_\alpha \otimes \mathbf{u}_\alpha),$$
$$\mathbf{q} = \sum_{\alpha=1}^{n}\left\{\rho_\alpha\left(\varepsilon_\alpha + \frac{1}{2}u_\alpha^2\right) + p_\alpha\right\}\mathbf{u}_\alpha, \tag{16.19}$$

where p_α is the partial pressure of the α-constituent and

$$p = \sum_{\alpha=1}^{n} p_\alpha \tag{16.20}$$

is the total pressure.

16.4.1 Entropy Principle and Its Restrictions

The existence of the linear operator (16.10) confirms the Galilean invariance of field equation (16.6) and determines the velocity dependence of the production terms (16.11). Another important restriction comes from the entropy inequality (2.8).

In the present case,

$$h^0 = \rho S = \sum_{\alpha=1}^{n} \rho_\alpha s_\alpha, \tag{16.21}$$

where S is the total entropy density and s_α are the entropy densities of each constituent.

Statement 7 *The entropy density h^0 (16.21) of the mixture is a convex function with respect to the densities $\mathbf{u} \equiv (\rho_\alpha, \rho_\alpha \mathbf{v}_\alpha, \frac{1}{2}\rho_\alpha v_\alpha^2 + \rho_\alpha \varepsilon_\alpha)$.*

The proof of the statement is almost trivial: since for every α the entropy density $\rho_\alpha s_\alpha$ is a convex function of the densities of the corresponding α-fluid, then the entropy density of the mixture, being the sum of convex functions, is also a convex function of the whole densities $\mathbf{u}$ ($\boldsymbol{\alpha} = 1, \ldots n$).

Let us recall that the main field components have to satisfy the relation $(2.13)_1$. For the balance-law system (16.1) in the case of Euler fluids, it reads

$$dh^0 = d(\rho S) = \sum_{\alpha=1}^{n}\left\{\tilde{\Lambda}^{\rho_\alpha} d\rho_\alpha + \tilde{\boldsymbol{\Lambda}}^{\mathbf{v}_\alpha} d(\rho_\alpha \mathbf{v}_\alpha) + \tilde{\Lambda}^{\varepsilon_\alpha} d\left(\frac{1}{2}\rho_\alpha v_\alpha^2 + \rho_\alpha \varepsilon_\alpha\right)\right\}, \tag{16.22}$$

where

$$\tilde{\mathbf{u}}' = (\tilde{\Lambda}^{\rho_\alpha}, \tilde{\boldsymbol{\Lambda}}^{\mathbf{v}_\alpha}, \tilde{\Lambda}^{\varepsilon_\alpha}), \quad (\alpha = 1, \dots, n) \tag{16.23}$$

is the vector of the main field associated to the system (16.1).

Statement 8 *The main field components for the mixture of Euler fluids described by the system (16.1) have the form:*

$$\tilde{\Lambda}^{\rho_\alpha} = \frac{-g_\alpha + \frac{1}{2}v_\alpha^2}{T_\alpha}; \quad \tilde{\boldsymbol{\Lambda}}^{\mathbf{v}_\alpha} = -\frac{\mathbf{v}_\alpha}{T_\alpha}; \quad \tilde{\Lambda}^{\varepsilon_\alpha} = \frac{1}{T_\alpha}, \quad (\alpha = 1, \dots, n), \tag{16.24}$$

where

$$g_\alpha = \varepsilon_\alpha - T_\alpha S_\alpha + \frac{p_\alpha}{\rho_\alpha} \tag{16.25}$$

are the chemical potentials of the constituents.

This statement is a consequence of the fact that the system (16.1), for what concerns the main parts of the differential operators, is constituted by uncoupled systems of single fluid equations. Consequently, the Gibbs relation holds for each constituent and the main field components (16.24) coincide, for each species, with the ones of single fluid (2.18).

The main field components of the system (16.1) will be used for calculation of the main field of the equivalent system (16.6). Let us denote this main field as follows:

$$\mathbf{u}' = (\Lambda^\rho, \boldsymbol{\Lambda}^{\mathbf{v}}, \Lambda^\varepsilon, \Lambda^{\rho_b}, \boldsymbol{\Lambda}^{\mathbf{v}_b}, \Lambda^{\varepsilon_b}) \quad (b = 1, \dots, n-1). \tag{16.26}$$

Equation $(2.13)_1$ written in new variables reads

$$\begin{aligned} dh^0 = d(\rho S) &= \Lambda^\rho d\rho + \boldsymbol{\Lambda}^{\mathbf{v}} d(\rho \mathbf{v}) + \Lambda^\varepsilon d\left(\frac{1}{2}\rho v^2 + \rho\varepsilon\right) \\ &+ \sum_{b=1}^{n-1} \left\{ \Lambda^{\rho_b} d(\rho_b) + \boldsymbol{\Lambda}^{\mathbf{v}_b} d(\rho_b \mathbf{v}_b) + \Lambda^{\varepsilon_b} d\left(\frac{1}{2}\rho_b v_b^2 + \rho_b \varepsilon_b\right)\right\}. \end{aligned} \tag{16.27}$$

The expressions (16.22) and (16.27) should be equivalent to each other and this yields the relation between the main fields $\tilde{\mathbf{u}}'$ and $\mathbf{u}'$.

Statement 9 *The main field components for the mixture of Euler fluids described by the system (16.6) have the form:*

$$\begin{aligned} &\Lambda^\rho = \tilde{\Lambda}^{\rho_n}, \quad \boldsymbol{\Lambda}^{\mathbf{v}} = \tilde{\boldsymbol{\Lambda}}^{\mathbf{v}_n}, \quad \Lambda^\varepsilon = \tilde{\Lambda}^{\varepsilon_n}, \\ &\Lambda^{\rho_b} = \tilde{\Lambda}^{\rho_b} - \tilde{\Lambda}^{\rho_n}, \quad \boldsymbol{\Lambda}^{\mathbf{v}_b} = \tilde{\boldsymbol{\Lambda}}^{\mathbf{v}_b} - \tilde{\boldsymbol{\Lambda}}^{\mathbf{v}_n}, \quad \Lambda^{\varepsilon_b} = \tilde{\Lambda}^{\varepsilon_b} - \tilde{\Lambda}^{\varepsilon_n} \end{aligned} \tag{16.28}$$

for $b = 1, \ldots, n-1$, *i.e.,*

$$
\begin{aligned}
\Lambda^{\rho} &= -\frac{1}{T_n}\left(g_n - \frac{1}{2}(\mathbf{u}_n + \mathbf{v})^2\right), \\
\Lambda^{\mathbf{v}} &= -\frac{1}{T_n}(\mathbf{u}_n + \mathbf{v}), \\
\Lambda^{\varepsilon} &= \frac{1}{T_n}, \\
\Lambda^{\rho_b} &= -\frac{1}{T_b}\left(g_b - \frac{1}{2}(\mathbf{u}_b + \mathbf{v})^2\right) + \frac{1}{T_n}\left(g_n - \frac{1}{2}(\mathbf{u}_n + \mathbf{v})^2\right), \\
\Lambda^{\mathbf{v}_b} &= -\frac{\mathbf{u}_b}{T_b} + \frac{\mathbf{u}_n}{T_n} + \left(\frac{1}{T_b} - \frac{1}{T_n}\right)\mathbf{v}, \\
\Lambda^{\varepsilon_b} &= -\frac{1}{T_n} + \frac{1}{T_b}.
\end{aligned}
\tag{16.29}
$$

The main field permits to determine the production terms through the application of the residual inequality $(2.13)_2$. i.e.,

$$
\Sigma = \mathbf{u}' \cdot \mathbf{f} = \hat{\mathbf{u}}' \cdot \hat{\mathbf{f}} = \sum_{b=1}^{n-1}\left(\hat{\Lambda}^{\rho_b}\hat{\tau}_b + \hat{\boldsymbol{\Lambda}}^{\mathbf{v}_b} \cdot \hat{\mathbf{m}}_b + \hat{\Lambda}^{\varepsilon_b}\hat{e}_b\right) \geqslant 0,
$$

or explicitly

$$
\Sigma = \sum_{b=1}^{n-1}\left(-\frac{g_b - \frac{1}{2}u_b^2}{T_b} + \frac{g_n - \frac{1}{2}u_n^2}{T_n}\right)\hat{\tau}_b + \left(\frac{\mathbf{u}_b}{T_b} + \frac{\mathbf{u}_n}{T_n}\right) \cdot \hat{\mathbf{m}}_b + \left(\frac{1}{T_b} - \frac{1}{T_n}\right)\hat{e}_b \geqslant 0.
\tag{16.30}
$$

This inequality allows us to obtain the following structure of production terms.

Statement 10 *The internal (intrinsic) parts of the production terms* $(16.9)_2$ *are chosen in such a way that the residual inequality (16.30) is actually a quadratic form. In particular, in agreement with the kinetic theory [22],*

$$
\begin{aligned}
\hat{\tau}_b &= \sum_{c=1}^{n-1}\varphi_{bc}\left(\frac{g_n - \frac{1}{2}u_n^2}{T_n} - \frac{g_c - \frac{1}{2}u_c^2}{T_c}\right) + \sum_{c=1}^{n-1}\beta_{bc}\left(\frac{1}{T_c} - \frac{1}{T_n}\right), \\
\hat{\mathbf{m}}_b &= \sum_{c=1}^{n-1}\psi_{bc}\left(\frac{\mathbf{u}_n}{T_n} - \frac{\mathbf{u}_c}{T_c}\right), \\
\hat{e}_b &= \sum_{c=1}^{n-1}\theta_{bc}\left(\frac{1}{T_c} - \frac{1}{T_n}\right) + \sum_{c=1}^{n-1}\beta_{bc}\left(\frac{g_n - \frac{1}{2}u_n^2}{T_n} - \frac{g_c - \frac{1}{2}u_c^2}{T_c}\right),
\end{aligned}
\tag{16.31}
$$

where

$$\begin{bmatrix} \varphi_{bc} & \beta_{bc} \\ \beta_{bc} & \theta_{bc} \end{bmatrix}, \qquad \psi_{bc}$$

are phenomenological symmetric positive definite matrices ($b, c = 1, \dots, n-1$).

For processes not far from equilibrium, the previous matrices depend only on the equilibrium variables ρ_α and T. In the sequel, our analysis will be restricted to a model of non-reacting mixtures, for which $\tau_b = 0$.

16.4.2 Symmetric Hyperbolic System and Principal Subsystems

The main field components for the mixture of Euler fluids (16.29) symmetrize the system (16.6) according to Theorem 2.1.

Concerning the principal subsystems (see Sect. 2.4), taking into account (16.29) and (16.6), we can recognize the following interesting principal subsystems:

Case 1 *The single-temperature model is a principal subsystem of the multi-temperature*. Let us suppose that $\Lambda^{\varepsilon_b} = 0$ for $b = 1, \dots, n-1$, then

$$T_1 = \ldots = T_n = T.$$

This principal subsystem contains only the energy conservation equation for the mixture, while energy balance equations for the constituents are dropped. Thus, one may conclude that single-temperature model naturally appears as a principal subsystem of the multi-temperature system.

Case 2 *The equilibrium subsystem*. If we set

$$\Lambda^{\varepsilon_b} = \boldsymbol{\Lambda}^{\mathbf{v}_b} = \boldsymbol{\Lambda}^{\rho_b} = 0 \qquad \forall\, b = 1, \dots, n-1,$$

i.e.,

$$T_b = T, \;\; \mathbf{u}_b = 0, \;\; g_b = g \qquad \forall\, b = 1, \dots, n-1,$$

we have the equilibrium Euler subsystem (a single fluid system) with concentrations c_b being solutions of $g_1 = g_2 = \ldots = g_n$.

16.4.3 Characteristic Velocities and Their Upper Bound in the ST Model

The characteristic velocities for the *MT* model are simple to evaluate. Since, for each constituent, they are the same as the ones of a single fluid, i.e.,

$$\lambda_\alpha^{(1)} = v_{\alpha n} - c_{s\alpha}; \quad \lambda_\alpha^{(2,3,4)} = v_{\alpha n}; \quad \lambda_\alpha^{(5)} = v_{\alpha n} + c_{s\alpha}, \tag{16.32}$$

where $v_{\alpha n} = \mathbf{v}_\alpha \cdot \mathbf{n}$ are the normal component of the velocities at the wave front and

$$c_{s\alpha} = \sqrt{\left(\frac{\partial p_\alpha}{\partial \rho_\alpha}\right)_{s_\alpha}}$$

are the sound velocities. For an ideal gas, for example, we have

$$p_\alpha = \frac{k_B}{m_\alpha}\rho_\alpha T_\alpha, \quad \varepsilon_\alpha = c_v^{(\alpha)} T_\alpha, \quad c_v^{(\alpha)} = \frac{k_B}{m_\alpha(\gamma_\alpha - 1)} \tag{16.33}$$

and

$$c_{s\alpha} = \sqrt{\frac{k_B}{m_\alpha}\gamma_\alpha T_\alpha}. \tag{16.34}$$

Instead, in the case of the *ST* model, the evaluation of the velocities is very difficult even in an equilibrium state due to the fact that the characteristic polynomial is, in general, irreducible (see e.g. [10, 26]). But thanks to the subcharacteristic property (2.25) of principal subsystems, we are able to establish the following lower and upper bounds for the characteristic velocities of the ST model:

$$\min_\alpha (v_{\alpha n} - c_{s\alpha}^*) \leqslant \lambda_{\min}^{ST}; \qquad \max_\alpha (v_{\alpha n} + c_{s\alpha}^*) \geqslant \lambda_{\max}^{ST},$$

where now

$$c_{s\alpha}^* = \sqrt{\frac{k_B \gamma_\alpha}{m_\alpha} T}.$$

16.4.4 Qualitative Analysis and K-Condition in the Mixture Theories

We made, in Sect. 2.6, the qualitative analysis for a system of hyperbolic type that is composed of two groups, as is usual in RET: one group is formed by conservation laws and the other group by balance laws. In this case, the coupling K-condition given in Sect. 2.6.2 plays an important role.

For *ST* theory without chemical reactions, it was proved [27, 28] that the K-condition is violated for some genuinely nonlinear eigenvalues. Therefore, from the results in [29], global smooth solutions can not exist even though initial data are small enough. Instead, for a *MT* system, it is possible to verify that the K-condition is satisfied for all eigenvalues. Therefore we can conclude [16]:

Statement 11 *If the initial data of the MT model are perturbations of equilibrium state, smooth solutions exist for all time and tend to an equilibrium constant state.*

This statement also shows clearly that the *MT* model is more realistic than the *ST* model.

16.5 Average Temperature

The *MT*-mixture theory explained above is the most realistic theory, and is consistent with the kinetic theory [30]. It is also a necessary theory in physics, in particular, in plasma physics [22]. Nevertheless, from a theoretical point of view, a serious problem still remains in it: How to measure the temperature of each constituent?

In this section, instead of studying this difficult problem directly, we study the problem about the macroscopic average temperature of a mixture. Let us consider the definition of the average temperature firstly proposed by Ruggeri and co-workers in [17, 18, 31, 32]. The main idea is to investigate the definition of internal energy so as to introduce the (average) temperature T as a state variable of the mixture. Then the intrinsic internal energy ε_I (see $(16.3)_3$) of the *MT* mixture resembles the structure of the intrinsic internal energy of a *ST* mixture.

Therefore, the following implicit definition of an average temperature is adopted:

Definition 16.1 The average temperature T is the one that corresponds to the barycentric intrinsic internal energy, i.e., it is defined through the relation:

$$\rho\varepsilon_I(\rho_\beta, T) = \sum_{\alpha=1}^{n} \rho_\alpha \varepsilon_\alpha(\rho_\alpha, T) = \sum_{\alpha=1}^{n} \rho_\alpha \varepsilon_\alpha(\rho_\alpha, T_\alpha). \tag{16.35}$$

By expanding this relation in the neighborhood of the average temperature we have

$$T = \frac{\sum_{\alpha=1}^{n} \rho_\alpha c_v^{(\alpha)} T_\alpha}{\sum_{\alpha=1}^{n} \rho_\alpha c_v^{(\alpha)}}, \tag{16.36}$$

where

$$c_v^{(\alpha)} = \left.\frac{\partial \varepsilon_\alpha(\rho_\alpha, T_\alpha)}{\partial T_\alpha}\right|_{T_\alpha = T}$$

is the specific heat at constant volume of constituent α. We observe that (16.36) gives the exact value of the average temperature in the case of the mixture of ideal gases for which the $c_v^{(\alpha)}$ are constant.

This definition of the average temperature has some advantages over usual ones used frequently in the literature. Firstly, as a consequence of the definition, the conservation law of energy of a mixture $(16.6)_3$ becomes an evolution equation for the average temperature T as in the cases of *ST* and *CT* mixtures. So, in the case of spatial homogeneous solution of the differential system (16.6) (solutions depend only on time), T is constant and the nonequilibrium temperature of each constituent T_α converges to T for large time, as shown below.

The second advantage is related to the entropy of the whole mixture that, thanks to the introduction of this average temperature, reaches its maximum value when $T_\alpha = T$. In fact, Ruggeri and Simić introduced the *diffusion temperature flux*

$$\Theta_\alpha = T_\alpha - T, \tag{16.37}$$

and they proved in [18], using the Gibbs equation for each constituent, that the entropy density near equilibrium becomes a negative definite quadratic form with respect to the nonequilibrium variables Θ_α:

$$\rho S = \sum_{\alpha=1}^{n} \rho_\alpha s_\alpha(\rho_\alpha, T_\alpha) = \sum_{\alpha=1}^{n} \rho_\alpha s_\alpha(\rho_\alpha, T) - \frac{1}{2T^2}\sum_{\alpha=1}^{n} \rho_\alpha c_v^{(\alpha)} \Theta_\alpha^2 + O(\Theta_\alpha^3). \tag{16.38}$$

Therefore in the present case the nonequilibrium entropy k (5.18) is given by

$$k = -\frac{1}{2\rho T^2}\sum_{\alpha=1}^{n} \rho_\alpha c_v^{(\alpha)} \Theta_\alpha^2,$$

and the entropy density S has a maximum in equilibrium.

16.5.1 Alternative Form of the Differential System

It is convenient in the following analysis to rewrite the system (16.6) using the material derivatives:

$$\frac{d}{dt} = \frac{\partial}{\partial t} + \mathbf{v}\cdot\nabla, \quad \frac{d_b}{dt} = \frac{\partial}{\partial t} + \mathbf{v}_b\cdot\nabla$$

for $b = 1, \cdots n-1$. Taking into account the definition of the average temperature given by (16.35), we have

$$\begin{cases} \frac{d\rho}{dt} + \rho \operatorname{div} \mathbf{v} = 0, \\ \rho \frac{d\mathbf{v}}{dt} - \operatorname{div}\mathbf{t} = 0, \\ \rho \frac{\partial \varepsilon_I}{\partial T}\frac{dT}{dt} = \rho^2 \frac{\partial \varepsilon_I}{\partial \rho} \operatorname{div} \mathbf{v} + \sum_{b=1}^{n-1} \frac{\partial \varepsilon_I}{\partial c_b} \operatorname{div} \mathbf{J}_b + \mathbf{t} \operatorname{grad} \mathbf{v} - \operatorname{div} \mathbf{q}, \\ \frac{d_b \rho_b}{dt} + \rho_b \operatorname{div} \mathbf{v}_b = 0, \\ \rho_b \frac{d_b \mathbf{v}_b}{dt} - \operatorname{div}\mathbf{t}_b = \hat{\mathbf{m}}_b, \\ \rho_b \frac{d_b \varepsilon_b}{dt} - \mathbf{t}_b \cdot \nabla \mathbf{v}_b + \operatorname{div} \mathbf{q}_b = \hat{e}_b. \end{cases} \tag{16.39}$$

The differential system of equations governs the evolution of ρ, $\mathbf{v}$, T, ρ_b, $\mathbf{J}_b$ and Θ_b, provided that we assign the constitutive equations of p_α, ε_α and, for dissipative fluids, also the heat fluxes $\mathbf{q}_\alpha$ and the viscous stress tensors $\boldsymbol{\sigma}_\alpha$.

16.6 Examples of Spatially Homogeneous Mixture and Static Heat Conduction

In this section two simple examples are provided in order to support the previous theoretical considerations, and to stress the main features of the multi-temperature approach and the role of the average temperature.

16.6.1 Solution of a Spatially Homogenous Mixture

First we consider a non-reacting mixture of gases in spatially homogeneous fields, i.e., a mixture with field variables depending solely on time [18, 33]. The governing equation (16.39) can be written in the following form:

$$\frac{d\rho}{dt} = 0, \quad \frac{d\mathbf{v}}{dt} = \mathbf{0}, \quad \frac{dT}{dt} = 0, \tag{16.40}$$

$$\frac{d\rho_b}{dt} = 0, \quad \rho_b \frac{d\mathbf{v}_b}{dt} = \hat{\mathbf{m}}_b, \quad \rho_b \frac{d\varepsilon_b}{dt} = \hat{e}_b \tag{16.41}$$

where now $d/dt = \partial/\partial t$. From (16.40), (16.41), it is easy to conclude that

$$\rho = \text{const.}, \quad \mathbf{v} = \text{const.}, \quad T = \text{const.},$$
$$\rho_b = \text{const.}, \quad (b = 1, \dots, n),$$

and, due to the Galilean invariance, we may choose $\mathbf{v} = \mathbf{v}_0 = \mathbf{0}$ without any loss of generality. It is also remarkable that the average temperature of the mixture remains constant during the process: $T(t) = T_0$.

In the sequel we shall regard only small perturbations from an equilibrium state, $\mathbf{v}_\alpha = \mathbf{v}_0 = \mathbf{0}$, $T_\alpha = T_0$, $(\alpha = 1, \dots, n)$, and analyze their behavior. Therefore, the right-hand side of $(16.41)_{2,3}$ could be linearized in the neighborhood of equilibrium. Taking into account (16.31), we obtain

$$\rho_b \frac{d\mathbf{v}_b}{dt} = -\sum_{c=1}^{n-1} \frac{\psi_{bc}^0}{T_0} (\mathbf{v}_c - \mathbf{v}_n), \tag{16.42}$$

$$\rho_b c_v^{(b)} \frac{dT_b}{dt} = -\sum_{c=1}^{n-1} \frac{\theta_{bc}^0}{T_0^2} (T_c - T_n), \tag{16.43}$$

where ψ_{bc}^0 and θ_{bc}^0 are entries of positive definite matrices evaluated in equilibrium. Note that $\mathbf{v}_b - \mathbf{v}_n = \mathbf{u}_b - \mathbf{u}_n$ and $T_b - T_n = \Theta_b - \Theta_n$.

In the particular case of a binary mixture, the explicit solution of equations (16.42) and (16.43) can be obtained as

$$\mathbf{v}_1(t) = \mathbf{v}_1(0) e^{-\frac{t}{\tau_v}}, \qquad T_1(t) = T_0 + (T_1(0) - T_0) e^{-\frac{t}{\tau_T}},$$

where τ_v and τ_T represent the relaxation times, which, for an ideal gas, assume the expression:

$$\tau_v = \frac{\rho_1 \rho_2 T_0}{\psi_{11}^0 \rho}, \qquad \tau_T = \frac{k \rho_1 \rho_2 T_0^2}{\theta_{11}^0 (\rho_1 m_2 (\gamma_2 - 1) + \rho_2 m_1 (\gamma_1 - 1))}. \tag{16.44}$$

Starting from these solutions, other field variables can be obtained by using the following equations:

$$\rho_1 \mathbf{v}_1 + \rho_2 \mathbf{v}_2 = \rho \mathbf{v} = \mathbf{0}, \tag{16.45}$$

$$\begin{aligned} \rho_1 c_v^{(1)} T_1 + \rho_2 c_v^{(2)} T_2 &= (\rho_1 c_v^{(1)} + \rho_2 c_v^{(2)}) T \\ &= (\rho_1 c_v^{(1)} + \rho_2 c_v^{(2)}) T_0. \end{aligned} \tag{16.46}$$

It is obvious that, due to dissipative character of the system, *all the non-equilibrium variables exponentially decay and converge to their equilibrium values*. In order to compare the values of τ_v and τ_T for ideal gases, and also to compute the actual

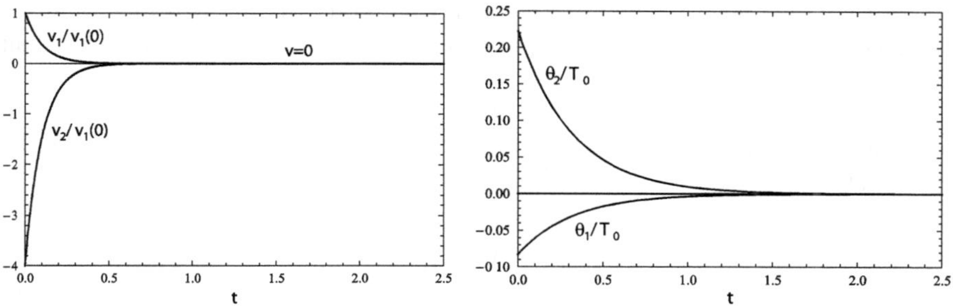

Fig. 16.1 Dimensionless velocities and diffusion temperature fluxes of the constituents versus time

values of variables in numerical examples, the relations from kinetic theory has to be recalled [22]:

$$\theta_{11}^{0} = \frac{3m_1 m_2}{(m_1 + m_2)^2} k T_0^2 \Gamma_{12}'; \quad \psi_{11}^{0} = \frac{2m_1 m_2}{m_1 + m_2} T_0 \Gamma_{12}', \tag{16.47}$$

where Γ_{12}' represents volumetric collision frequency, and the following estimate can be obtained:

$$\frac{\tau_T}{\tau_v} = \frac{2}{3} \frac{\rho(m_1 + m_2)}{\rho_1 m_2(\gamma_2 - 1) + \rho_2 m_1(\gamma_1 - 1)} > \frac{2}{3(\gamma_{\max} - 1)} \geqslant 1, \tag{16.48}$$

($\gamma_{\max} = \max\{\gamma_1, \gamma_2\} \leqslant 5/3$).

In Fig. 16.1, we present the graphs of normalized velocities and diffusion temperature fluxes [18]. It can be observed that, due to inequality (16.48), *the mechanical diffusion vanishes more rapidly than the thermal one.* This is in sharp contrast with the widely adopted approach that ignores the influence of the multiple temperature of each constituent of the mixture.

16.6.2 Solution of Static Heat Conduction

Another simple example is the one-dimensional mixture of gases at rest ($\mathbf{v}_\alpha = 0$) without chemical reactions ($\tau_\alpha = 0$) between two walls $0 \leqslant x \leqslant L$, maintained at two different temperatures $T(0) = T_0$, $T(L) = T_L$ [32].

In both *CT* and *ST* models, the static field equation reduces to the global energy equation $(16.39)_3$ that reads div $\mathbf{q} = 0$. In the one-dimensional case, this equation, combined with the Fourier law with constant heat conductivity, yields the classical result of a linear temperature profile as for a single fluid:

$$T'' = 0 \iff T = (T_L - T_0)\xi + T_0$$

where $\xi = x/L$ and $'$ denotes $d/d\xi$. For what concerns the densities they are obtained by the conditions that the pressure of each constituent must be constant due to the momentum equations.

In the *MT* model, the situation is quite different. Let us consider the simple case of a binary mixture ($n = 2$). In the linear case, by taking into account (16.31), system (16.1) reduces to

$$\begin{cases} \dfrac{dp_1}{dx} = 0, \ \dfrac{dp_2}{dx} = 0, \\ \dfrac{dq_1}{dx} = \beta\left(T_2 - T_1\right), \\ \dfrac{dq_2}{dx} = \beta\left(T_1 - T_2\right), \end{cases} \tag{16.49}$$

where $\beta = \theta_{11}/T_0^2$. By using the Fourier law, $(16.49)_{2,3}$ can be rewritten as

$$\begin{cases} T_1'' = \nu_1(T_1 - T_2), \\ T_2'' = \nu_2(T_2 - T_1), \end{cases} \tag{16.50}$$

where we assume that the dimensionless quantities,

$$\nu_1 = \frac{\beta L^2}{\chi_1}, \qquad \nu_2 = \frac{\beta L^2}{\chi_2}, \tag{16.51}$$

are constant. The system (16.50) is equivalent to

$$\hat{T}'' = 0, \qquad \Theta'' - \omega^2\,\Theta = 0$$

with $\hat{T} = \nu T_1 + (1-\nu)T_2, \ \ \Theta = T_2 - T_1$ and

$$\nu = \frac{\nu_2}{\nu_1 + \nu_2} = \frac{\chi_1}{\chi_1 + \chi_2}, \qquad \omega = \sqrt{\nu_1 + \nu_2}.$$

Consequently, we get the solution in the form:

$$T_1 = \hat{T} - (1-\nu)\Theta, \quad T_2 = \hat{T} + \nu\Theta \tag{16.52}$$

where

$$\hat{T} = A\,\xi + B, \qquad \Theta = \frac{1}{\sinh(\omega)}\left\{\Theta_L \sinh(\omega\,\xi) + \Theta_0\, \sinh(\omega(1-\xi))\right\}, \tag{16.53}$$

and A, B, Θ_0, Θ_L are constants of integration. In the case of ideal gases, equations $(16.49)_1$ and (16.33) yield the constant internal energy density of each constituent:

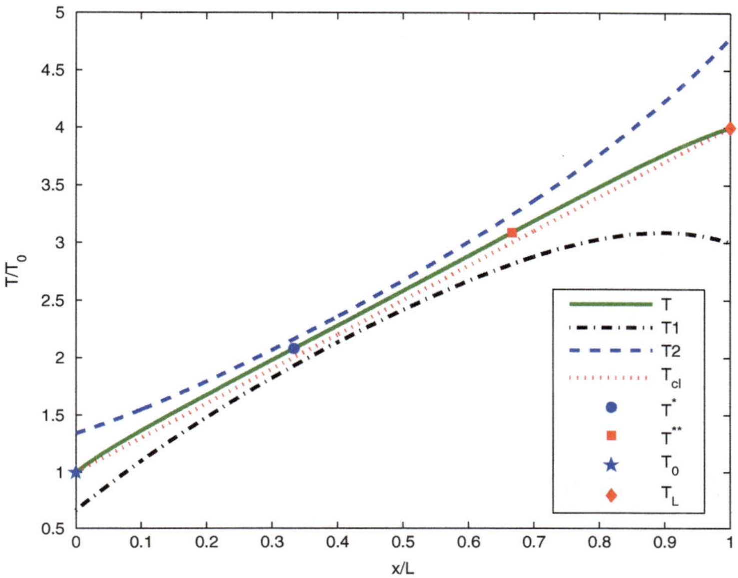

Fig. 16.2 Graphs of the average temperature T and constituent temperatures T_1, T_2 in terms of the dimensionless distance x/L. T_{cl} represents the classical straight line solution. T_0 is the temperature unit

$$\rho_\alpha T_\alpha c_v^{(\alpha)} = P_\alpha = Const., \quad (\alpha = 1, 2). \tag{16.54}$$

And (16.36) yields the average temperature:

$$\frac{1}{T} = \frac{\pi}{T_1} + \frac{1-\pi}{T_2}, \quad \text{with } \pi = \frac{P_1}{P_1 + P_2}. \tag{16.55}$$

The constant π belongs to $[0, 1]$. It is interesting to observe that the *coldness* $1/T$ (inverse of the average temperature) belongs to the convex envelope of the component coldnesses $1/T_1$ and $1/T_2$. Equations (16.52), (16.53) and (16.55) give the explicit solution of T_1, T_2 and T as the function of ξ and five constants of integration: $(A, B, \Theta_0, \Theta_L, \pi)$. We observe the behavior of T is not a straight line in contrast to the classical case of *CT* or *ST* theories; the multi-temperature effect is that the temperature is not a linear function of x (see Fig. 16.2). Due to (16.49)–(16.51), when $\epsilon = 1/\beta$ tends towards zero, the solution of (16.52), (16.53) and (16.55) converges towards the classical solution $T_1 = T_2 = T = \hat{T}$ for any $\xi \in]0, 1[$. This result is true also at the boundary when Θ_0 and Θ_1 are of same order as ϵ.

Let us introduce the concentration $c = \rho_1/(\rho_1 + \rho_2)$ with $c(0) = c_0$. Then (16.54), (16.52), and (16.53) imply

$$c = \frac{c_0}{c_0 + \Omega(1 - c_0)}, \quad \text{with } \Omega = \frac{T_1}{T_2}\frac{T_{20}}{T_{10}}$$

and

$$T_{10} = B - (1 - \nu)\Theta_0, \quad T_{20} = B + \nu\Theta_0.$$

The concentration is a function of the position x whereas, in the classical case, $\Omega = 1$ and $c = c_0$. Ruggeri and Lou [32] studied the method how to determine in a unique way the constants of integration. They proved that, for a mixture of n constituents, the measurement of the average temperature at $2(n-1)$ points allows to determine the temperature of each constituent in all points.

16.7 Maxwellian Iteration

In order to understand a connection among the extended models of *MT*, *ST* and the classical model *CT*, we use the *Maxwellian iteration* (for more detail see Chap. 18, Sects. 18.2.1 and 18.3). In the present case, the Maxwellian iteration is carried out as follows: put the zeroth iterates, i.e., the values of quantities evaluated in an equilibrium state into the left-hand side (l.h.s.) of the system $(16.39)_{5,6}$, then we obtain the first iterates from the right-hand side (r.h.s.) of the system.

Taking into account the fact that in zero-*th* iteration $\mathbf{v}_\alpha^{(0)} = \mathbf{v}$ and consequently

$$\frac{d_b^{(0)}}{dt} = \frac{d}{dt}, \qquad \mathbf{J}_b^{(0)} = \mathbf{u}_b^{(0)} = \mathbf{0}$$

and moreover

$$T_\alpha^{(0)} = T, \quad \mathbf{q}^{(0)} = \mathbf{q}_b^{(0)} = 0, \quad \mathbf{t}^{(0)} = -p^{(0)}\mathbf{I} = -p_0\mathbf{I}, \quad t_b^{(0)} = -p_b^{(0)}\mathbf{I},$$

we obtain:

$$\begin{aligned} &\rho_b \left(\frac{d\mathbf{v}}{dt}\right)^{(0)} + \operatorname{grad} p_b^{(0)} = \hat{\mathbf{m}}_b^{(1)}, \\ &\rho_b \left\{ \left(\frac{\partial \varepsilon_b}{\partial \rho_b}\right)^{(0)} \left(\frac{d\rho_b}{dt}\right)^{(0)} + \left(\frac{\partial \varepsilon_b}{\partial T_b}\right)^{(0)} \left(\frac{dT}{dt}\right)^{(0)} \right\} + p_b^{(0)} \operatorname{div}\mathbf{v} = \hat{e}_b^{(1)}. \end{aligned} \tag{16.56}$$

On the other hand, from the zero-*th* order of $(16.39)_{2,3,4}$, we have

$$\rho \left(\frac{d\mathbf{v}}{dt}\right)^{(0)} = -\operatorname{grad} p_0,$$

$$\rho \frac{\partial \varepsilon_I}{\partial T} \left(\frac{dT}{dt} \right)^{(0)} = \left(\rho^2 \frac{\partial \varepsilon_I}{\partial \rho} - p_0 \right) \operatorname{div}\mathbf{v}, \tag{16.57}$$

$$\left(\frac{d\rho_b}{dt} \right)^{(0)} = -\rho_b \operatorname{div}\mathbf{v},$$

and therefore, inserting (16.57) into (16.56), we obtain

$$-\frac{\rho_b}{\rho} \operatorname{grad} p_0 + \operatorname{grad} p_b^{(0)} = \hat{\mathbf{m}}_b^{(1)}, \tag{16.58}$$

$$\Omega_b \operatorname{div}\mathbf{v} = \hat{e}_b^{(1)}, \tag{16.59}$$

where

$$\Omega_b = p_b^{(0)} + \rho_b \left\{ -\rho_b \left(\frac{\partial \varepsilon_b}{\partial \rho_b} \right)^{(0)} + \left(\rho^2 \frac{\partial \varepsilon_I}{\partial \rho} - p_0 \right) \frac{\left(\dfrac{\partial \varepsilon_b}{\partial T_b} \right)^{(0)}}{\rho \dfrac{\partial \varepsilon_I}{\partial T}} \right\}. \tag{16.60}$$

Taking into account the expressions of the productions (16.31) and the definitions of the diffusion flux $\mathbf{J}_a$ (16.14) and of the thermal diffusion Θ_a (16.37), after some arrangement (an interested reader can consult the details in the original paper [18]), we obtain the fact that the approximation of the momentum equation of each species (16.58) gives the Fick law $(16.16)_3$:

$$\mathbf{J}_a^{(1)} = \tilde{L}_a \operatorname{grad}\left(\frac{1}{T} \right) - \sum_{b=1}^{n-1} L_{ab} \operatorname{grad}\left(\frac{g_b - g_n}{T} \right), \tag{16.61}$$

while for what concerns the approximation of energy equation (16.59) we obtain new constitutive equations:

$$\Theta_a^{(1)} = -k_a \operatorname{div}\mathbf{v}, \tag{16.62}$$

where k_a is a linear combination of Ω_b given in (16.60).

The equation (16.62), *obtained by means of the Maxwellian iteration, gives the temperature of each species as a constitutive equation, in a similar way as the Fick law gives the velocity of each species.*

It is possible to prove [18] that for a mixture of ideal gases

$$\Omega_b = \rho_b T c_v^{(b)} \frac{\sum_{\alpha=1}^{n} \rho_\alpha c_v^{(\alpha)} (\gamma_b - \gamma_\alpha)}{\sum_{\alpha=1}^{n} \rho_\alpha c_v^{(\alpha)}}.$$

In the next section we will deduce and justify (16.62) using a classical TIP approach.

Equation (16.62) cannot be obtained in the classical theory, and it is interesting to observe that this is probably due to the fact that, if the gas has constituents with the same degrees of freedom, i.e., the same γ, then $\Omega_b = 0$, and consequently $k_a = 0$. And, from (16.62), $\Theta_b^{(1)} = 0$, and then all the temperatures are equal in the first Maxwellian iteration. In this last case, in order to reveal the temperature difference between constituents, it is necessary to go to the second order Maxwellian iteration. This is the subject of a paper in preparation by Ruggeri and Simić [34].

In conclusion of this section, we have proved the following important result:

Statement 12 *Using the Maxwellian iteration the Fick law (16.61) is obtained as the first approximation of the momentum balance equation of each species (16.39)$_5$ and the equivalent thermal diffusion law (16.62) is the first approximation of the energy balance equation of each constituent (16.39)$_6$.*

16.8 A Classical Approach to Multi-Temperature Mixtures

As seen above, by using the Maxwellian iteration, the *ST* model reduces to the classic model, but when we start from the *MT* model we obtain new constitutive equations (16.62) that do not exist in the classical theory of mixtures.

To justify the results of the Maxwellian iteration, Gouin and Ruggeri [17] constructed a classical theory of mixture with multi-temperature. The idea is to use the usual equations in the classical approach (16.13), but now we suppose that each constituent has its own temperature. In this approach, the role of the average temperature defined (16.35) is fundamental. In fact the multi-temperature effect appears through the pressure. In fact near equilibrium, we have (16.36) and therefore:

$$p = \sum_{\alpha=1}^{n} p_\alpha(\rho_\alpha, T_\alpha) = p_0 + \Pi_\theta,$$

where

$$p_0 = \sum_{\alpha=1}^{n} p_\alpha(\rho_\alpha, T), \quad \Pi_\theta = \sum_{b=1}^{n-1} r_b\, \Theta_b$$

and

$$r_b = \frac{1}{\rho_n c_v^{(n)}} \left\{ \rho_n c_v^{(n)} \frac{\partial p_b}{\partial T_b}(\rho_b, T) - \rho_b c_v^{(b)} \frac{\partial p_n}{\partial T_n}(\rho_n, T) \right\}. \tag{16.63}$$

Therefore, the total pressure p of a mixture is the sum of the equilibrium part p_0 depending on ρ_α and T and a new dynamic pressure part (as a nonequilibrium term) Π_θ due to the difference of temperatures between the constituents.

We assume the internal energy $\varepsilon(\rho, T, c_b)$ and the equilibrium pressure $p_0(\rho, T, c_b)$ satisfy the Gibbs equation:

$$TdS = d\varepsilon - \frac{p_0}{\rho^2}\, d\rho - \sum_{b=1}^{n-1} (g_b - g_n)\; dc_b\,. \tag{16.64}$$

The difference between (16.15) and (16.64) consists in the fact that, in (16.64), T means the average temperature when each component α has its own temperature T_α and p_0 takes the place of p. Let us consider first the case of a Stokesian fluid $\operatorname{tr}\boldsymbol{\sigma} = 0$. Taking into account that, from (16.38), ρS depends only on the average temperature up to the first order expansion and using (16.13) to eliminate the time derivatives, we obtain from (16.64) the following entropy balance [18]:

$$\rho\frac{dS}{dt} + \operatorname{div}\left\{\frac{1}{T}\left(\mathbf{q} - \sum_{b=1}^{n-1}(g_b - g_n)\,\mathbf{J}_b\right)\right\} =$$

$$\mathbf{q}\cdot\operatorname{grad}\left(\frac{1}{T}\right) - \sum_{b=1}^{n-1}\mathbf{J}_b\cdot\operatorname{grad}\left(\frac{g_b - g_n}{T}\right) + \frac{1}{T}\operatorname{tr}\left(\mathbf{J}_{mech}\,\mathbf{D}\right), \tag{16.65}$$

where the mechanical flux is given by

$$\mathbf{J}_{mech} = \boldsymbol{\sigma} - \Pi_\theta\,\mathbf{I}. \tag{16.66}$$

Equation (16.65) can be interpreted as a balance of entropy, if we consider

$$\boldsymbol{\Phi} = \frac{1}{T}\left(\mathbf{q} - \sum_{b=1}^{n-1}(g_b - g_n)\,\mathbf{J}_b\right)$$

and

$$\Sigma = \mathbf{q}\cdot\operatorname{grad}\left(\frac{1}{T}\right) - \sum_{b=1}^{n-1}\mathbf{J}_b\cdot\operatorname{grad}\left(\frac{g_b - g_n}{T}\right) + \frac{1}{T}\operatorname{tr}\left(\mathbf{J}_{mech}\,\mathbf{D}\right) \tag{16.67}$$

as the entropy flux and the entropy production, respectively.

We observe that the entropy production is the sum of products of the following quantities:

Dissipative fluxes	**Thermodynamic forces**
heat flux $\mathbf{q}$	temperature gradient $\operatorname{grad}\left(\frac{1}{T}\right)$,
diffusion fluxes $\mathbf{J}_b$	chemical potential gradients $\operatorname{grad}\left(\frac{g_b - g_n}{T}\right)$,
mechanical flux $\mathbf{J}_{mech}$	velocity gradient $\mathbf{D}$.

In accordance with the case of a single temperature model [9, 10], which we have discussed in Sects. 1.1.2 and 1.1.3, the fluxes depend linearly on the associated forces in *TIP* near equilibrium, (see also for the general methodology of *TIP* [23–25, 35]):

- For the heat flux and the diffusion fluxes, we obtain the constitutive equations in the form of $(16.16)_{2,3}$.
- For Stokesian fluids, by taking into account (16.66), the last term of (16.67), corresponding to the mechanical production of the entropy, can be written in a separated form:

$$\frac{1}{T}\,\mathrm{tr}\,(\mathbf{J}_{mech}\,\mathbf{D}) = \frac{1}{T}\,\mathrm{tr}\left(\boldsymbol{\sigma}\,\mathbf{D}^D\right) - \frac{1}{T}\,\Pi_\theta\,\mathrm{div}\,\mathbf{v}.$$

We obtain the constitutive equation of the viscous stress tensor in the form of $(16.16)_1$ (with $\nu = 0$), and the dynamic pressure part due to the difference of temperatures yields

$$\Pi_\theta = \sum_{b=1}^{n-1} r_b\,\Theta_b = -L_\pi\,\mathrm{div}\,\mathbf{v}, \tag{16.68}$$

where L_π is a scalar coefficient.

As the production of entropy must be non-negative, (16.67) and therefore the *phenomenological coefficients* must satisfy the inequalities (16.17) and

$$L_\pi \geqslant 0.$$

Taking into account that terms r_b given by (16.63) depend on (ρ_b, T), we deduce, from (16.68), that constitutive quantities Θ_a (depending a priori on $\nabla\mathbf{v}$) must be proportional to div $\mathbf{v}$:

$$\Theta_a = -k_a\,\mathrm{div}\,\mathbf{v} \quad (a = 1, \cdots, n-1).$$

This is in perfect agreement with (16.62) by the Maxwellian iteration procedure presented in the previous section.

Let $\|M_{ab}\|$ be the matrix such that $k_a = \sum_{b=1}^{n-1} M_{ab}\,r_b$, we have

$$\Theta_a = -\sum_{b=1}^{n-1} M_{ab}\,r_b\,\mathrm{div}\,\mathbf{v} \quad (a = 1, \cdots, n-1). \tag{16.69}$$

Introducing expression (16.69) into (16.68), we obtain

$$L_\pi = \sum_{a,b=1}^{n-1} M_{ab}\,r_a r_b \geqslant 0.$$

And assuming the Onsager symmetry property, $M_{ab} = M_{ba}$ $(a, b = 1, \cdots, n-1)$, we deduce that coefficients M_{ab} are *associated with a positive definite quadratic form.*

In conclusion, the results are the same as in the classical theory, but moreover we have obtained new constitutive equation (16.69) for the difference of temperatures.

We have considered the simple case of Stokes fluids. If the fluid is non Stokesian, the Navier-Stokes stress tensor of viscosity is given by $(16.16)_1$ where ν is the bulk viscosity. The stress tensor $\mathbf{t}$ becomes

$$\mathbf{t} = -(p_0 + \Pi_\theta)\,\mathbf{I} + \boldsymbol{\sigma} = -p\,\mathbf{I} + 2\,\mu\,\mathbf{D}^D,$$

with

$$p = p_0 + \Pi_\theta + \Pi_\sigma.$$

The nonequilibrium pressure $p - p_0$ is separated into two different parts. The first one $\Pi_\sigma = -\nu \operatorname{div} \mathbf{v}$ is related to the bulk viscosity and the second one $\Pi_\theta = -L_\pi \operatorname{div} \mathbf{v}$ is related to the multi-temperature effect between components.

Due to the non-zero dynamic pressure even for Stokes fluids, we conclude that the model of multi-temperature mixtures of fluids has a great importance. Perhaps such a model may be useful to analyze the evolution of the early universe in which the dynamic pressure seems to be essential [36, 37].

In the present survey, we did not mention a reactive mixture, which is particularly interesting. We here only recall that thermodynamic properties of multi-component reactive mixtures of gases have been widely investigated by Giovangigli and Massot in [38] (see also [39–41]).

References

1. S. Chapman, T.G. Cowling, *The Mathematical Theory of Non-uniform Gases* (Cambridge University Press, Cambridge, 1991)
2. J.M. Burgers, *Flow Equations for Composite Gases* (Academic, New York, 1969)
3. D. Kannappan, T.K. Bose, Transport properties of a two-temperature argon plasma. Phys. Fluids **20**(1), 1668 (1977)
4. D. Kannappan, T.K. Bose, Transport properties of a two-temperature helium plasma. Phys. Fluids **23**(7), 1473 (1980)
5. T.K. Bose, R.V. Seeniraj, Two-temperature noble gas plasmas. I. Thermodynamics and transport coefficients. J. Indian Inst. Sci. **64**(10), 181 (1983)
6. A. Sellitto, V.A. Cimmelli, D. Jou, Influence of electron and phonon temperature on the efficiency of thermoelectric conversion. Int. J. Heat Mass Transf. **80**, 344 (2015)
7. K. Xu, X. He, C. Cai, Multiple temperature kinetic model and gas-kinetic method for hypersonic non-equilibrium flow computations. J. Comput. Phys. **227**, 6779 (2008)
8. C. Truesdell, *Rational Thermodynamics* (McGraw-Hill, New York, 1969)
9. I. Müller, *Thermodynamics* (Pitman, Boston/London, 1985). ISBN: 0-273-08577-8

10. I. Müller, T. Ruggeri, *Rational Extended Thermodynamics*, 2nd edn. (Springer, New York, 1998)
11. I. Müller, A thermodynamic theory of mixtures of fluids. Arch. Ration. Mech. Anal. **28**, 1 (1968)
12. K. Hutter, *Continuum Methods of Physical Modeling* (Springer, New York, 2004)
13. K.R. Rajagopal, L. Tao, *Mechanics of Mixtures* (World Scientific, Singapore, 1995)
14. K. Wilmanski, *Continuum Thermodynamics—Part I: Foundations* (World Scientific, Singapore, 2008)
15. R.J. Atkin, R.E. Craine, Continuum theories of mixtures: basic theory and historical development. Q. J. Mech. Appl. Math. **29**, 209 (1976)
16. T. Ruggeri, S. Simić, On the hyperbolic system of a mixture of eulerian fluids: a comparison between single and multi-temperature models. Math. Methods Appl. Sci. **30**, 827 (2007)
17. H. Gouin, T. Ruggeri, Identification of an average temperature and a dynamical pressure in a multitemperature mixture of fluids. Phys. Rev. E **78**, 016303 (2008)
18. T. Ruggeri, S. Simić, Average temperature and maxwellian iteration in multitemperature mixtures of fluids. Phys. Rev. E **80**, 026317 (2009)
19. T. Ruggeri, Multi-temperature mixture of fluids. Theor. Appl. Mech. **36**, 207 (2009)
20. T. Ruggeri, Some recent results on multi-temperature mixture of fluids, in *Continuous Media with Microstructure*, ed. by B. Albers (Springer, Berlin/Heidelberg, 2010), pp. 39-57. ISBN: 978-3-642-11444-1
21. S. Simić, M. Pavić-Čolić, D. Madjarević, Non-equilibrium mixtures of gases: modelling and computation. Riv. Mat. Univ. Parma 6–1 (2015, in press)
22. T.K. Bose, *High-Temperature Gas Dynamics* (Springer, Berlin, 2003)
23. S.R. de Groot, P. Mazur, *Non-equilibrium Thermodynamics* (North-Holland, Amsterdam, 1963)
24. C. Eckart, The thermodynamics of irreversible processes. II. Fluid mixtures. Phys. Rev. **58**, 269 (1940)
25. L. Onsager, Reciprocal relations in irreversible processes. I. Phys. Rev. **37**, 405 (1931); Reciprocal relations in irreversible processes. II. Phys. Rev. **38**, 2265 (1931)
26. T. Ruggeri, S. Simić, Nonlinear wave propagation in binary mixtures of Euler fluids. Contin. Mech. Thermodyn. **16**, 125 (2004)
27. T. Ruggeri, Global existence, stability and non linear wave propagation in binary mixtures, in *Proceedings of the International Meeting in honour of the Salvatore Rionero 70th Birthday, Napoli 2003*, ed. by P. Fergola, F. Capone, M. Gentile, G. Guerriero (World Scientific, Singapore, 2004), pp. 205–214
28. T. Ruggeri, Some recent mathematical results in mixtures theory of euler fluids, in *Proceedings WASCOM 2003*, ed. by R. Monaco, S. Pennisi, S. Rionero, T. Ruggeri (World Scientific, Singapore, 2004), pp. 441–454
29. J. Lou, T. Ruggeri, Acceleration waves and weak Shizuta-Kawashima condition. Suppl. Rend. Circ. Mat. Palermo "Non Linear Hyperbolic Fields and Waves. A tribute to Guy Boillat" **78**, 187 (2006)
30. E. Ikenberry, C. Truesdell, On the pressure and the flux of energy in a gas according to Maxwell's kinetic theory. J. Ration. Mech. Anal. **5**, 1 (1956)
31. T. Ruggeri, S. Simić, in *Proceedings Mathematical Physics Models and Engineering Sciences* (Liguori Editore, Napoli, 2008), p. 455
32. T. Ruggeri, J. Lou, Heat conduction in multi-temperature mixtures of fluids: the role of the average temperature. Phys. Lett. A **373**, 3052 (2009)
33. T. Ruggeri, S. Simić, in *Asymptotic Methods in Non Linear Wave Phenomena*, ed. by T. Ruggeri, M. Sammartino (World Scientific, Singapore, 2007), p. 186
34. T. Ruggeri, S. Simić, Non-equilibrium temperatures in the mixture of gases via maxwellian iteration (in preparation)
35. I. Gyarmati, *Non-equilibrium Thermodynamics. Field Theory and Variational Principles* (Springer, Berlin, 1970)

36. S.R. de Groot, W.A. van Leeuwen, Ch.G. van Weert, *Relativistic Kinetic Theory* (North-Holland, Amsterdam, 1980)
37. S. Weinberg, Entropy generation and the survival of protogalaxies in an expanding universe. Astrophys. J. **168**, 175 (1971)
38. V. Giovangigli, M. Massot, Entropic structure of multicomponent reactive flows with partial equilibrium reduced chemistry. Math. Methods Appl. Sci. **27**, 739 (2004)
39. F. Conforto, R. Monaco, F. Schürrer, I. Ziegler, Steady detonation waves via the Boltzmann equation for a reacting mixture. J. Phys. A **36**, 5381 (2003)
40. M. Groppi, G. Spiga, Kinetic theory of a chemically reacting gas with inelastic transitions. TTSP **30**(4–6), 305 (2001)
41. F. Conforto, A. Jannelli, R. Monaco, T. Ruggeri, On the Riemann problem for a system of balance laws modelling a reactive gas mixture. Physica A **373**, 67 (2007)

Chapter 17
Shock Structure and Temperature Overshoot in Macroscopic Model of Mixtures

Abstract In this chapter, we study the shock structure in a mixture on the basis of the model of multi-temperature mixtures explained in the previous Chap. 16. For simplicity, the study is restricted to weak and moderately strong shocks in a binary mixture of ideal gases without viscosity and heat conductivity. The model predicts the existence of the temperature overshoot of the heavier constituent, which was also predicted by other sophisticated approaches. This phenomenon is a consequence of weak energy exchange between the constituents, either due to large mass difference, or large rarefaction of the mixture. In the range of small Mach number, it is also shown that the shock thickness (or equivalently, the inverse of Knudsen number) decreases with the increase of the Mach number: a behavior similar to a single fluid.

17.1 Introduction

The aim of this chapter is to present a survey of a recent paper of Madjarević et al. [1] that gives a systematic analysis of the shock structure problem in a binary MT mixture of Euler fluids.

Several studies [2–4] indicated that for certain values of parameters (Mach number, mass fraction, mass ratio, and diameter ratio of the atoms) there appears a temperature overshoot of a heavier constituent—a region within a shock profile where the temperature raises above the terminal temperature of the mixture. The complexity of numerical schemes has prevented massive calculations and detailed study of this aspect of the shock structure.

Even though the model of a mixture of Euler fluids can be regarded as simplified since viscosity and heat conductivity were neglected, the results of the shock structure in Helium-Argon mixture [5] obtained by using the hyperbolic MT model are in good agreement with available experimental data [6]. This put the macroscopic MT model at the same level of accuracy as more sophisticated models of the kinetic theory [7], or of DSMC [8], at least for weak shocks. The main advantage of the MT model is its tractability with only moderate numerical efforts and with the use of standard numerical packages. These facts have stimulated further systematic study of the problem.

T. Ruggeri, M. Sugiyama, *Rational Extended Thermodynamics beyond the Monatomic Gas*, DOI 10.1007/978-3-319-13341-6_17

The main results in this chapter are concerned with the shock thickness and the temperature overshoot of a heavier constituent described in terms of three parameters—mass ratio, Mach number and mass concentration.

17.2 Binary Mixture of Euler Fluids

We consider a binary mixture of ideal monatomic gases without viscosity and heat conductivity. The characteristic velocities are given in (16.32) with sound velocity (16.34) and $\gamma_1 = \gamma_2 = \gamma = 5/3$.

The average temperature of the MT mixture (16.35) becomes, in the present case,

$$(\rho_1 c_{V_1} + \rho_2 c_{V_2})T = \rho_1 c_{V_1} T_1 + \rho_2 c_{V_2} T_2. \tag{17.1}$$

The total pressure and the intrinsic part of the internal energy are in the same form as in the case of a single-component gas:

$$p = p_1 + p_2 = \rho \frac{k_B}{m} T, \qquad \rho\varepsilon_I = \rho_1\varepsilon_1 + \rho_2\varepsilon_2 = \rho \frac{k_B}{(\gamma - 1)m} T,$$

provided that we introduce the average mass $m = m(c)$ and the average temperature T of the mixture in the following form:

$$\frac{1}{m(c)} = \frac{c}{m_1} + \frac{1-c}{m_2}, \qquad T = c\frac{m(c)}{m_1}T_1 + (1-c)\frac{m(c)}{m_2}T_2, \tag{17.2}$$

where c is the concentration related to the mass densities:

$$\rho_1 = \rho c, \quad \rho_2 = \rho(1-c).$$

As in the previous chapter, we introduce the difference of the temperatures $\Theta = T_2 - T_1$, so-called diffusion temperature, and the ratio of the masses of the constituents,

$$\mu = \frac{m_1}{m_2}, \quad 0 < \mu \leqslant 1,$$

where we have assumed $m_1 \leqslant m_2$. The temperatures of the constituents can now be expressed in terms of the new variables T, Θ and c, using μ as a parameter:

$$T_1 = T - f(c)\Theta, \quad T_2 = T + (1 - f(c))\Theta,$$

where auxiliary function $f(c)$ has the following form:

$$f(c) = \frac{\mu(1-c)}{c + \mu(1-c)}.$$

Finally, we also use the diffusion flux $\mathbf{J}$ of the constituent, instead of its velocity $\mathbf{v}_1$. Since relative velocities obey the relation $\rho_1\mathbf{u}_1 + \rho_2\mathbf{u}_2 = 0$, due to (16.4), the diffusion flux of constituent 1 is defined as

$$\mathbf{J} = \rho_1\mathbf{u}_1 = -\rho_2\mathbf{u}_2.$$

Therefore, we take the change of variables

$$(\rho, \mathbf{v}, T, \rho_1, \mathbf{v}_1, T_1) \to (\rho, \mathbf{v}, T, c, \mathbf{J}, \Theta),$$

and rewrite the system of governing equations as if it were a single fluid with extended fields. This is the idea of Ruggeri when he tried to unify the second sound phenomena between gas and crystals [9]. We will see the explicit form in the next section when we write the shock structure equations.

The source terms satisfy the Galilean invariance (16.11):

$$\mathbf{m}_1 = \hat{\mathbf{m}}_1, \quad e_1 = \hat{e}_1 + \hat{\mathbf{m}}_1 \cdot \mathbf{v}, \tag{17.3}$$

and (16.31) gives

$$\hat{\mathbf{m}}_1 = -\psi_{11}\left(\frac{\mathbf{u}_1}{T_1} - \frac{\mathbf{u}_2}{T_2}\right), \quad \hat{e}_1 = -\theta_{11}\left(-\frac{1}{T_1} + \frac{1}{T_2}\right), \tag{17.4}$$

where ψ_{11} and θ_{11} are positive phenomenological coefficients. They can be related to state variables and relaxation times for diffusion τ_v and temperature τ_T (16.44) and (16.47). For a monatomic gas, the ratio between relaxation times (16.48) becomes

$$\frac{\tau_T}{\tau_v} = \frac{m_1 + m_2}{c\,m_2 + (1-c)\,m_1} > 1. \tag{17.5}$$

17.3 Shock Structure Problem

The shock structure, which we discussed in Sect. 3.4, is a continuous solution with steep gradients of state variables in the neighborhood of singular surface—the shock wave—which is diffused due to dissipative mechanisms taken into account. Our attention will be restricted to the shock structure related to plane shocks, moving at constant speed s in the direction orthogonal to the singular surface. Consequently, one space variable, say x, is enough for the problem, and the shock structure is described as a traveling wave solution, depending on a single variable $\varphi = x - st$, which asymptotically connects equilibrium states in front and behind the shock wave. These assumptions transform the model into a set of ordinary differential equations where the velocity, the diffusion flux and the source term (momentum exchange) are described by a single component, i.e., $\mathbf{v} = (v, 0, 0)$, $\mathbf{J} = (J, 0, 0)$ and $\hat{\mathbf{m}}_1 = (\hat{m}_1, 0, 0)$, in Cartesian coordinates.

17.3.1 Dimensionless Shock Structure Equations

The problem of the shock structure is studied in dimensionless form. For this purpose, we introduce the dimensionless variables by scaling the state variables and independent variable φ with appropriate upstream (unperturbed) equilibrium values, indicated by the subscript 0:

$$\tilde{\rho} = \frac{\rho}{\rho_0}, \; \tilde{u} = \frac{u}{a_0}, \; \tilde{T} = \frac{T}{T_0}, \; \tilde{J} = \frac{J}{\rho_0 a_0}, \; \tilde{\Theta} = \frac{\Theta}{T_0}, \; \tilde{\varphi} = \frac{\varphi}{l_0}, \; M_0 = \frac{u_0}{a_0}, \tag{17.6}$$

where $u = v - s$ is the relative velocity of a mixture with respect to the shock wave, l_0 is the upstream reference length and $a_0 = \{\gamma (k_B/m_0) T_0\}^{1/2}$ is the upstream speed of sound; $m_0 = m(c_0)$ is the equilibrium average mass of the mixture and M_0 is the upstream Mach number. For the sake of simplicity, tilde will be dropped in the sequel.

The upstream reference length l_0 is usually taken as the mean free path of the atoms. The average mean free path in the mixture will be expressed in terms of other more primitive properties of the constituents [10]:

$$l_0 = \frac{n_1}{n}(l_1)_0 + \frac{n_2}{n}(l_2)_0, \tag{17.7}$$

$$(l_\alpha)_0 = \frac{1}{\pi d_{12}^2}\left[n_1\left(1 + \frac{m_\alpha}{m_1}\right)^{1/2} + n_2\left(1 + \frac{m_\alpha}{m_2}\right)^{1/2}\right]^{-1},$$

where n_α are number densities of the constituents, and $n = n_1 + n_2$ is the mixture number density.

Using the scaled variables (17.6), we obtain the following set of dimensionless equations:

$$\begin{aligned}
&\frac{d}{d\varphi}(\rho u) = 0, \\
&\frac{d}{d\varphi}\left(\rho u^2 + \frac{1}{\gamma}\frac{m_0}{m}\rho T + \frac{J^2}{\rho c(1-c)}\right) = 0, \\
&\frac{d}{d\varphi}\left\{\left(\frac{1}{2}\rho u^2 + \frac{1}{\gamma - 1}\frac{m_0}{m}\rho T + \frac{J^2}{2\rho c(1-c)}\right)u + \left(\frac{uJ}{\rho c(1-c)} + \frac{1}{\beta}\right)J\right\} = 0, \\
&\frac{d}{d\varphi}(\rho c u + J) = 0,
\end{aligned} \tag{17.8}$$

$$\frac{d}{d\varphi}\left\{\rho c u^2 + \frac{J^2}{\rho c} + 2uJ + \frac{1}{\gamma}\frac{m_0}{m_1}\rho c\,(T - f(c)\Theta)\right\} = -\frac{l_0}{\tau_D a_0} m_\mu(T, c, \Theta) J,$$

$$\frac{d}{d\varphi}\left\{\left(\frac{1}{2}\rho c\left(u + \frac{J}{\rho c}\right)^2 + \frac{1}{\gamma - 1}\frac{m_0}{m_1}\rho c\,(T - f(c)\Theta)\right)\left(u + \frac{J}{\rho c}\right)\right\}$$

$$= -\frac{l_0}{\tau_D a_0} m_\mu(T, c, \Theta) J u + \frac{l_0}{\tau_T a_0} e_\mu(\rho, T, c, \Theta)\Theta.$$

Equation $(17.8)_{1-3}$ represent conservation laws of mass, momentum and energy of the mixture, while $(17.8)_{4-6}$ are the balance laws of mass, momentum and energy of the constituent 1. Auxiliary functions in source terms read

$$m_\mu(T, c, \Theta) = \frac{T + [1 - c - f(c)]\Theta}{[T - f(c)\Theta][T + (1 - f(c))\Theta]} T,$$

$$e_\mu(\rho, T, c, \Theta) = \frac{1}{\gamma(\gamma - 1)}\frac{m_0}{m_1}\frac{m}{m_2}\frac{\rho c(1 - c)T^2}{[T - f(c)\Theta][T + (1 - f(c))\Theta]}$$

and ratios of masses can be expressed as

$$\frac{m_0}{m} = \frac{c + \mu(1 - c)}{c_0 + \mu(1 - c_0)}, \quad \frac{m_0}{m_1} = \frac{1}{c_0 + \mu(1 - c_0)}, \quad \frac{m}{m_2} = \frac{\mu}{c + \mu(1 - c)}.$$

In (17.8) we have also used abbreviation $1/\beta$ for the following expression:

$$\frac{1}{\beta} = \frac{1}{\gamma - 1}\left[\frac{m_0}{m_1}(1 - \mu)T - \frac{m_0}{m_1}\frac{m}{m_2}\Theta\right] + \frac{J^2}{2\rho^2}\left[\frac{1}{c^2} - \frac{1}{(1 - c)^2}\right],$$

where β can be interpreted as thermal inertia [11].

17.3.2 Boundary Conditions and Numerical Procedure

By using the procedure described in Sect. 3.4, the shock structure problem can be approached with the use of dynamical-system theory. The system (17.8) can be seen formally in the form (3.38), where $\mathbf{u} = (\rho, u, T, c, J, \Theta)^T$ is the column vector of state variables.

Mere observation of the source terms in (17.8) yields that the diffusion flux and the diffusion temperature vanish both in upstream and downstream stationary points, i.e., $J_0 = J_1 = 0$, $\Theta_0 = \Theta_1 = 0$. Thus, for a given upstream equilibrium state $\mathbf{u}_0$, one may determine the downstream equilibrium state $\mathbf{u}_1$ by integration of the conservative part of the system $(17.8)_{1-4}$. Nontrivial solution in dimensionless form reads

$$\mathbf{u}_0 = \begin{bmatrix} \rho_0 \\ u_0 \\ T_0 \\ c_0 \\ J_0 \\ \Theta_0 \end{bmatrix} = \begin{bmatrix} 1 \\ M_0 \\ 1 \\ c_0 \\ 0 \\ 0 \end{bmatrix}, \quad \mathbf{u}_1 = \begin{bmatrix} \rho_1 \\ u_1 \\ T_1 \\ c_1 \\ J_1 \\ \Theta_1 \end{bmatrix} = \begin{bmatrix} \frac{4M_0^2}{3+M_0^2} \\ \frac{3+M_0^2}{4M_0} \\ \frac{1}{16}\left(14 - \frac{3}{M_0^2} + 5M_0^2\right) \\ c_0 \\ 0 \\ 0 \end{bmatrix}. \tag{17.9}$$

Note that relations between mixture state variables ρ_1, u_1, T_1 and ρ_0, u_0, T_0, correspond to the solution of the usual Rankine-Hugoniot relations between the state variables at the shock wave for a single fluid. On the other hand, constituent-related state variables c, J and Θ have the same equilibrium values in front and behind the shock. Since the diffusion flux J and the diffusion temperature Θ vanish in equilibrium, they can be regarded as genuine nonequilibrium variables. Also, concentration is the same in both equilibrium states, $c_1 = c_0$, and in the sequel it will be termed equilibrium concentration, without special regard to upstream or downstream state. Finally, downstream equilibrium can be regarded as a one-parameter family of states parametrized by the Mach number (i.e., shock speed: see (3.49)), $\mathbf{u}_1 = \mathbf{u}_1(\mathbf{u}_0, M_0)$. In this way, the third parameter—Mach number—is naturally introduced in the model through the boundary conditions.

The problem can be approached numerically in two different ways: as a boundary-value problem, or as an initial-value problem. We adopt the latter way.

The numerical procedure has a few crucial points. First, it was shown [12] that stationary points change their character when the shock parameter—Mach number—crosses certain critical value. The critical value, $M_0 = 1$, corresponds to the highest characteristic speed of the equilibrium subsystem, here consisted of the conservation laws for the mixture. Actually, if the highest characteristic speed is genuinely nonlinear (3.17), it was shown [13] that there is a single eigenvalue of the system (17.8), linearized at stationary point, which changes the sign. Furthermore, upstream equilibrium $\mathbf{u}_0$ is a saddle point (in a generalized sense), while downstream equilibrium $\mathbf{u}_1$ is a stable node. Thus, one may "follow" the direction of the eigenvector corresponding to a positive eigenvalue in $\mathbf{u}_0$ and asymptotically reach the stable stationary point $\mathbf{u}_1$. Since the domain on which the heteroclinic orbit is defined is infinite, $-\infty < \varphi < \infty$, the integration should be performed on a truncated domain, $\varphi \in [\varphi_0, \varphi_1]$. This domain should be sufficiently large so as to secure the condition that terminal values $\mathbf{u}(\varphi_0)$ and $\mathbf{u}(\varphi_1)$ lie in small neighborhoods of the stationary points $\mathbf{u}_0$ and $\mathbf{u}_1$, respectively.

17.3.3 Profile of Shock Structure

The procedure described above can be applied if the solution of the shock structure equations exists. To analyze the existence of solution in binary mixture we write down the characteristic velocity (16.32) of the full system in equilibrium, where both constituents have common velocity v and temperature T. They read (in increasing order)

$$\lambda^{(1)} = v - \sqrt{\gamma \frac{k_B}{m_1} T}, \quad \lambda^{(2)} = v - \sqrt{\gamma \frac{k_B}{m_2} T},$$
$$\lambda^{(3)} = \lambda^{(4)} = 0,$$
$$\lambda^{(5)} = v + \sqrt{\gamma \frac{k_B}{m_2} T}, \quad \lambda^{(6)} = v + \sqrt{\gamma \frac{k_B}{m_1} T},$$

while characteristic speeds of equilibrium subsystem are (in increasing order)

$$\mu^{(1)} = v - \sqrt{\gamma \frac{k_B}{m_0} T}, \quad \mu^{(2)} = 0, \quad \mu^{(3)} = v + \sqrt{\gamma \frac{k_B}{m_0} T}.$$

By the Theorem 2.3 [14], the subcharacteristic conditions hold

$$\lambda^{(1)} < \mu^{(1)}, \quad \mu^{(3)} < \lambda^{(6)}.$$

Admissible equilibrium states (17.9) are determined by the Lax condition $\mu_1^{(1)} < s < \mu_0^{(1)}$ (or $\mu_0^{(3)} < s < \mu_1^{(3)}$), which can be equivalently expressed as $M_0 > 1$ (or $M_0 < -1$). However, non-existence of the smooth shock structure (Theorem 3.1) is related to the condition $s < \lambda_0^{(1)}$ (or $s > \lambda_0^{(6)}$). Non-existence condition then reads

$$s < \lambda_0^{(1)} \quad \Leftrightarrow \quad M_0 > M_{0\text{crit}}^{(0)} = \sqrt{\frac{1}{c_0 + (1 - c_0)\mu}},$$

which corresponds to regions II and III in parameter space shown in Fig. 17.1. The following situation may occur:

$$\lambda_0^{(1)} < s < \lambda_1^{(2)} \quad \Leftrightarrow \quad M_{0\text{crit}}^{(0)} > M_0 > M_{0\text{crit}}^{(1)} = \sqrt{\frac{4\mu + 3(1-\mu)c_0}{4\mu - (1-\mu)c_0}},$$

which corresponds to region IV in Fig. 17.1. In this case the singularity which appears in downstream equilibrium moves into the domain as M_0 is increased, but does not prevent the existence of the smooth shock structure since it is a regular

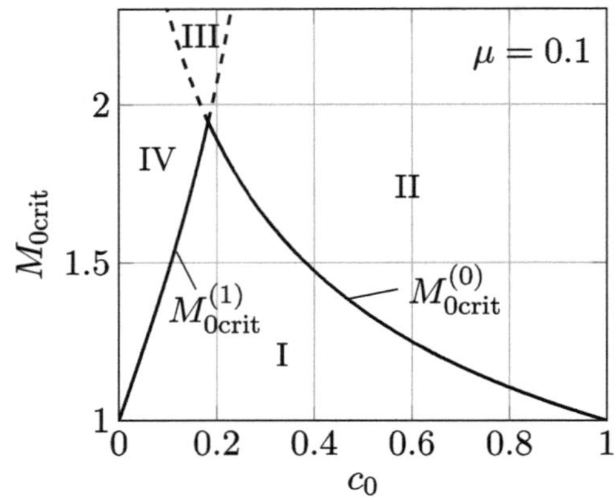

Fig. 17.1 Critical values of Mach number M_0 and unperturbed concentration c_0: possible regions

singularity [15]. The proof of this statement in the case of mixtures is neither at our disposal at the moment, nor it can be proved numerically by the initial value strategy described above.

In this study we restrict our attention to cases in which $M_0 < M_{0\text{crit}} = \min\{M^{(0)}_{0\text{crit}}, M^{(1)}_{0\text{crit}}\}$ and numerical solution can be obtained as the solution of initial value problem in the different regions evaluated in Fig. 17.1.

17.4 Shock Structure and Temperature Overshoot

Temperature overshoot is one of the peculiarities of the shock structure in mixtures whose constituents have disparate masses. It manifests through existence of the region of non-zero width where the temperature of one constituent raises above the terminal, i.e., downstream equilibrium temperature of the mixture. This phenomenon was observed in numerical calculations based on Boltzmann equations for mixtures [2, 3, 16] and DSMC [8]. Available experimental data do not provide enough evidence to support numerical simulations, although Harnett and Muntz [6] regard the overshoot of the parallel temperature of Argon as the onset of the overshoot of its mean temperature.

In the studies mentioned above, it was emphasized that the temperature overshoot is the most significant in the case of small molar fraction of heavier component. Abe and Oguchi [7] offered a physical explanation of this phenomenon. They stated that, in the case of vanishingly small mole fraction of heavier component, the main structure of the shock wave is determined by the lighter one. This causes the deceleration of heavier component and, at the same time, conversion of kinetic into thermal energy. However, dissipation through conduction is slow process which cannot diffuse thermal energy gained by deceleration. As a consequence, the internal energy (temperature) of heavier component is raised above the terminal one. In our model, momentum and energy transfer through viscosity and heat conduction are neglected. We are focused on dissipation caused by mutual exchange of momentum and energy between the constituents, where the most prominent role is played by their mass ratio μ. Thus, we examine the temperature overshoot from this perspective, analyzing its dependence on the mass ratio μ, as well as upstream Mach number M_0 and equilibrium concentration c_0.

The numerical solution of the present model predicts the temperature overshoot. Figure 17.2 is an example of this effect. It has an outstanding feature, not reported in previous studies, that the temperature overshoot varies non-monotonically with mass ratio. Namely, there exists a value μ^* of the mass ratio which determines the local minimum of temperature overshoot. Since other studies were based on limited number of numerical simulations, which provided information on certain particular cases only, this phenomenon remained unobserved thus far.

The outstanding feature of non-monotonic behavior of the temperature overshoot can be understood as follows: (1) For $\mu < \mu^*$, the temperature overshoot is increased due to large mass difference and low Knudsen number (Kn). The flow is between hydrodynamic and slip flow regime, but the mass ratio is too small to yield sufficient exchange of energy between the constituents (2) For $\mu > \mu^*$, Kn is increased, which puts the flow into transition regime. Although the masses of the constituents become comparable, the exchange of energy is prevented by rarefaction of the mixture, i.e., small number of cross-collisions which could cause it. Consequently, the temperature of heavier constituent cannot be attenuated, and temperature overshoot is increased.

Therefore, in a simplified model of MT mixtures, where viscosity and heat conductivity are neglected, small mutual exchange of energy between the constituents can be pointed out as main physical reason for the increase of temperature overshoot. It can occur for two reasons: (a) large mass discrepancy between the constituents (small μ), and (b) more rarefaction of a mixture.

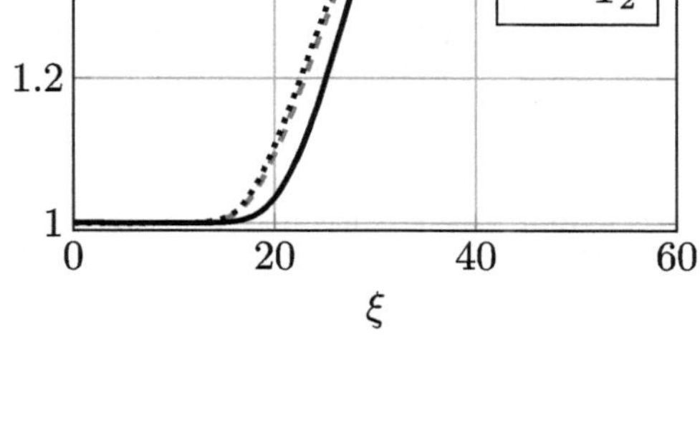

Fig. 17.2 Temperature profiles in the shock structure (T—average temperature of the mixture, T_1—temperature of the lighter constituent, T_2—temperature of the heavier constituent): $M_0 = 1.6$, $c_0 = 0.21$, $\mu = 0.1$

17.5 Shock Thickness and the Knudsen Number

One of the parameters which describe the shock structure globally is the shock thickness. It is usually defined as follows:

$$\delta = \left| \frac{v_1 - v_0}{(dv/d\varphi)_{\max}} \right| .$$

It is important to notice that the dimensionless shock thickness is equal to the reciprocal of the Knudsen number:

$$\tilde{\delta} = \frac{\delta}{l_0} = \frac{1}{\mathrm{Kn}}, \tag{17.10}$$

whose value helps to distinguish between different flow regimes. In view of (17.10), shock thickness will carry also the information about the flow regime. In this section we analyze the shock thickness (and Kn) in terms of mass ratio, Mach number and upstream concentration.

The dependence of Kn on the Mach number is monotonous for fixed mass ratio and upstream concentration. It increases with the increase of the Mach number, which amounts to a decrease of the shock thickness. This is rather expected result which is similar to the behavior of a single-component gas. However, experimental facts about shock structure in a single fluid, as well as comparative study based on Navies-Stokes Fourier model, reveals that this tendency seem to be opposite for larger Mach numbers (see [17]). Since our calculations are confined to small Mach number flows, at most $M_0 \leqslant 2.0$, the results obtained here are in agreement with the single fluid model in this range.

More details can be found in the original paper [1]. In this paper, a comparison between the single and multi-temperature models is also made.

References

1. D. Madjarević, T. Ruggeri, S. Simić, Shock structure and temperature overshoot in macroscopic multi-temperature model of mixtures. Phys. Fluids **26**, 106102 (2014)
2. S. Kosuge, K. Aoki, S. Takata, Shock-wave structure for a binary gas mixture: finite-difference analysis of the boltzmann equation for hard sphere molecules. Eur. J. Mech. B Fluids **17**, 87 (2001)
3. A. Raines, Study of a shock wave structure in gas mixtures on the basis of the boltzmann equation. Eur. J. Mech. B/Fluids **21**, 599 (2002)
4. G.A. Bird, Aspects of the structure of strong shock waves. Phys. Fluids **13**, 1172 (1970)
5. D. Madjarević, S. Simić, Shock structure in helium-argon mixture—a comparison of hyperbolic multi-temperature model with experiment. Europhys. Lett. **102**, 44002 (2013)
6. L.N. Harnet, E. Muntz, Experimental investigation of normal shock wave velocity distribution functions in mixtures of argon and helium. Phys. Fluids **10**, 565 (1972)

7. K. Abe, H. Oguchi, An analysis of shock waves in binary gas mixtures with special regard to temperature overshoot, Report No. 511. Institute of Space and Aeronautical Science, University of Tokyo (1974)
8. G. Bird, The structure of normal shock waves in a binary gas mixture. J. Fluid Mech. **31**, 657 (1968)
9. T. Ruggeri, The binary mixtures of euler fluids: a unified theory of second sound phenomena, in *Continuum Mechanics and Applications in Geophysics and the Environment*, ed. by B. Straughan, R. Greve, H. Ehrentraut, Y. Wang (Springer, Berlin, 2001), pp. 79–91
10. G. Bird, *Molecular Gas Dynamics and the Direct Simulation of Gas Flows* (Clarendon, Oxford, 1994)
11. T. Ruggeri, S. Simić, On the hyperbolic system of a mixture of eulerian fluids: a comparison between single and multi-temperature models. Math. Methods Appl. Sci. **30**, 827 (2007)
12. S. Simić, Shock structure in continuum models of gas dynamics: stability and bifurcation analysis. Nonlinearity **22**, 1337 (2009)
13. S. Simić, Shock structure in the mixture of gases: stability and bifurcation of equilibria, in *Thermodynamics – Kinetics of Dynamic Systems*, Chap. 8, ed. by J.C.M. Pirajãn (InTech, Rijeka, 2011)
14. G. Boillat, T. Ruggeri, Hyperbolic principal subsystems: entropy convexity and subcharacteristic conditions. Arch. Ration. Mech. Anal. **137**, 305 (1997)
15. W. Weiss, Continuous shock structure in extended thermodynamics. Phys. Rev. E **52**, R5760 (1995)
16. R. Monaco, Shock-wave propagation in gas mixtures by means of a discrete velocity model of the boltzmann equation. Acta Mech. **55**, 239 (1985)
17. I. Müller, T. Ruggeri, *Rational Extended Thermodynamics*, 2nd edn. (Springer, New York, 1998)

Part VIII
Maxwellian Iteration and Objectivity

Chapter 18
Hyperbolic Parabolic Limit, Maxwellian Iteration and Objectivity

Abstract In this chapter, we discuss the parabolic limit of extended thermodynamics via the Maxwellian iteration, and we observe that the usual constitutive equations, which are nonlocal in space, are approximations of some balance laws of ET when some relaxation times are negligible. An important consequence is that these equations need not satisfy the objectivity principle. To avoid misunderstanding, we should mention that the principle still continues to be valid for constitutive equations.

We also discuss the point that, under suitable assumptions, the conditions dictated by the entropy principle in the hyperbolic case guarantee the validity of the entropy principle also in the parabolic limit. Lastly we express our opinion concerning the limitation of the parabolic regularized version of ET theories.

18.1 Different Constitutive Equations

We have seen that the physical laws in continuum theories are expressed by the *balance laws*, which, under regularity conditions, assume the form (1.2).

In order to have a closed system, we need *constitutive equations*. A very rough mathematical definition of constitutive equations may be considered as the equations that are necessary to close the system. That is, choosing an independent field $\mathbf{u} \in R^N$, we have to give relations between the $5N$ components of the vectors $\mathbf{F}^0, \mathbf{F}^i, \mathbf{f}$ and the N components of the unknown vector $\mathbf{u}$. But of course, as we will see below, this definition has no physical meaning because the additional equations must represent the real constitutive properties of the material.

For a long time, constitutive equations have been made in an empirical way. They belong substantially to the one of the following three big classes:

- *Local constitutive equations* Examples are:
  - stress-strain relation in non-linear elasticity: $\mathbf{t} \equiv \mathbf{t}(\mathbf{E})$ (Hooke's law in the linear case).

T. Ruggeri, M. Sugiyama, *Rational Extended Thermodynamics beyond the Monatomic Gas*, DOI 10.1007/978-3-319-13341-6_18

- The caloric and thermal equations of state in Euler fluids that express the internal energy and the pressure as functions of the mass density and the temperature:

$$\varepsilon \equiv \varepsilon(\rho, T), \quad p \equiv p(\rho, T).$$

- All rational extended thermodynamic theories in which we assume the local dependence of the vectors of the density, fluxes and productions (2.1).

Introducing such constitutive equations into the balance laws, we obtain a differential system which, in general, is hyperbolic.

- *Non-local type (in space)* In the case of a one-component dissipative fluid, examples are:

  - Fourier's law: $(1.10)_3$.
  - Navier-Stokes' law: $(1.10)_{1,2}$.
  - In the case of a mixture of dissipative fluids with n constituents, well-known examples are the Navier-Stokes', Fourier's, and Fick's laws (16.16).

  When we introduce such constitutive equations into the balance laws, we obtain a system of differential equations where some spatial derivatives are of second order and the time derivatives are of first order. These differential systems have a parabolic structure.

- *Non-local type (in time)* Examples are as follows: visco-elastic materials or, in general, all materials in which the stress depends not only on the present deformation but also on the history of the deformation (constitutive equations with memory). Except for the case of exponential memory kernel, the mathematical structure of such systems is of integro-differential type.

18.2 Frame-Dependence of the Heat Flux

In the modern constitutive theory, all the constitutive equations must obey two universal principles that are the first two in the Axioms of RET (see Sect. 2.2): *objectivity principle* and *entropy principle*.

A long debate came out in the literature after Ingo Müller published a famous paper [1] in which he proved that the Fourier and Navier-Stokes "constitutive" equations (1.10) violate the objectivity principle. At that time Müller was convinced that his result indicates that the objectivity principle is not a valid principle. And then a huge literature appeared between supporter and non-supporter of the objectivity principle. Several authors added artificial time derivatives to try to recover the objectivity for the heat equation and for the stress. Here we record observations on the subject made independently by Bressan [2] and Ruggeri [3]. They observed that a possible interpretation of Müller's result is that the objectivity principle is indeed universal, but the Fourier and Navier-Stokes laws are not "true" constitutive equations.

The precise and convincing answer was presented by rational extended thermodynamics. In fact we have seen in Sect. 5.3.6 that the Navier-Stokes Fourier laws are approximations of the balance laws of 14 moment, and therefore these are not constitutive equations. It is not necessary for the laws to satisfy the frame indifference principle.

18.2.1 Maxwellian Iteration and the Parabolic Limit

To reveal the relationship between extended and classical models, a formal iterative scheme known as the Maxwellian iteration is applied [4, 5]. In general, the first iterates are obtained from the right-hand sides of balance laws by putting the "zero-*th*" iterates—equilibrium values—into the left-hand sides. The second iterates are obtained from the right-hand sides by putting the first iterates into the left-hand sides, and so on. Therefore the Maxwellian iteration is substantially composed of (i) an identification of the relaxation times and (ii) a formal power expansion of the solution in terms of the relaxation times: a sort of Chapman–Enskog procedure at macroscopic level.

We proved, in Sect. 5.3.6 (5.61), that:

Statement 13 *The Fourier and Navier Stokes laws (1.10) are the first order approximation of the Maxwellian iteration of the ET balance law system (5.59). Therefore they are not true constitutive equations and need not satisfy the objectivity principle.*

A similar situation exists in the case of a mixture of fluids holding the Fick law for the mass diffusion. In fact as we have seen in Sect. 16.7 that

Statement 14 *The Fick law (16.61) is the first approximation in the Maxwellian iteration of momentum balance equation of each species* $(16.39)_5$.

Moreover

Statement 15 *The new non-local "constitutive equation" for the temperature differences (16.62) is the first approximation obtained by the Maxwellian iteration of the energy balance of each constituent* $(16.39)_6$.

Another simple example in the context of the mixture theory is well-known Darcy's law for porous media saying that the relative velocity between the fluid part $\mathbf{v}_F$ and the solid one $\mathbf{v}_S$ is proportional to the pressure gradient in the fluid (see, e.g. [6]):

$$\nabla p_F = -\frac{k}{\mu}\left(\mathbf{v}_S - \mathbf{v}_F\right), \tag{18.1}$$

where k and μ are, respectively, the permeability and the viscosity.

Statement 16 *Darcy's law (18.1) is an approximation to the balance law of the linear momentum for the fluid flowing through the porous solid being treated as a rigid body, i.e., (18.1) is a limit case of (see [7]):*

$$\rho_F \, \dot{\mathbf{v}}_F + \nabla \, p_F = -\frac{k}{\mu} \left(\mathbf{v}_S - \mathbf{v}_F \right).$$

We have seen that, using the Maxwellian iteration, we can obtain from extended thermodynamics—at least formally—the usual non-local constitutive equations of the classical theory. Therefore the parabolic systems of classical theories appear, from a physical point of view, to be approximations of the corresponding hyperbolic systems when some relaxation times are negligible:

- Navier Stokes' and Fourier's laws as a limit case of momentum and energy balance equations in extended thermodynamics.
- Fick's law as a limit case of momentum equations of each species in a mixture with single-temperature.
- The new diffusion equation for the difference of the temperatures in mixture with multi-temperature as a limit case of the energy balance equation of each constituent.
- Darcy's law for porous material is a limit case of momentum equation.

Although the previous non-local equations are not constitutive equations but approximations of balance laws, the non-local equations have been very useful. In many applications, the relaxation times are sometimes negligibly small and the non-local equations are relevant in such situations. The advantage of the non-local approximation is that, in this limit, we are able to measure non-observable quantities like heat flux, viscous stress and, in particular, velocity and temperature of each species in a mixture of fluids using the classical constitutive equations.

An interesting analysis on constitutive equations can be read on the recent book [8].

18.3 Maxwellian Iteration and the Entropy Principle

Clearly a major open problem in this framework is the rigorous proof of the convergence of the solutions via Maxwellian iterations. To make a little step toward this proof, first of all, we have to focus our attention on another very subtle point: is the entropy principle preserved in the Maxwellian iteration scheme? In other words: if the "full" hyperbolic theory satisfies the entropy inequality, are we sure that the corresponding parabolic limit satisfies automatically a suitable entropy inequality?

Here we give some results due to Ruggeri [9] in one space-dimension (but the results remain valid in any space-dimension):

Theorem 18.1 *If the system of balance laws of ET is endowed with a convex entropy and the processes are not far from equilibrium, the entropy principle is preserved in the Maxwellian iteration. And as a consequence, if the original* *hyperbolic system is entropic, the parabolic limiting system is also entropic. Instead, for processes far from equilibrium, in general, this is not true.*

Proof Let us consider, as in all RET theories, that the system is split into two blocks of M conservation equations (2.26) and of $N - M$ balance laws (2.27). And, with the notation $\mathbf{u}' \equiv (\mathbf{v}, \mathbf{w})$ (we omit the prime), it was proved in Sect. 2.5 that the equilibrium manifold in the main field components is the hyperplane $\mathbf{w} = 0$ [see (2.29)].

In extended thermodynamics, there are many interesting cases where the processes are not far from equilibrium. In this case, we have

$$h' = h'^0 = \frac{1}{2}\mathbf{K}(\mathbf{v})\mathbf{w} \cdot \mathbf{w} + h'_{eq}(\mathbf{v}), \tag{18.2}$$

and therefore if we consider one space-dimension, the symmetric system (2.11), which is linear in $\mathbf{w}$, assumes the form:

$$\mathbf{H}\, \partial_t \mathbf{v} + \mathbf{A}\, \partial_x \mathbf{v} + \mathbf{B}\, \partial_x \mathbf{w} = 0, \tag{18.3}$$

$$\mathbf{K}\, \partial_t \mathbf{w} + \mathbf{B}^T \partial_x \mathbf{v} + \mathbf{C}\, \partial_x \mathbf{w} = -\mathbf{L}\mathbf{w}, \tag{18.4}$$

where $\mathbf{v} \in R^M$, $\mathbf{w} \in R^N$, $\mathbf{H} \equiv h'_{\mathbf{vv}}$ $(M \times M) \in Sym^+$ (symmetric positive definite matrix), $\mathbf{A} \equiv k_{\mathbf{vv}}$ $(M \times M) \in Sym$ (symmetric matrices), $\mathbf{B} \equiv k_{\mathbf{vw}}$ $(M \times N)$, $\mathbf{C} \equiv k_{\mathbf{ww}}$ $(N \times N) \in Sym$ and $k = h'^1(\mathbf{v}, \mathbf{w})$. Moreover $\mathbf{L} \equiv \mathbf{L}(\mathbf{v})$ $(N \times N) \in Sym^+$ because of the residual inequality $\Sigma = -\mathbf{L}\mathbf{w} \cdot \mathbf{w} \leqslant 0$.

According with the Maxwellian iteration procedure Sect. 18.2.1, the first iterates are obtained from the right-hand sides of balance laws by putting the "zero-*th*" iterates—equilibrium values—on the left-hand sides, i.e., putting $\mathbf{w} = 0$ on the left side of (18.4) we have

$$\bar{\mathbf{B}}^T \partial_x \mathbf{v} = -\mathbf{L}\mathbf{w}^{(1)},$$

where $\bar{\mathbf{B}} \equiv \mathbf{B}(\mathbf{v}, \mathbf{0})$. Then substituting this into (18.3) we obtain as the first Maxwellian iteration (taking only the principal part, i.e., we do not consider first-order spatial derivative):

$$\mathbf{H}\, \partial_t \mathbf{v} \simeq \mathbf{D}\, \partial_{xx} \mathbf{v}, \tag{18.5}$$

where the diffusion matrix $\mathbf{D}$ is given by

$$\mathbf{D} = \bar{\mathbf{B}}\mathbf{L}^{-1}\bar{\mathbf{B}}^T \in Sym^+. \tag{18.6}$$

Consequently the entropy principle is preserved in the passage from the hyperbolic system to the parabolic limit.

In the general case, however, the entropy principle is not necessarily preserved. In this case, instead of (18.3) and (18.4), we have

$$\mathbf{H}\,\partial_t \mathbf{v} + \mathbf{G}\,\partial_t \mathbf{w} + \mathbf{A}\,\partial_x \mathbf{v} + \mathbf{B}\,\partial_x \mathbf{w} = 0, \tag{18.7}$$

$$\mathbf{G}^T\,\partial_t \mathbf{v} + \mathbf{K}\partial_t w + \mathbf{B}^T \partial_x \mathbf{v} + \mathbf{C}\,\partial_x \mathbf{w} = -\mathbf{L}\mathbf{w} \tag{18.8}$$

with $\mathbf{G} \equiv h'_{\mathbf{vw}}$ $(M \times N)$ and $\mathbf{K} \equiv h'_{\mathbf{ww}}$ $(N \times N) \in Sym^+$. After the Maxwellian iteration, we obtain a parabolic system similar to (18.5) but with a diffusion matrix $\mathbf{D}$:

$$\mathbf{D} = \bar{\mathbf{B}}\mathbf{L}^{-1}\bar{\mathbf{B}}^T - \bar{\mathbf{B}}\mathbf{L}^{-1}\bar{\mathbf{G}}^T\bar{\mathbf{H}}^{-1}\bar{\mathbf{A}}. \tag{18.9}$$

This matrix, in general, does not have a definite sign. Therefore the most general system for which the entropy principle is preserved in the parabolic limit is the following one: The matrix $\mathbf{D}$ of the system given by (18.9) belongs to Sym^+. We call these special systems *entropy-principle preserving systems.*

A simple entropy-principle preserving system is the one for which h' is given by the sum of two functions, one is the function of only $\mathbf{v}$ and the other is the function of only $\mathbf{w}$. An interesting case of the processes not far from equilibrium (18.2) is a special class of this system. In fact in this case we have $\mathbf{G} \equiv 0$ and the diffusion matrix (18.9) reduces to (18.6) and becomes symmetric positive definite. This result seems to indicate that the parabolic classical theories have a limiting validity only near equilibrium according with the assumption of local equilibrium!

18.4 Regularized System and Non-subshock Formation

In the previous section we considered the full Maxwellian iteration for all relaxation times assuming implicitly that all of them have the same order of magnitude. In this section, we consider relaxation times with different order of magnitude.

An interesting idea is the one proposed by Torrilon and Struchtrup [10] who have constructed a sophisticated method to obtain parabolic extended systems called *regularized* systems from the hyperbolic systems of ET. Philosophy in its construction, however, seems to be similar to the Maxwellian iteration. By this method, they were able to derive, in the case of monatomic gases, a regularized 13 Grad system that is a natural parabolic extension of the NSF system of equations. As the regularized system contains first-order derivatives in time and, in some equations, second-order derivatives in space, it is of parabolic type. This fact is in sharp contrast to the hyperbolic 13 Grad system. To obtain the regularized 13 Grad system, the authors start from the ET system with 23 moments. Therefore we can regard the regularized parabolic Grad system as a sort of approximation of the hyperbolic ET23. By the same procedure, the 14-moment system (5.75) for rarefied

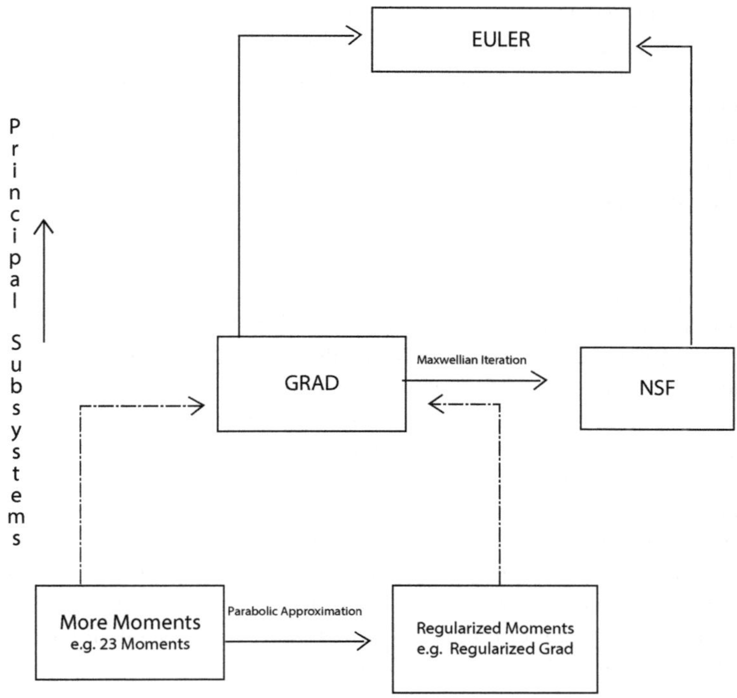

Fig. 18.1 Systems and sub-systems and the parabolic limit

polyatomic gases was regularized by Rahimi and Struchtrup in [11] starting from ET with 36 moments.

In Fig. 18.1, the relationship between the different models is sketched: We can see that the Euler-fluid sub-system is effectively a particular case of the Navier-Stokes Fourier system. Nevertheless the Euler-fluid subsystem is also a principal sub-system of the Grad system. And moreover the Navier-Stokes Fourier system itself is a particular case of the Grad system when the Maxwellian iteration is used. This situation is also valid if we take the ET theory with many moments. When we adopt the regularization approximation, we have, as parabolic counterpart, the regularized moment-equations. In any case, given an N-moment hyperbolic system, all the previous systems of moments are principal sub-systems according to the general *nesting structure* due to Boillat and Ruggeri [12]. Moreover we can see that all the parabolic limits can be considered as approximations of the hyperbolic system as sketched in Fig. 18.1.

The regularized extended system appears very interesting. But one of the main assertions that the parabolization of a system of equations avoids the formation of sub-shock in the shock structure solution, in the present authors' opinion, leads to a misunderstanding. For example, if we accept that the NSF theory is an approximation of the 14-moment hyperbolic system, the validity of the theory cannot exceed the validity of the 14-moments system. Therefore, from the Table 10.2, we notice that, for 14 moments, there appears a sub-shock at the critical Mach number depending on D but in the range $1.34 \leqslant M_{crit} \leqslant 1.65$. The upper and lower bounds correspond to $D \to \infty$ and $D \to 3$, respectively. Therefore we cannot expect that the NSF theory can be valid for any Mach number but we can expect at most until the critical Mach number of the 14-moment theory even though no sub-shock formation exists in the theory. The same is true for the regularized theory. Its validity cannot exceed the critical Mach number of the starting hyperbolic ET theory.

We have a clear explanation to the no-evidence of the sub-shock in experiments thanks to the mathematical properties of the nesting theories of RET. In fact, for a fixed truncation order N, the validity of the ET theory, as far as shock waves are concerned, is up to the Mach number where the shock velocity s reaches the maximum characteristic velocity evaluated in equilibrium. Beyond this limitation, from the Boillat and Ruggeri Theorem 3.1, a sub-shock emerges, and the model is no longer valid. We need to increase the number of truncation N taking more fields into the model. In this case, according to the properties of the principal subsystem, i.e., Theorem 2.3 and inequality (10.45), the maximum characteristic velocity increases. Therefore we have now a sub-shock formation with larger critical Mach number. In other words, we need more moments in order to let a theory be valid for larger Mach numbers. In the limit of infinite Mach number, we substantially deal with the Boltzmann equation itself to predict smooth shock structure!

18.5 Conclusion

In conclusion, extended thermodynamics seems to indicate in clear manner that non-local relations are not constitutive equations but approximations of balance laws. The true constitutive equations are in local form and they obey the material frame difference. The physical systems are hyperbolic in agreement with the relativity principle that any disturbance propagates with finite speed. Nevertheless, non-local equations such as the usual Fourier, Navier-Stokes, Fick, Darcy laws and others are useful to measure non-observable quantities and they are good approximations in many practical problems. The Maxwellian iteration preserves the entropy principle at least for processes not far from equilibrium. Hyperbolic systems with dissipation (balance laws with production terms) can have global smooth solutions provided that the initial data are small.

References

1. I. Müller, On the frame dependence of stress and heat flux. Arch. Ration. Mech. Anal. **45**, 241 (1972)

2. A. Bressan, On relativistic heat conduction in the stationary and nonstationary cases, the objectivity principle and piezoelasticity. Lett. Nuovo Cimento **33**(4), 108 (1982)
3. T. Ruggeri, Generators of hyperbolic heat equation in nonlinear thermoelasticity. Rend. Sem. Mat. Padova **68**, 79 (1982)
4. I. Müller, T. Ruggeri, *Rational Extended Thermodynamics*, 2nd edn. (Springer, New York, 1998)
5. E. Ikenberry, C. Truesdell, On the pressure and the flux of energy in a gas according to Maxwell's kinetic theory. J. Ration. Mech. Anal. **5**, 1 (1956)
6. T. Ruggeri, *Maximum of Entropy Density in Equilibrium and Minimax Principle for an Hyperbolic System of Balance Laws Contributions to Continuum Theories, Anniversary Volume for Krzysztof Wilmanski*, ed. by B. Albers. WIAS-Report No. 18 (2000), pp. 207–214
7. K.R. Rajagopal, On a hierarchy of approximate models for flows of incompressible fluids through porous solids. Math. Models Methods Appl. Sci. **17**, 215 (2007)
8. D.A. Signer, *Stability of Non-linear Constitutive Formulations for Viscoelastic Fluids*. Springer Briefs in Applied Sciences and Technology (Springer, Heidelberg, 2014), pp. 9–88
9. T. Ruggeri, Can constitutive relations be represented by non-local equations? Q. Appl. Math. **70**, 597 (2012)
10. M. Torrilon, H. Struchtrup, Regularized 13 moment equations: shock structure calculations and comparison to Burnett models. J. Fluid Mech. **513**, 171 (2004)
11. B. Rahimi, H. Struchtrup, Capturing non-equilibrium phenomena in rarefied polyatomic gases: a high-order macroscopic model. Phys. Fluids **26**, 052001 (2014)
12. G. Boillat, T. Ruggeri, Hyperbolic principal subsystems: entropy convexity and subcharacteristic conditions. Arch. Ration. Mech. Anal. **137**, 305 (1997)

Chapter 19
Open Problems

Abstract We list up some open problems and try to discuss the perspective on the ET theory.

19.1 Open Problems

There are still many open problems that may be interesting subjects to be studied in the next future. Some of them are listed as follows:

- **ET of dense gases and the convexity of the entropy density**: The ideas of the binary hierarchy and of the independence between the internal energy and the pressure give, in our opinion, the correct direction for exploring rarefied polyatomic gases. In fact, we have seen in the previous chapters that, in the case of 14 fields, the closures via phenomenological approach, via MEP, and via Grad procedure produce, give the same differential equations. Moreover the system of field equations for a monatomic gas becomes a (singular) limiting case of the system of a polyatomic gas. In this case, if the gas is polytropic, the convexity of the entropy in equilibrium is automatically satisfied under the usual stability requirement for the thermal and caloric equations of state and the condition $D > 3$. The excellent agreements of the ET theory with experimental data also support strongly this assertion.

  However, the delicate question still remains: To what extent the present model is valid for a dense gas? As seen in Chaps. 5 and 6, the binary hierarchy is a natural generalization of the single hierarchy of fields for monatomic gases. This is perfectly consistent with the moment theory in the kinetic theory for polyatomic gases, although we know that the range of the validity of the kinetic theory basing on the Boltzmann equation is restricted to rarefied gases. Therefore the binary hierarchy is fully reliable in the case of rarefied polyatomic gases.

  On the other hand, we have seen in Chap. 5, Sect. 5.5 that the convexity of the entropy of a dense gas is satisfied only in a bounded domain. And we have found that, in the case of hard-sphere systems, the domain reduces to a point when D approaches to the monatomic limit (see Fig. 5.1). In the case of van der Waals fluids, the situation is similar.

T. Ruggeri, M. Sugiyama, *Rational Extended Thermodynamics beyond the Monatomic Gas*, DOI 10.1007/978-3-319-13341-6_19

These observations seem to indicate that our theory is valid only for moderately dense gases. We need to improve the present ET theory in order to have a wider range of applicability, in particular, near the monatomic limit.

- **ET of relativistic gas**: In the case of monatomic gases, a successful relativistic theory was obtained as we have presented briefly in Chap. 4, Sect. 4.7. As we have seen in the relativistic hierarchy (4.43), as usual in relativistic analysis, the energy and momentum are combined together into the energy-momentum tensor. In the case of the binary hierarchy (10.1), it is still not so clear what is the natural relativistic counterpart. This is also a problem from the kinetic-theoretical viewpoint. This is, in our opinion, a very interesting and challenging problem that we want to investigate in near future.
- **Non-linear closure for more than six fields**: We have seen in Chaps. 11 and 12 that, in ET6, it is possible to have a theory that is valid even far from equilibrium. Interesting challenge is to obtain general (N, M)-systems for which a similar analysis to that of non-linear ET6 can be made.
- **Boundary values of ET**: This is still an open problem not only for a monatomic gas as we discussed in Chap. 4, Sect. 4.2.2 but also for a polyatomic gas. Some preliminary results are given in Chap. 9 but, in the case of many fields, this problem is still difficult to solve.
- **Applications of ET to various practical subjects in engineering, biology etc.**: ET is expected to make a contribution to many fields like nano-technology, physical biology. But no systematic applications are made up to now at least in the context of RET methodology. We believe there are huge possibilities for such studies.

The previous problems are probably the most fundamental and intriguing ones in this field. There are, instead, some works now in progress. These are not included in the present book but will soon appear:

- **Dissipative mixture:** In Chap. 16, we presented only mixtures of non-dissipative Eulerian fluids. A more realistic case is to consider a mixture of dissipative fluids in which every species obeys the requirements of ET14.
- **Shock wave with phase transition:** As mentioned in Sect. 3.3, it is interesting for us to study the problematic of shock and phase transition generalizing the previous study on the basis of the Euler fluid. The work is important also for practical applications. Moreover a complete analysis of Riemann problem is now in preparation.
- **Acceleration wave analyzed by the ET14 theory:** In Chap. 14, we studied acceleration waves by the ET6 theory. Similar study by the ET14 theory is interesting because, in this case, we will have the D-dependence (dependence on the degrees of freedom of a molecule) in wave propagation phenomena. Study of the interaction between acceleration wave and shock wave is also in preparation.

Author Index

T. Ruggeri, M. Sugiyama, *Rational Extended Thermodynamics beyond the Monatomic Gas*, DOI 10.1007/978-3-319-13341-6

Subject Index

T. Ruggeri, M. Sugiyama, *Rational Extended Thermodynamics beyond the Monatomic Gas*, DOI 10.1007/978-3-319-13341-6

MIX
Papier aus verantwortungsvollen Quellen
Paper from responsible sources
FSC® C105338

If you have any concerns about our products,
you can contact us on
ProductSafety@springernature.com

In case Publisher is established outside the EU,
the EU authorized representative is:
Springer Nature Customer Service Center GmbH
Europaplatz 3, 69115 Heidelberg, Germany

Printed by Libri Plureos GmbH
in Hamburg, Germany